Current Topics in Microbiology 109 and Immunology

Editors

The Molecular Biology of Adenoviruses 1

30 Years of Adenovirus Research 1953–1983

Edited by Walter Doerfler

With 69 Figures

Springer-Verlag
Berlin Heidelberg NewYork Tokyo 1983

Professor Dr. WALTER DOERFLER
Institut für Genetik
der Universität zu Köln
Weyertal 121
D-5000 Köln 41, FRG

ISBN-13:978-3-642-69462-2 e-ISBN-13:978-3-642-69460-8
DOI: 10.1007/978-3-642-69460-8

Softcover reprint of the hardcover 1st edition 1983

Library of Congress Catalog Card Number 15-12910

2123/3130-543210

The series "The Molecular Biology of Adenoviruses" is dedicated to Wallace Rowe (4, July 1983†) in memory of his approach to science

Preface

A puzzling epidemiological problem was the driving force behind the discovery of human adenoviruses by Wallace Rowe and his colleagues 30 years ago. The development of a plaque assay for poliomyelitis virus in 1953 led us to the threshold of quantitative virology, and in the same year the double-helical structure of DNA was discovered and became a cornerstone of molecular biology.

The potential of adenoviruses as research tools in the molecular and cellular biology of eukaryotic cells was recognized as early as the late 1950s and early 1960s by several investigators. Structural and biochemical studies dominated the early years. In 1962, some of the adenoviruses were the first human viruses shown to be oncogenic in experimental animals. Thus adenovirology offered the investigator the entire gamut of host cell interactions, productive and abortive, as well as transformed and tumor cell systems. The possibilities that adenoviruses afforded for the study of the molecular biology and genetics of eukaryotic cells were fully realized in the late 1960s and the 1970s.

Over many years, adenoviruses have proved to be a very successful model for research in molecular biology, facilitating the recognition and development of new principles in biology that have turned out to be generally applicable. Work on the icosahedral structure of the virion, the functional organization of the viral genome, a novel mode of DNA replication, problems of viral (foreign) DNA insertion into the host chromosome, the concept of transforming genes, the splicing of RNA, inverse correlations between DNA methylation and gene expression, and many other aspects attest to the viability of this model system. Recently, effects of adenovirus type 12 transformation on the class I major histocompatibility system of the host have been documented. Considering the complexity of the tumor problem, important parameters still remain to be discovered, and the model of transforming viral or cellular genes will perhaps have to be refined and modified. The effects of adenovirus infection on amplifications and rearrangements of host genes are just beginning to be recognized. Recently discovered nucleotide sequence homologies between the E1A region of adenovirus type 12 and certain oncogenes raise tantalizing questions.

The role of adenoviruses in studies on the molecular biology of eukaryotes has occasionally been compared with that of bacteriophage lambda in investigations of prokaryotes. Some of the basic features of the organization and expression of the viral genome are still unknown, although the entire nucleotide sequence of human adenovirus type 2 has become available.

Adenoviruses will be used in the future to work out the detailed mechanisms and controls of some of the main reactions in molecular biology, and in that important role may indeed resemble that of bacteriophage lambda.

The main topics of adenovirus research have been repeatedly summarized in many excellent reviews. The three volumes in this series, The Molecular Biology of Adenoviruses, provide summaries of current research as well as more formal reviews.

I thank the editors of the series Current Topics in Microbiology and Immunology for inviting me to be guest editor, and I am indebted to all the contributors to the three volumes for submitting their manuscripts on time. I am particularly grateful to Prof. DIETRICH GÖTZE and MARGA BOTSCH at Springer-Verlag, Heidelberg, and to PETRA BÖHM and BIRGIT KIERSPEL in Köln for their careful and painstaking work.

Köln, December 1983 WALTER DOERFLER

Table of Contents

Indexed in Current Contents

List of Contributors

Ackerman, S., The Wistar Institute of Anatomy and Biology, Philadelphia, PA 19104, USA

Akusjärvi, G., Departments of Medical Genetics and Microbiology, The Biomedical Center, Box 589, S-75123 Uppsala

Brackmann, K.H., Institute for Molecular Virology, St. Louis University Medical Center, 3681 Park Avenue, St. Louis, MO 63110, USA

Bunick, D., Biology Department, University of North Carolina, Chapel Hill, NC 27514, USA

Concino, M., The Wistar Institute of Anatomy and Biology, Philadelphia, PA 19104, USA

Deuring, R., Center for Cancer Research, Massachusetts Institute of Technology, Cambridge, HA 02139, USA

Doerfler, W., Institute of Genetics, University of Cologne, Weyertal 121, D-5000 Cologne 41

Eick, D., Institut für Virologie, Universität Freiburg, Hermann-Herder-Str. 11, D-7800 Freiburg

Gahlmann, R., Institute of Genetics, University of Cologne, Weyertal 121, D-5000 Cologne 41

Green, M., Institute for Molecular Virology, St. Louis University Medical Center, 3681 Park Avenue, St. Louis, MO 63110, USA

Leisten, R., Institute of Genetics, University of Cologne, Weyertal 121, D-5000 Cologne 41

Lichtenberg, U., Institute of Genetics, University of Cologne, Weyertal 121, D-5000 Cologne 41

Lucher, L.A., Institute for Molecular Virology, St. Louis University Medical Center, 3681 Park Avenue, St. Louis, MO 63110, USA

Perricaudet, M., Institut de Recherches, Scientifiques sur le Cancer, 7, Rue Guy-Mocquet, F-94800 Villejuif

Pettersson, U., Departments of Medical Genetics and Microbiology, The Biomedical Center, Box 589, S-75123 Uppsala

PHILIPSON, L., EMBO Laboratory, Meyerhofstr. 1, D-6900 Heidelberg

RICHARDSON, W.D., National Institute for Medical Research, The Ridgeway, Mill Hill, London NW7 1AA, United Kingdom

SALAS, M., Centro de Biología Molecular (CSIC-UAM), Universidad Autónoma, Canto Blanco, E-Madrid 34

SCHULZ, M., Institute of Genetics, University of Cologne, Weyertal 121, D-5000 Cologne 41

STABEL, S., EMBO Laboratory, Meyerhofstr. 1, D-6900 Heidelberg

STILLMAN, B.W., Cold Spring Harbor Laboratory, P.O. Box 100, Cold Spring Harbor, NY 11724, USA

SUSSENBACH, J.S., Laboratory for Physiological Chemistry, State University of Utrecht, NL-Utrecht

SYMINGTON, J.S., Institute for Molecular Virology, St. Louis University Medical Center, 3681 Park Avenue, St. Louis, MO 63110, USA

TAMANOI, F., Cold Spring Harbor Laboratory, P.O. Box 100, Cold Spring Harbor, NY 11724, USA

VAN DER VLIET, P.C., Laboratory for Physiological Chemistry, State University of Utrecht, NL-Utrecht

VIRTANEN, A., Departments of Medical Genetics and Microbiology, The Biomedical Center, Box 589, S-75123 Uppsala

WEINMANN, R., The Wistar Institute of Anatomy and Biology, Philadelphia, PA 19104, USA

WESTPHAL, H., Laboratory of Molecular Genetics, National Institutes of Health, Bethesda, MD 20205, USA

ZANDOMENI, R., The Wistar Institute of Anatomy and Biology, Philadelphia, PA 19104, USA

Structure and Assembly of Adenoviruses

LENNART PHILIPSON

Department of Microbiology, University of Uppsala, The Biomedical Center, Box 581, S-751 23 Uppsala
Present address: European Molecular Biology Laboratory, Postfach 102209, D-6900 Heidelberg, FRG

Current Topics in Microbiology and Immunology, Vol. 109

1 Introduction

Adenoviruses, this year celebrating the 30th anniversary of their discovery by man, are still in the foreground of scientific interest. With their multiple genes they are in fact one of the outstanding model systems for studying gene expression and its control in mammalian cells, as evidenced in numerous reviews (Philipson et al. 1975; Doerfler 1977; Ginsberg 1979; Philipson 1979; Nevins and Chen-Kiang 1981; Flint and Broker 1981). The advances in the biochemistry and molecular biology of these viruses have not yet been balanced by equal progress in ultrastructural research of the virion itself and its assembly in the cell, although the architectural problems have recently received attention (Pereira and Wrigley 1974; Everitt et al. 1975; Brown et al. 1975; Corden et al. 1976; Burnett et al. 1978; Nermut 1980). Although adenoviruses are recovered from several hosts, including among others humans, monkeys, mice, and birds, with distinct but small differences in single polypeptides, the virion itself is of remarkably uniform morphology (Fig. 1). They all have an icosahedral shell and the capsid is built up from 252 capsomers, comprising 240 hexons, each surrounded by six neighboring capsomers, and 12 pentons at the vertices of the icosahedron each surrounded by five peripentonal hexons (Valentine and Pereira 1965; Ginsberg et al. 1966). Inside the capsid is a nuclear core containing the viral DNA and at least three viral proteins. The adenoviruses have been classified as viruses with an icosahedral nucleocapsid, but since the viral DNA is associated with viruscoded core proteins, the term nucleocapsid should probably be reserved for the core, and in analogy with other viruses the outer shell should be called the capsid. The human adenoviruses have for obvious reasons been investigated in more detail than those from other species. On the basis of biochemical, immunological, morphological and biological criteria they have been divided into seven subgenera, previously called subgroups. Each subgenus contains one or several serotypes, now referred to as species, separated on the basis of neutralization with type-specific antisera. Table 1 shows the basis for the current classification.

2 Architecture of the Virion

The icosahedral shape of the capsid leads to some confusion in the estimates given for the size of the virion. Diameter values of 60–70 nm have been published but it is not always clear whether these refer to the edge-to-edge distance or to the vertex-to-vertex distance. The dimensions of an icosahedron can always be calculated from the length of its edge (Mattern 1969). The size for the edge of the capsid has been measured to be 43 nm and the edge-to-edge distance then corresponds to 70 nm and the vertex-to-vertex distance to 82 nm. These calculations give a diameter of the virion along the fivefold symmetry axis of around 73 nm (Nermut 1975).

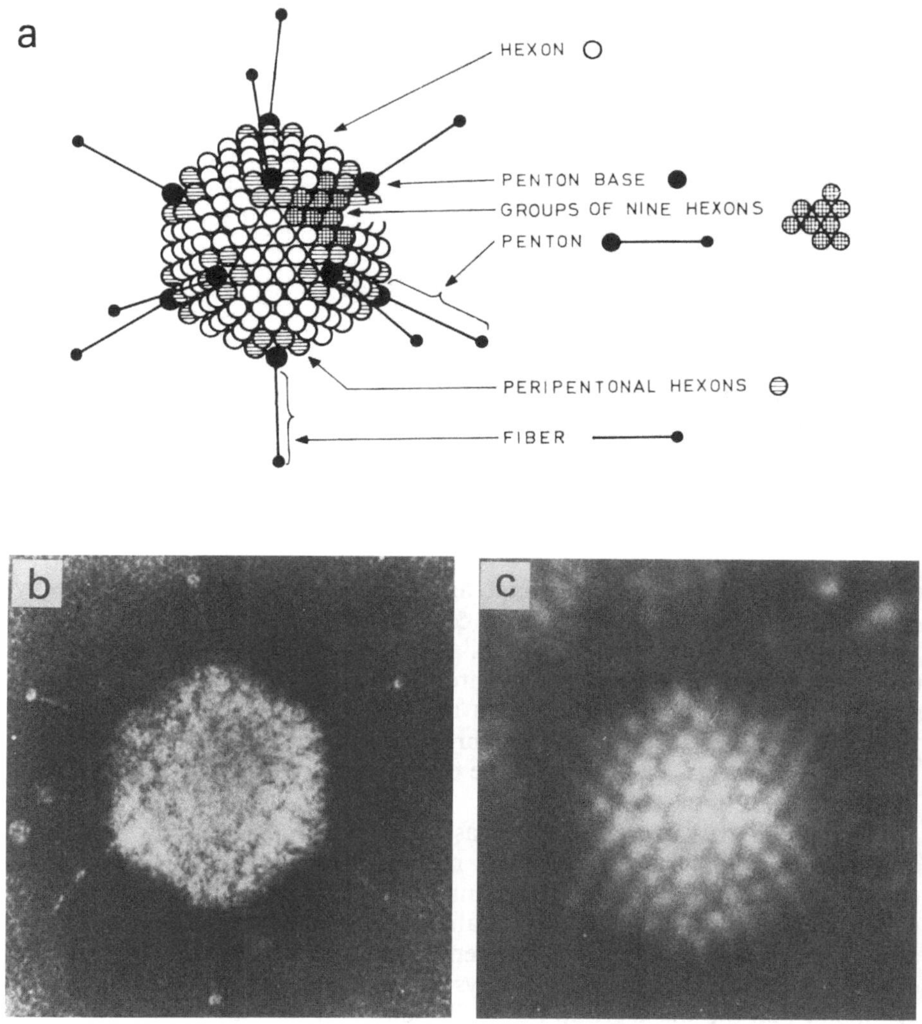

Fig. 1a–c. Structure of the adenovirus capsid. **a** The icosahedral outline of the adenovirus capsid and the location of various components. [PHILIPSON and PETTERSSON (1973)]. **b.** Electron micrograph showing the Ad5 virion contrasted with sodium silicotungstate (VALENTINE and PEREIRA 1965). Note the antenna-like fiber protruding from each vertex. Figure kindly provided by Dr. W. RUSSELL, Mill Hill, London). **c** Electron micrograph of Ad5 virus contrasted with sodium silicotungstate, showing the regular icosahedral symmetry of adenoviruses. (HORNE et al. 1959)

The major subunit of the icosahedral shell of the virion is the hexon, which has attracted extensive attention because it is produced in large amounts in the infected cell, is soluble in the native form, and can be examined in the electron microscope. Electron microscopy (VALENTINE and PEREIRA 1965) and low-angle X-ray diffraction (TEJG-JENSEN et al. 1972) of

Table 1. Classification of human adenoviruses. [Adapted from WIGAND et al. (1982)]

Subgenus[a] (subgroup)	Species (serotypes)	GC in DNA (%)	Hemag-gluti-nation subgroup[b]	Length of fibers (nm)	Onco-genicity in vivo[c]
A	12, 18, 31	47–49	IV	28–31	High
B	3, 7, 11, 14, 16, 21, 34, 35	50–52	I	9–11	Weak
C	1, 2, 5, 6	57–59	III	23–31	None
D	8, 9, 10, 13, 15, 17, 19, 20, 22–30, 32, 33, 36, 37	57–60	II	12–13	None
E	1	57	III	17	None

[a] Two additional subgenera, F and G, have recently been described (WADELL et al. 1980)
[b] Group I agglutinates monkey erythrocytes and group II rat erythrocytes in a tight lattice forming a complete pattern. Groups III and IV agglutinate rat erythrocytes in an incomplete pattern
[c] Almost all nononcogenic serotypes have been shown to transform rodent cells in vitro

the hexons defined their basic shape and dimensions. They were originally described as spherical bodies with a central hole (VALENTINE and PEREIRA 1965) or hexagons (PETTERSSON et al. 1967). Biochemical work established that each hexon of Ad2 contains three identical polypeptides, each with a molecular weight of 108000 (MAIZEL et al. 1968a; GRÜTTER and FRANKLIN 1974; JÖRNVALL et al. 1981) and the threefold symmetry was also established by analyzing groups of hexons in the electron microscope (CROWTHER and FRANKLIN 1972).

The adenovirus hexon protein was the first animal virus protein to be crystallized. Following the original method described by PEREIRA et al. (1968), purified hexons dialyzed against 0.8 M KH_2PO_4 form a precipitate which is gradually converted into tetrahedral crystals. FRANKLIN et al. (1971) and CORNICK et al. (1971) used different conditions and obtained pyramidal crystals. Structural studies of both crystal types revealed that they had cubic symmetry and the space group $P2_13$. The length of the cubic cell is 10.99 nm and each unit cell contains four hexons. Since there are three asymmetrical units in the cell there must be three crystallographic asymmetrical units per hexon. The cylinder axis of the hexon is parallel to the axis of threefold symmetry. Consequently, there are three structural units symmetrically arranged around the cylinder axis of the hexon, which was also confirmed by electron microscopy (CROWTHER and FRANKLIN 1972). Attempts to reconstruct the hexon from electron-microscopic images indicated that different portions of the hexons have different symmetries (NERMUT 1975; NERMUT and PERKINS 1979).

Recent structural studies by X-ray crystallography (BURNETT et al. 1978; BERGER et al. 1978; BURNETT, 1983) have provided a more detailed 3D model of the hexon suggesting that it is a holotriangular prism (Fig. 2) with the following principal features: the lowest 1-nm portion of the hexon facing

a

Fig. 2a, b. A model of hexon based on an electron density map at 0.6 nm resolution. **a** The intact trimer, illustrating the dense base and the more open top. **b** The model has been split open to emphasize the different features of the top and base. The triangular top (*left*), seen from above, shows the three towers surrounding a small hole and reveals the true threefold symmetry of the hexon trimer. The array of towers gives the adenovirus its spiky appearance since they are on the outside of the capsid. The base (*right*), seen from below, reveals the pseudohexagonal symmetry which allows the hexons to pack close together in the capsid. The large cavity in the base, which narrows in the middle to emerge between the towers at the top, is also seen. Figure kindly provided by Dr. R. BURNETT, Columbia University, New York)

b

the DNA core measures 7.5 nm in diameter and has an axial hole with a diameter of 3.5 nm. The middle portion from 1–5.2 nm is hexagonal with an 8.9-nm side, and the top portion from 5.2–11.6 nm facing the outside of the virion is triangular with a 7.5 nm side. The top triangle is twisted by around 30° relative to the middle hexagon. The centre of the channel is very narrow at the middle of the hexon but widens in both directions, especially toward the base. The height of the hexon is 11.6 nm. The hexons can be released from the virus in groups of nine, nonamers, each corresponding to one of the triangular facets of the icosahedron (Fig. 3). These nonamers have been useful in clarifying the polarity of the hexon, the majority showing a left-handed structure on hydrophobic carbon films but a right-handed structure on hydrophilic positively charged carbon films (NERMUT 1980). Thus the internal surface of the hexon is predominantly hydrophobic,

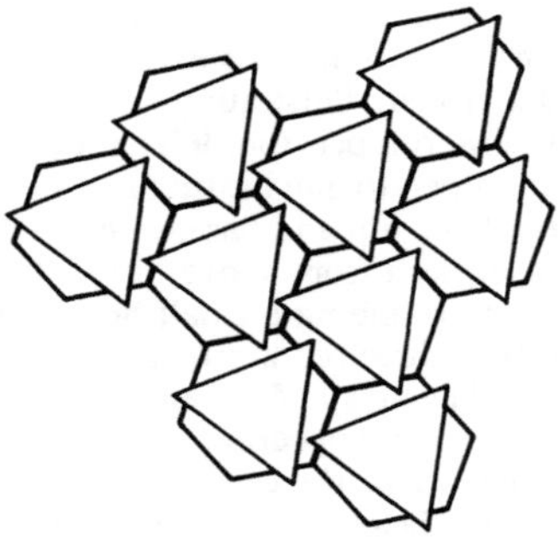

Fig. 3. A schematic model for a nonamer, showing the asymmetric hexon-hexon interactions. Note the offset of the triangular apices, representing the towers, at the top. This gives a large central cavity between the towers from three different hexon molecules at the centre of the nonamer. This is the arrangement seen on the capsid facet, as distinct from the other possible arrangement where the smaller cavity is at the centre. (Figure kindly provided by Dr. R.M. BURNETT, Columbia University, New York)

whereas the top is negatively charged at physiological pH, as would be expected if the hexon interacted with the core on the inside and gave the particle a negative charge on the outside.

2.1 Organization of the Virus Capsid

The structure of animal viruses was originally thought to be guided by two principles, one physical, dependent on minimum free energy, and one biological, dependent on natural selection (CASPAR 1966). The icosahedral capsid in adenovirus follows the requirements for the 5′3′2′ symmetries of an icosahedron, and therefore maximum bonding strength between the capsomers should give stability to the capsid. The hexon would then have six identical binding sites and hopefully six structural subunits. However, the hexon has only three polypeptides and shows threefold rotational symmetry (GRÜTTER and FRANKLIN 1974). The present model solves this dilemma; the hexon appears to have a pseudohexagonal base and a triangular top. The determination of pseudohexagonal symmetry leading to a close-packing model and the precise orientation of the triangles was established with crystallographic techniques (BURNETT, personal communication).

The interactions between the nonamers and the peripentonal hexons surrounding each of the pentons are, however, weaker than those within the nonamers. The penton-hexon interaction is obviously electrostatic, since dialysis at pH 6.3–6.5 causes release of pentons. Recent physicochemical data suggest that the penton base, like the hexon, contains only three polypeptides (DEVAUX et al. 1982). The minimum energy principle cannot therefore apply in the peripentonal region, and an additional linker protein between the penton base and the peripentonal hexons must provide additional stability. Protein IIIa may be a candidate for this linker, since it is probably present in five copies per vertex region (EVERITT et al. 1975; BOUDIN et al. 1980).

The fiber is highly asymmetric, consisting of a rod with a terminal knob (Fig. 1). The diameter of the rod is around 2 nm and the diameter of the knob around 4 nm. The length of the fiber varies among adenovirus subgenera (Table 1), and is 9–13 nm for subgenera B and D and 23–31 nm for subgenera A and C (NORRBY 1969b). Ad4 in subgenus E has 17-nm-long fibers. Avian adenoviruses, except for one species, have two fibers of different lengths extending from different sites on the penton base (LAVER et al. 1971; GELDERBLOM and MAICHLE-LAUPPE 1982). Dimers of fibers have been encountered but it is unclear whether they have a physiological role (WADELL and NORRBY 1969a; NORRBY et al. 1969b). Fibers from Ad5 have been crystallized (MAUTNER and PEREIRA 1971) and structural studies are currently under way (GREEN et al. 1983; DEVAUX et al. 1983). Conformational studies suggested that the native fiber is mainly in a β-pleated sheet configuration (BOULANGER and LOUCHEUX 1972). Inspection of the amino acid sequence revealed a repeating motif of 15 residues organized in two short β-strands and β-bends (GREEN et al. 1983). The fiber unit has been thought to consist of three probably identical polypeptide chains (SUNDQUIST et al. 1973a) but recent data suggest a dimer structure (DEVAUX et al. 1983). The fiber obviously functions as the attachment organelle for the virus at the plasma membrane of the cell at the initial phase of infection (PHILIPSON et al. 1968).

The general principle for building the adenovirus capsid therefore involves close packing of proteins, with morphological symmetry greater than the true molecular symmetry, in a capsid with icosahedral symmetry. Several additional proteins are, however, required in building the capsid of the adenoviruses, as became evident from the detailed analysis of the protein composition of the particles.

2.2 The Core

The adenovirus genome, which is a linear double-stranded DNA of around $20–23 \times 10^6$ daltons (GREEN et al. 1967; VAN DER EB et al. 1969), is packaged within the icosahedral capsid in a nucleoprotein structure called the core. The cores first identified by electron microscopy (EPSTEIN 1959) can be released from the outer shell by exposure of purified virions to one of several denaturants such as heat, desoxycholate, and pyridine (RUSSELL et al. 1971; PRAGE et al. 1970). After freeze fracturing the particles often fracture between the capsid and the core (BROWN et al. 1975; NERMUT 1978). If so, the edge of the cores can be measured, and values of 34 nm have been reported (NERMUT 1978). When the cores are isolated in solution without the capsid, they are not rigid enough to maintain an icosahedral shape. Freeze-fracture studies also suggest that the nucleoprotein is surrounded by a protein shell, but the protein covering the cores has not been identified. It has tentatively been suggested that protein V, the so-called minor core protein, might form a protein shell around the nucleoprotein containing the major core protein VII (NERMUT 1980). The adenovirus nucleoproteins do not contain cellular histones as do papova virions (ROBLIN et al. 1971; FREARSON and CRAWFORD 1972); the viral DNA is instead associated with two virus-encoded basic proteins V and VII. A third arginine-rich core protein called μ with a molecular weight of around 4000 has recently been described (Hosakawa and Sung 1976). Neither the precise location nor the role of this protein in the core has been determined.

2.3 Organization of the Nucleocapsid

The core-associated polypeptides probably mediate the packaging of adenovirus DNA within the virions, but the adenovirus DNA in core particles appears not to be protected from nuclease attack in a way similar to cellular DNA packaged in nucleosomes. Corden et al. (1976) reported that digestion of adenovirus cores with micrococcal nuclease yields discrete DNA products in a ladder arrangement from 200 to 1800 bp. Others have failed to reproduce this result, observing only protected viral DNA which was smaller than that obtained from cellular chromatin (Tate and Philipson 1979; Brown and Weber 1980). Native core preparations are, however, compact, with the structure of a thick fiber of around 30 nm in diameter (Vayda et al. 1983) which displays a morphology similar to that of chromatin fibers. These thick fibers contain polypeptides V and VII and also polypeptide μ, and all three proteins sediment together with released cores. When the cores are exposed to 0.5 *M* NaCl, proteins V and μ are released and the cores become at the same time less resistant to nuclease digestion and appear in the electron microscope as a beaded string similar to cellular nucleosomes. Although there is disagreement concerning a ladder arrangement of the protected DNA sequences at low concentrations of nucleases (Corden et al. 1976; Tate and Philipson 1979; Vayda et al. 1983), a consensus exists that polypeptide VII protects around 100–150 bp of viral DNA. The virus protein VII may therefore form a protein particle about which the DNA is wound in a similar fashion to nucleosomes. The diameter of such a filament would be about 9 nm provided there are three protein subunits of 2.5–3 nm in diameter per turn. Protein VII would then provide a helical core structure which is packed in a more condensed form with the protein V and the μ polypeptide. Finally, a terminal DNA binding protein is covalently attached to each 5′ end of the DNA but this cannot influence the structure since it is only present in two copies per viral genome.

3 Protein Composition

The adenovirus particle was originally claimed to be composed of two kinds of proteins, hexons and pentons, in addition to the core, but when the adenovirus proteins were analyzed by SDS-PAGE (Maizel et al. 1968a; Maizel 1971), a considerable complexity was revealed and it is now agreed that the particle contains at least nine unique polypeptides (II–IX and IIIa). In addition minor bands have been observed, some of which represent unique viral polypeptides present in only a few copies per virion. Other components such as polypeptides XI–XII have not yet been identified as virion proteins, but may represent the additional core protein μ (Hosakawa and Sung 1976) or cleavage products which remain when capsid polypeptides are cleaved during virion assembly. Figure 4 shows a schematic model of the adenovirus particle indicating the location of all polypeptides in the virion.

A protein kinase activity is released from purified adenovirus particles upon disintegration (Akusjärvi et al. 1978; Blair and Russel 1978). It has been solubilized and partially purified from Ad2 and Ad5 virions, but the enzymatic activity does not copurify with any of the known virion polypeptides (Akusjärvi et al. 1978). Therefore it remains to be established that this is a viral enzyme. The γ-phosphate of ATP is transferred to serine and threonine residues, preferentially in polypeptide IIIa but also in polypeptides V, VI, VII, and X. Casein and phosvitin can serve as exogeneous acceptors for this protein kinase, which is independent of cyclic nucleotides for activity (Akusjärvi et al. 1978; Blair and Russell 1978).

4 The Major Capsid Proteins

The adenoviruses produce large quantities of soluble viral antigens during their multiplication (Huebner et al. 1954; Hilleman et al. 1955). Separation by immunoelectrophoresis (Pereira et al. 1959) or DEAE chromatography (Klemperer and Pereira 1959; Philipson 1960; Wilcox and Ginsberg 1961; Haruna et al. 1961) resolved them into three distinct components, each related to one of the major capsid proteins – hexon, penton, and fiber – both by antigenicity and by structure (Wilcox and Ginsberg 1963a; Wilcox et al. 1963; Valentine and Pereira 1965; Norrby 1966a).

During the late phase of adenovirus replication host cell protein synthesis is almost totally shut off (Ginsberg et al. 1967) and the infected cells are primarily engaged in the synthesis of large quantities of viral structural proteins and viral DNA. The nascent polypeptides assemble into the structural proteins during or shortly after their synthesis (Horwitz et al. 1969; Velicer and Ginsberg 1970), but only 1–5% of the fibers and penton bases and 20–30% of the hexons are used for assembly of virus particles (White et al. 1969; Everitt et al. 1971). The structural proteins of the virus have attracted attention, since unlike structural proteins of other animal viruses, they are soluble under nondenaturing conditions and are available in sufficient quantities for detailed immunological and biochemical studies. The antigens which are present in the excess pool have been used as the source of material for purification of adenovirus capsid components.

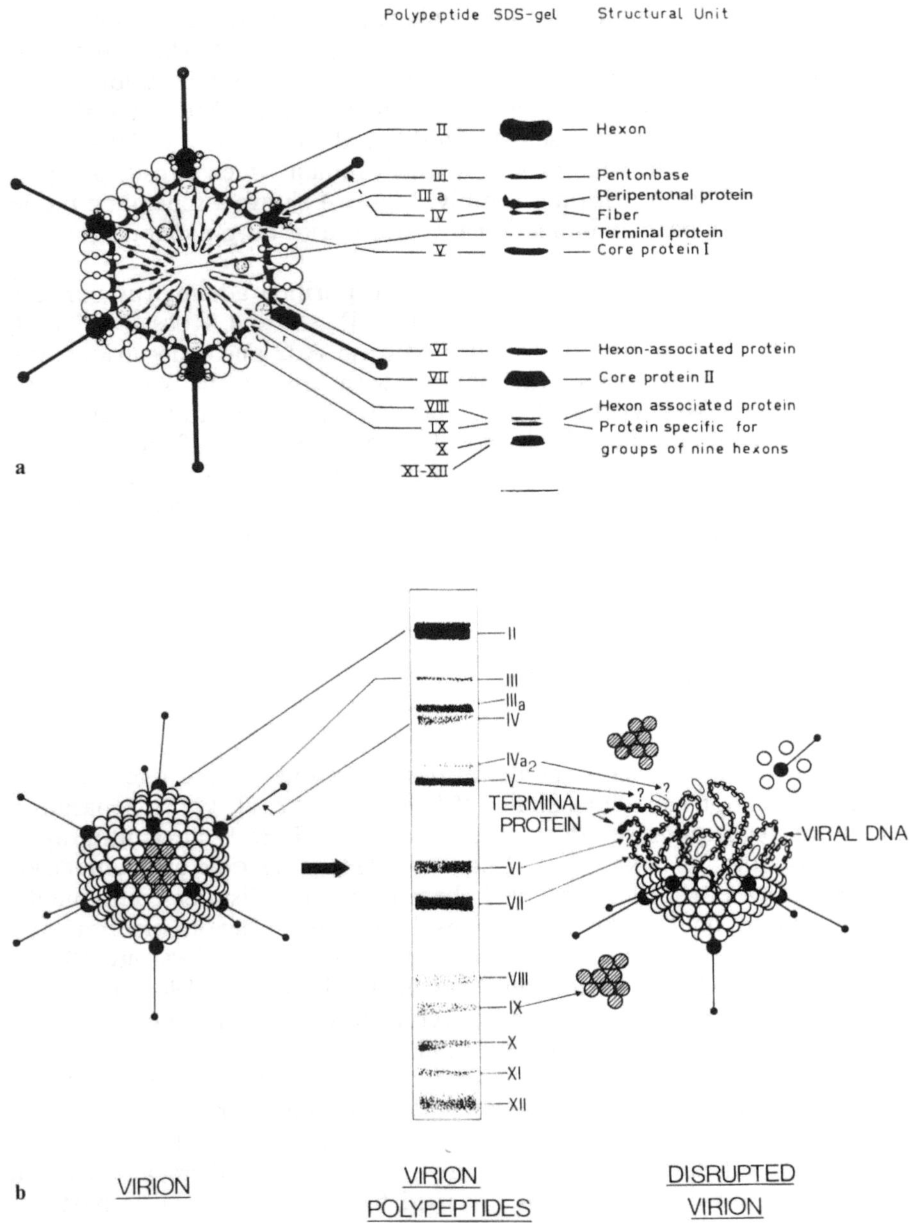

Fig. 4a, b. Two tentative models for the location of the proteins in the Ad2 virion. The polypeptide composition of the virion proteins are shown in a stained exponential 10–16% SDS–polyacrylamide gel, and the evidence for the location of the different protein moieties is summarized in the text. A covalently linked terminal protein is also indicated in the figure, but cannot be seen in a protein gel, and the corresponding polypeptide band is therefore indicated with a broken line. **a** A model based on protein-protein cross-linkage studies. [Modified from EVERITT et al. (1975)]. **b** A schematic model. [RUSSELL and PRECIOUS (1982)]

4.1 Hexon

Several methods for purification of adenovirus hexons from infected cells have been reported. KÖHLER (1965) used methanol precipitation at pH 4.0 and DEAE chromatography, whereas other investigators (VALENTINE and PEREIRA 1965; HOLLINSHEAD et al. 1967; RUSSELL et al. 1967; KJELLEN and PEREIRA 1968; MAIZEL et al. 1968b) used DEAE chromatography either alone or in combination with other types of chromatography (LEVINE and GINSBERG 1967) or electrophoresis (PETTERSSON et al. 1967; BOULANGER et al. 1969). A convenient method for large-scale purification was introduced by BOULANGER and PUVION (1973), giving high yields of purified proteins representing all the three major soluble antigens. After the discovery that hexons are crystallizable (PEREIRA et al. 1968), this approach was also introduced in purification and in combination with DEAE chromatography enough material was obtained to allow the determination of the complete amino acid sequence of the Ad2 hexon polypeptide (JÖRNVALL et al. 1981).

4.1.1 Biochemistry

Hexons from Ad2 sediment at 12.9 S (Table 2) (FRANKLIN et al. 1971) and their molecular weight is 324000 (JÖRNVALL et al. 1981). Several different estimates for the molecular weight of the hexon, ranging between 180000 and 400000, have been reported (VALENTINE and PEREIRA 1965; WASMUTH and TYTELL 1966; KÖHLER 1965; PETTERSSON et al. 1967; GRÜTTER and FRANKLIN 1974; DEVAUX et al. 1982) but it is now established that the Ad2 hexon consists of three identical polypeptide chains (HORWITZ et al. 1970; CORNICK et al. 1973; BOULANGER and PUVION 1973; GRÜTTER and FRANKLIN 1974; JÖRNVALL et al. 1974a), each with a molecular weight of 108K (JÖRNVALL et al. 1981). The precise molecular weight of the hexon polypeptide is derived from the complete amino acid sequence, which has so far only been determined for Ad2 hexons.

Ad2 hexons consist of 966 amino acid residues (JÖRNVALL et al. 1981) with a high level of dicarboxylic amino acids but no carbohydrates (PETTERSSON et al. 1967; BOULANGER et al. 1969). Seven cysteine residues are present; none of them appears to be engaged in disulphide bond formation (JÖRNVALL et al. 1974a; GRÜTTER and FRANKLIN 1974), but one is accessible to alkylation in the native molecule (GRÜTTER and FRANKLIN 1974; JÖRNVALL and PHILIPSON 1980).

The hexon contains an acetylated N-terminal amino acid (LAVER 1970; PETTERSSON 1971) followed by the sequence acetyl-Ala-Thr-Pro-Ser (JÖRNVALL et al. 1974b).

4.1.2 Primary Structure

The 966-residue sequence of the Ad2 hexon polypeptide exhibits some noteworthy features (Fig. 5) (JÖRNVALL et al. 1981). Proline is unevenly distributed along the hexon polypeptide chain (JÖRNVALL and PHILIPSON 1980);

Table 2. Physical and chemical properties of human Ad2 virions proteins. [Modified from LEMAY and BOULANGER (1980)]

Polypeptide	Capsid component	Number per virus particle	Molecular weight (K)		Subunit composition	Sedimentation coefficient	Isoelectric point
			Native	Subunit[a]			
II	Hexon	240	324[b]	108	Trimer	12.9	5.4–6.1
III	Penton base	12	246–256[c]	85	Trimer	9.5	5.8
III a	Vertex region	60	65	66	Monomer	6.0	6.0
IV	Fiber	12	156–207	62	Trimer or dimer	6.0	5.3–6.6
V	Core minor	180	47	48	Monomer	3.5	6.7
VI	Hexon-associated	450	50–72	24	Dimer or trimer	3.8	5.8
VII	Core major	1070	18–83	18.5	Monomer or tetramer	2.3	>7.5
VIII	Hexon-associated		13	13	Monomer	1.6	5.2
IX	Hexon nonamers	280	23	14	Dimer	2.3	5.9
III + IV	Penton	12	365[e]	85 + 62	Trimer + trimer	10.5	ND

[a] Determined by SDS-PAGE
[b] Data from JÖRNVALL et al. (1981)
[c] Data from DEVAUX et al. (1982)

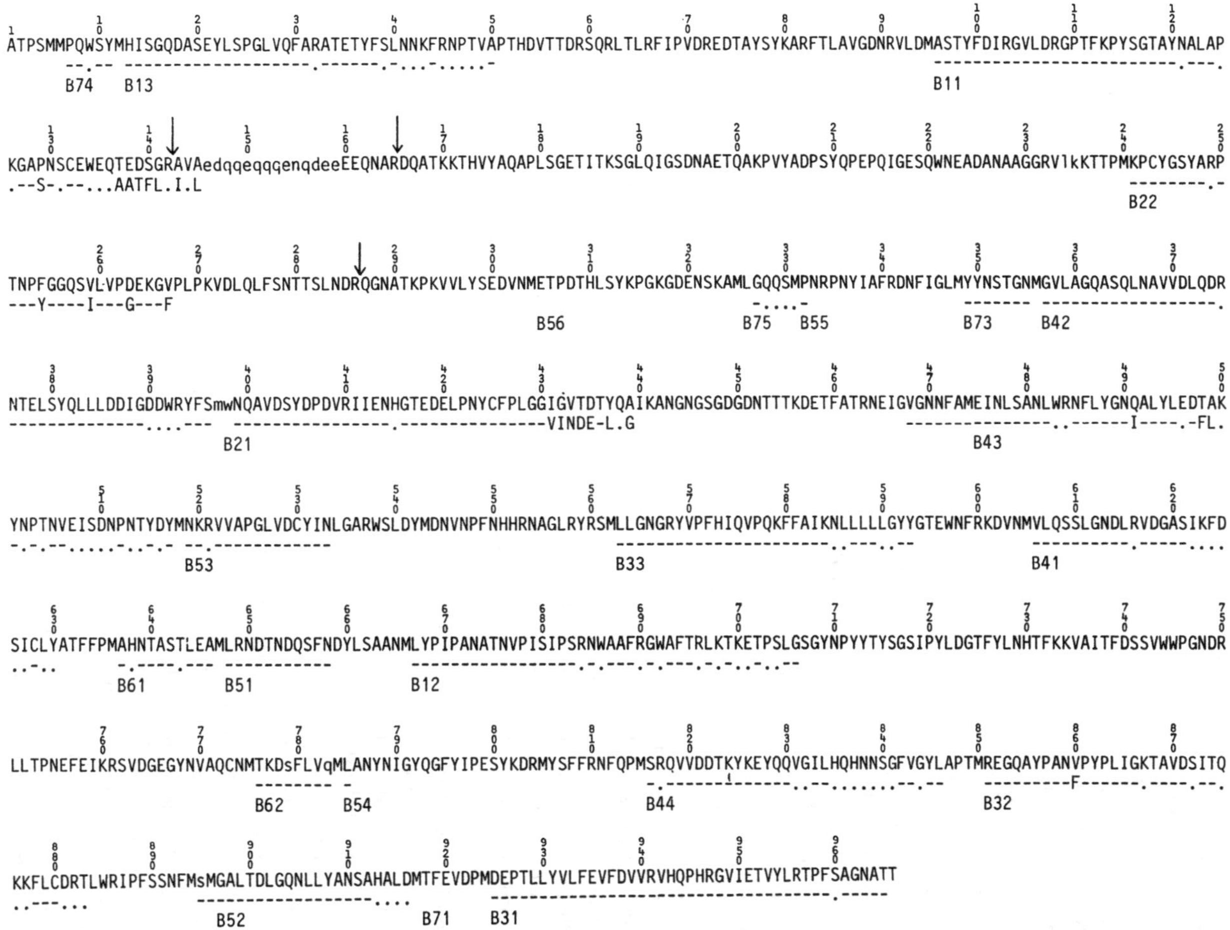

Fig. 5. Primary structure of Ad2 hexon (*upper line*) compared to that of Ad5 hexon (*lower line*). Positional numbers and peptide names are from the Ad2 hexon (JÖRNVALL et al. 1981). Amino acids are given in the one-letter abbreviations, noncapital letters show those residues not firmly established in the Ad2 hexon. In the Ad5 line, a *letter* shows a proved difference, a *dash* shows a proved identity, a *point* a tentative identity, and *lack of sign* a position not examined. The three peptide bonds sensitive to cleavage by trypsin in the native conformation of Ad2 hexon are indicated by *vertical arrows*

common in the first third of the sequence (27 of 335 residues), uncommon in the middle third (10/331), and common again in the C-terminal third (20/298). When they occur, the prolines are associated in pairs separated by one to five other amino acid residues. Clusters of hydrophobic residues are rare, and only in the pentaleucine sequence starting at position 587 can more than three aliphatic hydrophobic residues be found. Charged residues, however, are prevalent. The sequence contains 207 such residues (21%), with an excess of 11 acidic residues over the sum of the basic ones. Residues with positive and negative charges are intermingled in the primary structure, except in a highly acidic region which includes 15 dicarboxylic residues (positions 106–161) and a region with nine charged residues (positions 552–580). The highly acidic region is probably exposed on the surface of the protein in its native conformation, since it can be eliminated by limited tryptic digestion (JÖRNVALL and PHILIPSON 1980).

Three- and four-residue repeats are common in the hexon sequence, and 84 such repeats have been identified affecting a total of 542 residues, but no regions with long repeated sequences are apparent. Consequently, the many short regions with similar or identical amino acid combinations do probably reflect strict structural requirements on the molecule rather than remnants of ancestral duplications.

Combined, the sequence data suggest that three portions of the hexon polypeptide chain may be distinguished, with different conformations in the middle than at the two ends and with elements regularly repeated in all regions. It will be interesting to know whether these conclusions are confirmed at the level of tertiary structure.

Although the complete amino acid sequence has only been determined for the human Ad2 hexon, the amino acid compositions for hexons from other human adenoviruses, i.e., Ad5 and Ad3, show only small differences, suggesting that the hexon is a well-conserved entity. STINSKI and GINSBERG (1975) compared the CNBr peptides from Ad2, Ad5, and Ad3 and found that approximately two-thirds of the CNBr fragments from Ad2 and Ad5 have identical isoelectric points, whereas Ad3 hexons have fewer CNBr fragments in common with Ad2 hexons. VON BAHR-LINDSTRÖM et al. (1982) determined around 59% of the sequence (529 residues for the hexon polypeptide from Ad5, and it is clear that Ad2 and Ad5 hexons are extremely similar (Fig. 5). Only 23 residues (4%) were found to be different, in most cases due to single base pair mutations. The two molecular ends of the polypeptide were well conserved, but at least two deletions were found in the Ad5 sequence, rendering Ad5 hexons 5 K–7 K (approximately 60 residues) smaller. These deletions include the three trypsin-sensitive sites located between residues 100 and 290 in Ad2, which explains why native Ad5 hexon is resistant to cleavage with trypsin (VON BAHR-LINDSTRÖM et al. 1982).

BOURSNELL and MAUTNER (1981) used an indirect approach to identify heterologous regions in the hexon polypeptide, identifying unique restriction enzyme cleavage sites in the genes for Ad2 and Ad5 hexons. Their results confirm that the two ends of the hexon molecule are well conserved and suggest that the heterologous regions are located at the sites for trypsin

Table 3. Antigens associated with the major structural proteins. [adapted from PHILIPSON et al. (1975)]

Protein	Corresponding polypeptide	Antigens		
		Designation	Specificity	Remarks
Hexon	II	α	Genus	Oriented towards the inside of the virion
		–	Inter- and intrasubgenus	
		ε	Species	Available at the surface of the virion
Penton base	III	β	Genus Inter- and intrasubgenus	Carries toxin activity
Fiber	IV	γ	Species	Reacts with hemagglutination inhibition antibody
		–	Intersubgenus	
		δ	Intrasubgenus	Located at the proximal part of the fiber; only present in subgenera A, C, D, and E
Major core protein	VII	–	Genus and species	Only human/Ad2 and human/Ad3 have been examined
Other virion polypeptides	IIIa	–	Genus	
	IX	–	Genus and species	Only human/Ad2 and human/Ad5 have been examined

sensitivity in the Ad2 hexon polypeptide (residues 100 and 290). This region is thus likely to contain type-specific antigenic determinants.

4.1.3 Immunology

Hexons from all adenoviruses except the avian and some bovine species contain a common antigenic specificity α (PEREIRA et al. 1959; WILCOX and GINSBERG 1963a; BÜRKI et al. 1979; WIGAND et al. 1982). An additional type-specific epitope (ε) in the hexon has also been described (WILCOX and GINSBERG 1963b; KÖHLER 1965; PETTERSSON 1971 and Table 3). Using more sensitive techniques, intra- and intersubgeneric specificities were also revealed, as would be expected from a large polypeotide (NORRBY 1969a, b; NORRBY and WADELL 1969). Human Ad12, which is only distantly related to other human adenoviruses, has hexons which differ with regard to charge and which share few antigenic determinants with hexons from other serotypes (NORRBY and WADELL 1969). Hexons from Ad4, the only species in subgenus E, are closely related to hexons from subgenus B and in particular to those from Ad16 (NORRBY and WADELL 1969).

Species-specific hexon determinants are located on the outside of the virion (NORRBY et al. 1969a; WILLCOX and MAUTNER 1976). The group-specific determinant α, in contrast, is probably located internally, since only homologous antisera interact with the virus particles (NORRBY et al. 1969a). Viruses from subgenus C have apparently few determinants on the outside, since homologous hexon antibodies fail to aggregate the virions (WADELL 1972; PRAGE et al. 1970). Species-specific antibodies recognizing the hexon interfere with attachment of subgenera B and D viruses to erythrocytes, presumably because the antibodies attached to hexons in the peripentonal region block the short fiber interacting with the receptor (Table 1) (NORRBY 1969b; NORRBY and WADELL 1969). Subgenus C virions with long fibers (Table 1) attach in the presence of homologous hexon antibodies.

The topography of the hexon antigenic determinants has been studied by partial proteolysis. Treatment of purified Ad2 hexons with trypsin does not change the antigenic specificity (PETTERSSON 1971), but introduces breaks in the sequence at residues 102/103, 165/166, and 286/287 without dissociation of the peptides from the molecule (JÖRNVALL and PHILIPSON 1980), suggesting that the region between residues 100 and 290 is exposed on the surface of the hexon (VON BAHR-LINDSTRÖM et al. 1982). In contrast, digestion with chymotrypsin, papain, or subtilisin removes more than 50% of the hexon polypeptide and parts of the α determinant, but the ε determinant remains unaltered (PETTERSSON 1971). The species-specific region of the hexon polypeptide is obviously retained in a compact structure even when it has suffered several cuts in the sequence.

Monoclonal antibodies provided a new approach for studying antigenic determinants of the hexon. CEPKO et al. (1981) prepared group-reacting monoclonal antibodies which precipitate native hexons and hexon nonamers but fail to precipitate intact virions and nascent hexon polypeptide chains. The trimeric hexon molecule is thus recognized by different antibodies than nascent hexon polypeptides, confirming previous results (STINSKI and GINSBERG 1974; ÖBERG et al. 1975). A collection of monoclonal antibodies against Ad5 hexons identified five epitopes in the hexon, confirming the immunological complexity of the hexon molecule, previously demonstrated in cross-adsorption studies (RUSSELL et al. 1981; NORRBY and WADELL 1969). A map of all epitopes in the hexon might finally be achieved with monoclonal antibodies and chemically synthesized peptides from the hexon.

Two classes of hexons have been identified after purification by electrophoresis (PETTERSSON 1971; BOULANGER et al. 1978). The slowly migrating hexons contain a shorter polypeptide, lacking approximately 50 residues at the N-terminal, probably generated through proteolytic cleavage (BOULANGER et al. 1978). These hexons have a high activity against species-specific antibodies (BOULANGER et al. 1978) and also induce neutralizing antibodies more efficiently than the fast hexons (PETTERSSON 1971). However, since the slow hexons are absent in virions, their biological role is still undefined.

Hexons have been reported by many workers to carry the antigen responsible for adenovirus neutralization (WILCOX and GINSBERG 1963b; KJELLEN and PEREIRA 1968; PEREIRA and LAVER 1970; NORRBY et al. 1969a; HAASE and PEREIRA 1972), but one group (PETTERSSON et al. 1967) claimed that highly purified preparations of Ad2 hexons induce insignificant titers of neutralizing antibody. These conflicting results may arise either because hexons, when extensively purified, change their conformation and thereby fail to induce neutralizing antibodies, or because less pure hexon preparations contain small amounts of a contaminating protein which is the neutralizing antigen. Since free hexons in solution have different antigenic properties than hexons in the virus particle (STINSKI and GINSBERG 1975), conformational changes may occur. The observation that hexons lacking an N-terminal segment are more efficient in inducing neutralizing antibodies (BOULANGER et al. 1978) suggests that the conformation rather than the primary sequence accounts for the difference.

4.2 Fiber

Several methods have been used for the purification of fibers from the soluble antigen pool in cells infected with Ad2 or Ad5. DEAE chromatography in combination with other separation methods, such as isoelectric focusing, hydroxylapatite chromatography, or gel exclusion chromatography, yields a preparation suitable for biochemical studies. Fibers are only present in infected cells at one tenth of the concentrations of hexons, and this has hampered biochemical characterization.

4.2.1 Biochemistry

Fibers from Ad2 and Ad5 sediment at about 6S (Table 2) (LEVINE and GINSBERG 1967; PETTERSSON et al. 1968; WADELL et al. 1969; DEVAUX et al. 1982), and molecular weight determinations show that intact fibers from Ad2 and Ad5 have a value of 156K–207K (SUNDQUIST et al. 1973a; DORSETT and GINSBERG 1975; DEVAUX et al. 1982). The radius of gyration is 7.5 nm (DEVAUX et al. 1982). The fiber appears to be composed of three polypeptide chains each with a molecular weight of around 62K (PETTERSSON et al. 1968; MAIZEL et al. 1968b; SUNDQUIST et al. 1973a; DORSETT and GINSBERG 1975), but recent structural studies imply a dimer structure (DEVAUX et al. 1983; GREEN et al. 1983). Whether the polypeptides are identical is still unclear. SDS-PAGE resolves one single band (MAIZEL et al. 1968b; SUNDQUIST et al. 1973a) but DORSETT and GINSBERG (1975) suggested that a trimeric fiber consists of two copies of one polypeptide and one copy of another. Although it is difficult to envisage how identical polypeptide subunits could assemble to form the asymmetric fiber structure with a knob and a shaft (Fig. 4), the adenovirus genome contains only one fiber gene, which encodes a polypeptide chain with a predicted molecular weight of

62K. The purified fiber subunit has furthermore the same amino acid composition (PETTERSSON et al. 1968; DORSETT and GINSBERG 1975) as predicted from the DNA sequence of the gene. Thus if two populations of fiber polypeptides exist, the difference between them is probably caused by post-translational modifications of the polypeptide chain.

Fibers contain low amounts of arginine and are rich in hydroxyamino acids compared to hexons (PETTERSSON et al. 1968; DORSETT and GINSBERG 1975). ISHIBASHI and MAIZEL (1974a) discovered that the fiber is a glycoprotein containing two residues of *N*-acetyl glycosamine per polypeptide chain, and this has recently been confirmed (BOUDIN and BOULANGER 1982; CHENG CHEE-SHEUNG and GINSBERG 1982). A mutant of Ad5 (H5*ts*102) fails to glycosylate fibers at the nonpermissive temperature, but assembles fiber capsomers which obviously have a changed immunological specificity (CHENG CHEE-SHEUNG and GINSBERG 1982).

The nucleotide sequence (HERISSE et al. 1981) of the Ad2 fiber predicts a polypeptide of 581 amino acids with a molecular weight of 61.9K, in perfect agreement with the estimates by SDS-PAGE. The amino acid sequence contains three cysteine residues, but no disulphide bonds have been detected. Nine potential glycosylation sites can be recognized, but those glycosylated have not yet been identified.

Nearly all studies on adenovirus fibers have been carried out with fibers from subgenus C, and no data is yet available concerning the structure of fibers from other subgenera.

4.2.2 Immunology

The fiber contains one type-specific determinant γ which resides in the knob of the subunit, as revealed by electron microscopy (NORRBY et al. 1969a). Subgenus B species (Table 1) with short fibers contain only this determinant, but species in subgenera A, C, and D, with longer fibers, have in the rod-like portion another subgroup-specific determinant, δ. The longest fibers in subgenus C have in addition intersubgenus-specific determinants (NORRBY 1968, 1969b; PETTERSSON et al. 1968). The epitope can be demonstrated by hemagglutination inhibition and immunodiffusion (PEREIRA and DE FIGUEIREDO 1962; VALENTINE and PEREIRA 1965; NORRBY 1966b; PETTERSSON et al. 1968), and the antibodies prevent the attachment of virions to red cell (PEREIRA and DE FIGUEIREDO 1962) and KB and HeLa cell receptors (PHILIPSON et al. 1968). The determinant δ is probably located at the junction between the fiber and the penton base, since they are masked in the intact penton structure (PETTERSSON and HÖGLUND 1969; WADELL and NORRBY 1969b).

4.2.3 The Fiber Is the Attachment Unit

The fiber probably mediates the early contact between viruses and cells. Purified fiber in quantities 10- to 100-fold higher than the number of virus receptors (around 10^4 per cell) can inhibit virus attachment to the cell surface, and attachment can be prevented by antibodies against fiber (PHILIPSON et al. 1968). After penetration, intracellular virus appears to have lost the fiber together with other proteins in the vertex region (SUSSENBACH 1967; PHILIPSON et al. 1976).

Hemagglutination by adenoviruses was first demonstrated by ROSEN (1958). All human species can agglutinate erythrocytes (ROSEN 1960; ROSEN et al. 1962; BAUER and WIGAND 1963; SCHMIDT et al. 1965; NORRBY 1969b), and the species-specific determinant γ of the fiber interacts with the red cell surface (PEREIRA and FIGUEIREDO 1962; VALENTINE and PEREIRA 1965; NORRBY 1966b; PETTERSSON et al. 1968; NORRBY et al. 1969a). The receptors are present on the erythrocyte membrane, and an adenovirus receptor from monkey red cells has been solubilized (NEURATH et al. 1969). Since intact virions carry several fibers, they can establish a bridge between the erythrocytes and give a complete hemagglutination pattern. Aggregate or dimers of pentons and fibers can in the same way give rise to complete hemagglutination. In contrast, pentons or fibers alone can only establish a monovalent link with the cell, and agglutination is detected when antibodies are used to bridge the receptor-associated structural units (PEREIRA and FIGUEIREDO 1962; NORRBY 1966b; NORRBY 1969b). Heterospecific antibodies must be used to demonstrate hemagglutination by penton and fiber monomers, since the type-specific determinant of the fiber is responsible for the attachment to the erythrocyte receptor (PEREIRA and FIGUEIREDO 1962; NORRBY and SKAARET 1967). This method of revealing adenovirus hemagglutination by heterospecific antisera is called the *hemagglutination enhancement reaction*. The short fibers of subgenus B fail to hemagglutinate since they only contain a type-specific determinant (NORRBY and SKAARET 1967) and are therefore unable to form divalent aggregates after interaction with heterospecific antibodies. Antibodies directed against the γ specificity of the fiber can be assayed by *hemagglutination inhibition tests*. Based on their ability to agglutinate different red cells, ROSEN (1960) proposed a subgenus classification for human adenoviruses (see Table 1). Subgroup I viruses agglutinate monkey erythrocytes, whereas subgroup II viruses agglutinate rat erythrocytes. Subgroup III viruses also agglutinate rat erythrocytes, but with a partial pattern, in which a fraction of the erythrocytes sediment without being agglutinated and therefore form a ring in a less dense lattice. The partial agglutination is probably due to a competition between mono- and multivalent components for the receptors on the erythrocytes (WADELL 1969). Members of subgroup IV, which includes the highly oncogenic adenoviruses, were originally thought to lack hemagglutination capacity but were later shown to agglutinate rat erythrocytes with a partial pattern (SCHMIDT et al. 1965; NORRBY et al. 1969b). Further subdivisions have been suggested based on the types of erythrocytes agglutinated by different species within each subgenus (NORRBY 1968, 1969b).

No enzymatic effect has been associated with the fiber. Purified fibers from Ad1 and Ad2 exhibit an activity referred to as *erythrocyte receptor modifying effect* (ERM) leading to inactivation of the receptors on human erythrocytes (KASEL et al. 1961; KASEL and HUBER 1964). This effect appears to be caused by excessive binding of fibers to erythrocyte receptors rather than enzymatic modification of the receptor (WADELL 1969).

4.3 Penton

The pentons are composed of two structural units, the base and the fiber (Fig. 4). They are labile and difficult to purify in large quantities, but DEAE chromatography, either alone or in combination with hydroxylapatite or gel exclusion chromatography as well as electrophoresis, yields useful amounts of Ad2 and Ad5 pentons (VALENTINE and PEREIRA 1965; MAIZEL et al. 1968b; PETTERSSON and HÖGLUND 1969).

4.3.1 Biochemistry

Pentons from Ad2 and Ad5 sediment at around 11S (PETTERSSON and HÖGLUND 1969; WADELL et al. 1969). Widely different estimates of the molecular weight of pentons, from 250K (VALENTINE and PEREIRA 1965) through 400K (PETTERSSON and HÖGLUND 1969) to 485K–505K (WADELL 1970) have been reported. DEVAUX et al. (1982) arrived at a value of 365K for the complete penton structure using several different methods. The penton base is composed of identical polypeptide chains each having a molecular weight of 85K (ANDERSON et al. 1973; EVERITT et al. 1973; DEVAUX et al. 1982). Molecular weight determinations using neutron scattering have given a value of 246K for the base (DEVAUX et al. 1982), suggesting a trimeric structure. This is not predicted by the pentagonal symmetry at the vertices of the icosahedron, suggesting that other proteins are required for building the capsid. The fiber is joined to the penton base by noncovalent bonds which can be dissociated by treatment with guanidine hydrochloride (NORRBY and SKAARET 1967), formamide (NEURATH et al. 1968), pyridine (PETTERSSON and HÖGLUND 1969), or desoxycholate (BOUDIN et al. 1979). Penton bases dissociated from the fiber can be purified with retained antigenicity and other biological characteristics (PETTERSSON and HÖGLUND 1969).

Free vertex capsomers also exist in adenovirus-infected cells and can be purified directly from the soluble antigen pool (BOUDIN et al. 1979; WINTERS et al. 1970). Cells infected with *ts* mutants defective in fiber synthesis overproduce the penton base and provide a convenient source for purification of the base (BOUDIN et al. 1979). The free base sediments at 9.5S, and the polypeptide has a blocked N-terminal like other capsid components (BOUDIN et al. 1979). Penton bases and fibers assemble in vitro into pentons which are indistinguishable from those found in the native capsid (BOUDIN and BOULANGER 1982). The C-terminal end of the fiber is obviously involved in the attachment to the penton base (BOUDIN and BOULANGER 1982) and antibodies against determinant δ of the fiber (Table 3) cause dissociation of the penton, suggesting that antibodies can induce conformational changes in the fiber (BOUDIN and BOULANGER 1981).

The amino acid composition of pentons from Ad2 resembles that of hexons in overall composition, although there are significant differences with regard to tyrosine and the hydroxyamino acids. The penton bases are more sensitive to proteolytic enzymes than other adenovirus capsid proteins. At high enzyme concentrations they are destroyed, liberating free fibers, but the antigenic and morphological properties remain unchanged at low concentrations of trypsin (PETTERSSON and HÖGLUND 1969; WADELL and NORRBY 1969b).

4.3.2 Immunology

The penton base carries an antigenic determinant, known as β, which is common to pentons from all human adenovirus species (Table 3). In addition, inter- and intrasubgenus-specific determinants have been revealed (WADELL and NORRBY 1969b) and can be demonstrated in hemagglutination enhancement assays; immunodiffusion reveals primarily the intrasubgenus-specific antigen. Pentons from Ad12 carry fewer determinants on their surface than pentons from other serotypes (WADELL and NORRBY 1969b).

4.3.3 Biological Activity

Adenovirus infected cells show cytopathic changes involving a granular appearance of the cytoplasm, a rounded cell morphology, and detachment of the cells from the surface. These effects become apparent either 8–24 h after infection with high concentrations of crude virus or 7–20 days after infection during production of progeny (PEREIRA and KELLY 1957). Soluble proteins from infected cells, in particular the penton, were later identified as the cause of the early cytopathic effects (EVERETT and GINSBERG 1958; PEREIRA 1958; ROWE et al. 1958; VALENTINE and PEREIRA 1965). A penton concentration of 0.1 $\mu g/10^6$ cells is required to induce the effect, but after two decades the molecular mechanism is still unclear. The recently observed rapid entry of adenoviruses through endosomes during virus penetration may be a target for the penton-associated effect (FITZGERALD et al. 1983). The morphological changes are not dependent on macromolecular synthesis, and the effect is fully reversible after washing the cells (PEREIRA 1958; ROWE et al. 1958). Low concentrations of trypsin destroy the cytopathic effect without interfering with the antigenic properties of the penton base (PETTERSSON and HÖGLUND 1969; WADELL and NORRBY 1969b), and antipenton but not antifiber sera inhibit the effect (PETTERSSON and HÖGLUND 1969; WADELL and NORRBY 1969b), all suggesting that the penton base alone is responsible for the phenomenon. This is supported by the fact that isolated penton bases are able to induce the cytopathic changes (PETTERSSON and HÖGLUND 1969; WINTERS et al. 1970; BOUDIN et al. 1979).

An endonuclease activity has also been associated with pentons (BURLINGHAM et al. 1971). The endonuclease is detected in purified virions, but hexons and fibers lack activity. The enzyme is specific for DNA and cleaves double-stranded DNA 20 times faster than single-stranded DNA, especially in GC-rich regions (BURLINGHAM et al. 1971). A preference for single-stranded scissions has, however, also been reported (MARUSYK et al. 1975). The endonuclease can be disassociated from pentons by treatment with high concentrations of salt, suggesting that the enzyme is of cellular origin and remains tightly associated with pentons even after several purification steps (REIF et al. 1977a, b). The biological significance of this enzyme remains unknown. The activity has, however, been demonstrated in several independent studies (BURLINGHAM et al. 1971; MARUSYK et al. 1975; CAJEAN-FEROLDI et al. 1977).

5 The Core Proteins

The adenoviruses encode their own basic core proteins. Three polypeptides are associated with the cores, polypeptides V, VII, and the μ protein (MAIZEL et al. 1968a; LAVER et al. 1968; RUSSELL et al. 1971; EVERITT et al. 1973; ANDERSON et al. 1973; HOSAKAWA and SUNG 1976; VAYDA et al. 1983). The most abundant protein, VII, accounts for about 10% of the protein mass of the virion, corresponding to around 1100 copies in each particle. The weight ratio of VII to DNA is approximately 1:1 (CORDEN et al. 1976; SUNG et al. 1977) in both Ad2 and Ad3 cores (LAVER 1970; PRAGE and PETTERSSON 1971; EVERITT et al. 1973). The purification procedures include extraction at acidic pH in combination with PAGE. Separation of the core proteins can also be achieved by phosphocellulose and gel exclusion chromatography (SUNG et al. 1977). ANDERSON et al. (1973) discovered that protein VII with a molecular weight of 18.5K (MAIZEL 1971; EVERITT et al. 1973; ANDERSON et al. 1973) is synthesized as a precursor pVII of 20K, which is cleaved into the mature form during virion maturation.

The molecular weight of the *native* protein VII is also around 18K (PRAGE and PETTERSSON 1971), suggesting that the protein exists as a monomer, but it may also aggregate into tri- or tetramers (LEMAY and BOULANGER 1980) which may be required to form a matrix for a nucleosome-like structure in the core (NERMUT 1979; VAYDA et al. 1983). The gene for protein VII and its precursor has not yet been sequenced, nor is the complete amino acid sequence of the protein known. It contains, however, both tyrosine and tryptophan residues and has alanine at its N-terminal (PRAGE and PETTERSSON 1971; SUNG et al. 1977). The amino acid composition shows that the protein is basic, containing approximately 23% arginine. It resembles arginine-rich histones, especially H4 with its high arginine and alanine contents, but differs in containing tryptophan and low amounts of lysine. The arginine residues are probably evenly distributed along the polypeptide chain (LAVER 1970) except in the N-terminal portions where they are clustered (LISCHWE and SUNG 1977). Approximately 50% of the phosphate groups in the DNA can be neutralized by the basic residues present in protein VII (PRAGE and PETTERSSON 1971). The complex between DNA and protein VII is stable; treatment with 3 *M* NaCl, 1% sarcosyl, or 5% desoxycholate leaves it intact. The purified protein interacts with adenovirus DNA, forming a compact structure in which the DNA appears to be organized as in viral cores (BLACK and CENTER 1979). Protein VII, like other basic proteins, is weakly antigenic, and the proteins from Ad2 and Ad3 share immunological determinants, but antigenic differences are also apparent (Table 3) (PRAGE and PETTERSSON 1971).

Little is known about polypeptide V; it is moderately basic (Table 2), has an estimated molecular weight of 48.5K, and there are about 180 copies per virion. Unlike polypeptide VII, it seems to mature without proteolytic trimming and it is less tightly bound to the DNA than protein VII (BROWN et al. 1975; VAYDA et al. 1983).

HOSAKAWA and SUNG (1976) described a virion polypeptide, designated μ, associated with adenovirus cores. It has a molecular weight of 4–5K and is present in 125 copies per virion. It has an unusual amino acid composition, the basic residues accounting for 69% with 54% arginine, and in addition it contains as much as 13% histidine. This protein was recently shown to be part of native cores, but it could be extracted from the cores together with protein V with 0.5 *M* NaCl (VAYDA et al. 1983).

6 Other Virion Proteins

SDS-PAGE revealed a remarkable complexity of the polypeptides in the virion. In addition to the polypeptides of the well-known capsid components – hexons, pentons, fibers, and cores – several additional polypeptides were identified (MAIZEL et al. 1968a, b) and some of them have subsequently been purified and characterized.

6.1 Protein IIIa

There are approximately 60 copies of polypeptide IIIa in the virus particle (BOUDIN et al. 1980; DEVAUX et al. 1982). It is produced in excess during the infectious cycle and has been purified both from infected cells and from virions (LEMAY et al. 1980) by extraction in the presence of urea and applying QAE-Sephadex or hydroxylapatite chromatography, which yields preparations suitable for biochemical and immunological studies. Polypeptide IIIa, like most internal adenovirus capsid proteins, is proteolytically cleaved during virion maturation (BOUDIN et al. 1980) from a pIIIa form of 67K into the mature protein with a molecular weight of 66K. Polypeptide IIIa obviously plays an important role in maintaining the adenovirus capsid structure (Devaux et al. 1982) since it bridges the space between the trimeric penton and the peripentonal hexons.

Polypeptide IIIa sediments at 6.0S and the native molecule has an estimated molecular weight of 65.1K, in good agreement with the value of 66K determined by SDS-PAGE. The native protein thus probably exists as a monomer. The Stokes' radius was found to be 2.7 nm and the frictional ratio $f:f_0$ was around 1.0. Electrophoretic analysis suggests that it has an isoelectric point around 6.0 (Table 2), and the amino acid composition reveals that it, like hexons and fibers, is rich in dicarboxylic amino acids (Lemay et al. 1980). The contents of serine and glycine are high but the basic amino acids are underrepresented. The high isoelectric point suggests that most of the acidic amino acid residues are amidified. The mature protein IIIa appears to have glycine as the N-terminal amino acid, whereas the N-terminal of the precursor is blocked by acetylation (Lemay et al. 1980).

Polypeptide IIIa is a phosphoprotein, and a protein kinase activity which is associated with adenovirus particles appears to phosphorylate this polypeptide preferentially (Akusjärvi et al. 1978; Blair and Russell 1978).

Antisera prepared against purified polypeptide IIIa precipitate similar polypeptides from cells infected with Ad2, Ad3, or Ad5, and no species-specific determinants have been detected. Although it was suggested that antibodies against polypeptide IIIa may have neutralizing activity (Everitt and Philipson 1974), more recent data (Boudin et al. 1980) establish that antibodies against protein IIIa have no role in adenovirus neutralization.

6.2 Protein VI

Other adenovirus capsid components – VI, VIII, IX and X–XII – have received less attention than the major capsid components because they are only present in small amounts in both infected cells and virions. A polypeptide with a molecular weight of 24K was designated polypeptide VI due to its electrophoretic mobility relative to the other virion proteins (Maizel et al. 1968a; Everitt et al. 1973; Anderson et al. 1973). There are about 450 molecules of polypeptide VI per virion and it is synthesized as a precursor pVI with a molecular weight of 27K (Öberg et al. 1975; Edvardsson et al. 1976) which is cleaved to VI during virion maturation. The mature polypeptide VI can be extracted from virions by low concentrations of urea at an acidic pH and the protein has been purified by a multistep procedure involving gel and ion exchange chromatography. It exists as a dimer or a trimer with a molecular weight of 50K–72K (Everitt and Philipson 1974; Lemay and Boulanger 1980). Polypeptide VI was claimed to be located on the inside of the capsid in close connection with the hexons (Everitt et al. 1975) but recently it has been shown to act as a DNA-binding protein (Russell and Precious 1982). The latter observation points towards a relationship with the core and not with the outer capsid.

The nucleotide sequence of the gene for polypeptide pVI has been determined, so the amino acid sequence of polypeptides pVI and VI can be predicted. The precursor consists of 249 amino acids yielding a molecular weight of 27.0K (Akusjärvi and Persson 1981). During maturation, 33 amino acids are removed from the N-terminal end of the precursor between amino acid residues 33 (glycine) and 34 (alanine). The C-terminal end of polypeptide pVI resembles a transmembrane protein with a stretch of uncharged, predominantly hydrophobic residues followed by four basic amino acids.

6.3 Protein VIII

Polypeptide VIII has a molecular weight of 13K by SDS-PAGE (Maizel et al. 1968a; Andersson et al. 1973; Everitt et al. 1973). It is synthesized as a large precursor with a molecular weight of 26K (Öberg et al. 1975). The proteolytic cleavage during virion assembly eliminates half of the precursor. Polypeptide VIII has been extracted from virions with urea in high salt and purified by preparative PAGE and gel exclusion chromatography (Everitt and Philipson 1974). It probably exists as a monomer but little is known about its biochemical properties. It is associated with hexons, although less firmly than polypeptide IX, and the hexon nonamers lack this polypeptide (Everitt and Philipson 1974).

The nucleotide sequence of the gene for pVIII (Galibert et al. 1979; Herisse et al. 1980) predicts that it consists of 227 amino acids, with a molecular weight of 24.7K. The polypeptide is unusually rich in proline and arginine, but cysteine residues are absent and the contents of glycine and serine are remarkably high. It is not known where in the sequence the cleavage takes place to generate the mature polypeptide, but cleavage leads to a drastic change in the shape of the precursor (Lemay and Boulanger 1980).

6.4 Protein IX

Polypeptide IX is a constituent of the adenovirus particle (Maizel et al. 1968a; Everitt et al. 1973; Andersson et al. 1973). Unlike several other viral polypeptides, it does not mature through proteolytic trimming. The native protein has been purified from virus particles or hexon nonamers by urea extraction at low pH followed by ion exchange chromatography on QAE-Sephadex (Everitt and Philipson 1974; Boulanger et al. 1979). Large amounts of polypeptide IX are produced late after infection in vast excess of the amounts required for forming the virions. The gene has been sequenced from several serotypes and the complete amino acid sequence has been predicted from the Ad2 (Aleström et al. 1980; Andersson and

Lewis 1980), Ad5 (Maat et al. 1980), Ad3 (Engler 1981), Ad7 (Dijkema et al. 1981), and Ad12 (Bos et al. 1981) genomes. Polypeptide IX from Ad2 comprises 139 amino acids corresponding to a molecular weight of 14.3K. The N-terminal amino acid is acetylated, and the amino acid composition reveals high contents of serine, alanine, and leucine. The serine and alanine residues are clustered. There is only a single tryptophane, but no histidine or cysteine residues. Polypeptide IX can be divided into C-terminal and N-terminal domains which are joined by an alanine-rich spacer (Engler 1981). Species within the same subgenus possess molecules with similar sequences. Homologies are also found in the N-terminal domain between subgenera (Engler 1981). Antibodies reveal cross-reaction between species, but species-specific epitopes are also present (Boulanger et al 1979).

The gene for polypeptide IX differs from other structural genes in being expressed at intermediate as well as late times after infection, even before the onset of viral DNA replication (Persson et al. 1978). The gene for polypeptide IX is, furthermore, located in a separate transcription unit and the mRNA does not require splicing for its maturation (Aleström et al. 1980), all suggesting that polypeptide IX has a biological function different from its structural role. The observation of Colby and Shenk (1981) that adenovirus mutants which lack this gene are still viable speaks against this hypothesis.

6.5 Proteins X–XII

Maizel et al. (1968a) reported in their original study of adenovirus polypeptides that the virus contains a polypeptide, designated X, which migrates with the dye front on SDS-PAGE. Andersson et al. (1973) later resolved this component into three bands which range in molecular weight between 5K and 6.5K. Nothing is known about their function. One component may be associated with the viral core, the μ protein, but it has not been shown that they are unique translation products. They may also represent degradation products after proteolytic cleavage of other viral proteins. It is of high priority to determine if proteins X–XII are degradation products or encoded by unique viral mRNAs.

7 Protein Neighbor Analysis

The precise topography of the proteins in adenovirus particles, i.e., the identification of the protein neighbors in the virion, is of importance for understanding the virion structure and also for clarifying the assembly pathway. The internal relationship between the various polypeptides in the virion has been studied by methods devised to liberate morphological substructures and analyze them by SDS-polyacrylamide gel electrophoresis before or after cross-linkage of the proteins or the nucleoproteins. In vitro labeling and antibody accessibility has also been used to localize the proteins.

Treatment of purified virions with heat (RUSSELL et al. 1967) or dialysis against low ionic strength buffers at a pH between 6.0 and 6.8 (PRAGE et al. 1968, 1970; LAVER et al. 1969) removes the vertex regions, comprising the fiber, the penton base together with the 60 peripentonal hexons surrounding the 12 pentons at the vertices, and in addition the protein IIIa (Fig. 4) (EVERITT et al. 1973). Capsids devoid of the vertex region are stable (RUSSEL et al. 1967) and obviously held together by intermolecular bonds between the hexons. The perforated capsids still contain virion DNA, which is now accessible to deoxyribonuclease, in contrast to DNA in intact virions (PRAGE et al. 1970). Treatment of purified virions with pyridine (PRAGE et al. 1970) or desoxycholate and heat (RUSSELL et al. 1971) disintegrates the virion even further. The pentons and the peripentonal hexons are released as monomers and the main body of the capsid is also disrupted, so that the majority of the hexons are liberated in nonamers. Monomeric units of all hexons may be obtained by disrupting the particles with one of several different agents, such as SDS (SMITH et al. 1965), formamide (STANSY et al. 1968), or 8 *M* urea (SHORTRIDGE and BIDDLE 1970). These treatments will not dissociate the trimeric units into single polypeptide chains.

Hexons purified from infected cells contain only polypeptide II, whereas hexons obtained by dissociating virions with pyridine are bound to polypeptide VI (EVERITT et al. 1973). For each mole of hexon protein, two moles of polypeptide VI was recovered, suggesting that a dimer of polypeptide VI is associated with each hexon trimer. Polypeptide VI cannot, however, be labelled from the outside by iodination, suggesting that it is located on the inside of the hexon shell (EVERITT et al. 1975). Polypeptide IIIa is probably located in the peripentonal region since it is associated with peripentonal hexons in degraded virus preparations. In cross-linking experiments, polypeptide IIIa is linked to peripentonal hexons, the penton base, and the internal core polypeptide V. It therefore probably extends from the surface of the virion to its interior (EVERITT et al. 1975).

The location of polypeptide VIII in the virion is still unresolved, even though this protein is associated with hexons after disruption of viral particles by freezing and thawing (EVERITT et al. 1973). It has been assigned a position inside the hexon shell (Fig. 4) because of its basic isoelectric point.

Polypeptide IX can be recovered in the nonameric hexon units released with pyridine or desoxycholate. Around 15 copies of polypeptide IX have been recovered per unit, and from considerations of symmetry it has been suggested that it is located at the corner to edge contacts between the hexons in the nonamers (BOULANGER et al. 1979; BURNETT, personal communication). Polypeptide IX is not labelled from the outside by iodination, but its antigenic determinants are exposed on the exterior of the virions. It has therefore been suggested that this polypeptide plays a cementing role in the capsid structure (EVERITT et al. 1975). Recently it was shown with a deletion mutant of Ad5 (*dl*313) lacking the polypeptide IX gene that this protein is not essential for assembly of Ad5 virions (COLBY and SHENK 1981). The mutant virions, on the other hand, appear to be considerably more heat labile than wild-type particles. Mutant virus also fails to produce hexon nonamers upon disruption with pyridine. Thus polypeptide IX might not be essential, but provides a stabilizing function in the outer capsid.

After disruption of viruses with the ionic detergent Sarcosyl, the cores contain eight to ten discernable spherical subunits with two to four units masked by the negative stain in the electron microscope. It has therefore been suggested that each core might consist of 12 subunits, one subunit per vertex (BROWN et al. 1975). Only polypeptide VII is recovered from Sarcosyl-generated cores, but after pyridine or desoxycholate disintegration both polypeptides V and VII are associated with the cores (BROWN et al. 1975; PRAGE et al. 1970; RUSSELL et al. 1971). Cross-linking studies have shown that polypeptide V is in contact with the penton base and also the polypeptide VII–DNA complex, suggesting that polypeptide V is associated with each vertex forming a link between the outer capsid and the core (EVERITT et al. 1975; NERMUT 1979).

More detailed analysis of the adenovirus core has revealed that only polypeptide VII can be cross-linked to the DNA by ultraviolet irradiation (SATO and HOSOKAWA 1981), and it has been suggested that protein VII forms a helical core and the DNA is wound around it with a compaction factor of 5–6 and a protected fragment of DNA of around 100–150 bp per three to four units of protein VII. The virion protein V may form a thin pelliculum around the core and has therefore been called a core shell (NERMUT 1979, 1980).

The locations of the small polypeptides X, XI, and XII have not yet been determined, as mentioned above. The minor components, proteins IVa_1 and IVa_2, appear to be necessary for the assembly process, the trace amounts identified in purified virus may derive from an incompletely assembled fraction of the particles.

8 Structural Organization of the DNA

The adenovirus genome is a linear molecule of double-stranded DNA without single-stranded breaks (VAN DER EB and VAN KESTEREN 1966; GREEN et al. 1967; VAN DER EB et al. 1969). The genomes of the oncogenic species Ad12, Ad18, and Ad31 appear to be slightly smaller than those of the nononcogenic species Ad1, Ad2, and Ad5 (GREEN et al. 1967).

The DNA of Ad2, Ad5, Ad12, and CELO virus is not circularly permuted, since it yields unique denaturation patterns in the electron microscope (DOERFLER and KLEINSCHMIDT 1970; DOERFLER et al. 1972; YOUNGHUSBAND and BELLETT 1971; ELLENS et al. 1974). It also lacks sticky ends and does not have the terminal repetitions present in several bacteriophage DNAs, since when native DNA is annealed, with or without prior digestion with exonuclease III, cyclic structures are not detected (GREEN et al. 1967; YOUNGHUSBAND and BELLETT 1971; MURRAY and GREEN 1973). Instead, adenovirus DNA contains an inverted terminal repetition which results in full-length single-stranded circles after denaturation and renaturation of intact DNA (WOLFSON and DRESSLER 1972; GARON et al. 1972, 1975). Treatment of these circles with exonuclease III converts them to linear molecules, suggesting that a hydrogen-bonded panhandle is formed between the two single-stranded ends of viral DNA. It was originally thought that the inverted repetition extended over 350–1400 nucleotide pairs (GARON et al. 1972; WOLFSON and DRESSLER 1972), but after cleavage of DNA specifically labeled at the 3′ ends with different restriction endonucleases, the length of the inverted terminal repetition was found to be only 100–140 nucleotides (ARRAND and ROBERTS 1979). Direct sequencing of the ends of adenovirus DNA has established that the repetition only extends over 102 and 103 nucleotides for the DNAs of Ad2 and Ad5 respectively (STEENBERGH et al. 1977; ARRAND and ROBERTS 1979). This unusual structure has only been found in adenovirus DNA and in adenovirus-associated virus DNA; the latter is dependent on a helper adenovirus for its replication (KOZCOT et al. 1973; BERNS and KELLY 1974).

Regions of AT-rich sequences have been revealed electron microscopically in Ad2 DNA at positions 0–0.15, 0.5–0.6 and 0.8–1.0 (DOERFLER and KLEINSCHMIDT 1970). The latter two regions are located in the genes for the hexon and the fiber proteins which both contain species specific sequences. The intact strands of adenovirus DNA can be separated by equilibrium sedimentation in CsCl gradients in the presence of poly(UG) (KUBINSKI and ROSE 1967; LANDGRAF-LEURS and GREEN 1971; PATCH et al. 1972). Separation depends on differential affinity of the two strands for the ribopolymer; the strand that binds more copolymer has a higher density than the opposite strand and is therefore designated the heavy strand. The strands of Ad5 DNA can also be separated in alkaline CsCl gradients (SUSSENBACH et al. 1973). The strand with higher density under these conditions corresponds to the light strand in poly(UG) CsCl gradients (TIBBETTS et al. 1974; SHARP et al. 1975). To avoid confusion, the nomenclature of SHARP et al. (1975) has now been adopted; the strands are named r-strand (right) and l-strand (left) according to the direction in which they are transcribed.

Adenovirus DNA extracted from purified virions by treatment with proteolytic enzyme is a linear duplex molecule. When the DNA is extracted with guanidine hydrochloride and chloroform followed by dialysis, a fast-sedimenting fraction containing relaxed circles is observed (ROBINSON et al. 1973). The circles linearize when treated with a protease, suggesting that they are held together by a protein. The protein joining the two ends constitutes less than 0.4% of the total protein of the virion (ROBINSON and BELLETT 1975; KEEGSTRA et al. 1977). Within the virion this protein has a molecular weight of around 55K and it is covalently linked to each end of viral DNA (REKOSH et al. 1977). This terminal protein is encoded in the viral genome in the E2B region of the l-strand (map position 19.8–23.5). The precursor protein with a molecular weight of 80K–87K is cleaved during virus maturation by the protease encoded in the L3 region of the late transcription unit (CHALLBERG et al. 1980; STILLMAN et al. 1981) and is obviously covalently linked to the 5′-terminal cytidine residue of the adenovirus DNA via an *O*-serine phosphoester linkage (CARUSI 1977; KELLY and LECHNER 1979). The terminal protein is necessary for DNA replication both in vivo and in vitro (CHALLBERG and KELLY 1979), first forming a covalent bond with a cytidine nucleotide and then serving as a primer for extension of the entire DNA strand, and this supports the earlier findings that adenovirus DNA replication is initiated at both ends of the viral DNA molecule (SUSSENBACH and KUIJK 1977). Both viral and cellular products are required for viral DNA replication, which now can be reproduced in a cell-free system (LICHY et al. 1982; NAGATA et al. 1982).

Adenovirus DNA is infectious when assayed by the calcium phosphate precipitation technique (NICOLSON and MCALLISTER 1972; GRAHAM and VAN DER EB 1973). The specific infectivity was initially low, 10–20 pfu/μg DNA, but has improved, and around 10^3 pfu/μg DNA is now regularly obtained. The essential role of the terminal protein was also established when measuring the infectivity of viral DNA, since the specific infectivity of adenovirus DNA is enhanced 10- to 100-fold using a DNA protein complex compared with naked DNA (SHARP et al. 1976). Since in vitro replication can also initiate on viral cores (GODING and RUSSELL 1983), a nucleoprotein particle may be required to obtain higher specific infectivities. The specific infectivity of the virus particle is around 10^9 pfu/μg DNA.

9 Assembly of Capsomers

During the late phase of adenovirus productive infection, host cell protein synthesis is shut off and viral mRNA is preferentially translated. Most of the viral polypeptides are rapidly released from polyribosomes and transported to the nucleus with a lag period of only 3–6 min (HORWITZ et al. 1969; VELICER and GINSBERG 1970). During this short interval the monomeric structural polypeptides of the hexon, the penton base, and the fiber assemble into capsid units (VELICER and GINSBERG 1970). The first step in adenovirus assembly is therefore the formation of the major structural units of the capsid from monomeric polypeptide chains. Where this occurs in the cell is not yet clear but it probably takes place at or shortly after translation, since the monomeric polypeptide for all the three major capsid units is insoluble, but the multimeric form is freely soluble. The penton base and the fiber are then combined into the penton, which may be regarded as an intermediary step in capsomer assembly. Penton assembly occurs rapidly for around 25% of newly synthesized polypeptides, but it takes up to 24 h before this process is completed (HORWITZ et al. 1969); it may therefore be a partially reversible process.

9.1 Hexon

The product formed by in vitro translation of hexon mRNA in heterologous systems can only be detected with antibodies against denaturated hexon which recognize the nascent chains in infected cells (STINSKI and GINSBERG 1974; ÖBERG et al. 1975). When polysomes from infected cells are used for in vitro translation, however, the trimer is formed (PERSSON et al. 1977), suggesting that assembly of the hexon requires another viral product.

The first genetic evidence that assembly of hexons may require one additional gene function was provided by LEIBOWITZ and HORWITZ (1975) using two *ts* mutants, both from a different complementation group than the hexon gene. Hexon polypeptides synthesized in cells infected with H5*ts*17 and H5*ts*20 at the nonpermissive temperature could assemble after shift-down if the cells were kept for 15 min but not if they were kept for 2 h at the nonpermissive temperature. Since the size of the hexons made in cells infected by H5*ts*17, H5*ts*20, and wild-type Ad2 were the same, the authors suggested that a gene unrelated to the hexon gene is required for formation of the trimeric unit. It was later established by marker rescue experiments that these two *ts* mutants are located in the nonstructural 100K protein gene (FROST and WILLIAMS 1978). These results were recently confirmed with two other *ts* mutants located in the 100K protein gene, H5*ts*115 and H5*ts*116 (OOSTEROM-DRAGON and GINSBERG 1981). A direct association between the 100K protein and nascent hexon polypeptides was also demonstrated, using antiserum or monoclonal antibodies against the 100K protein or against denatured hexons (OOSTEROM-DRAGON and GINSBERG 1981; CEPKO and SHARP 1982). It is therefore tempting to suggest that the 100K protein, which is almost as abundant as the hexon, is required in stoichiometric quantity in order to polymerize the hexon polypeptide into the capsid unit with its complicated structure of species-specific determinants on the outside surface and internal genus-specific determinants (WILCOX and GINSBERG 1963b). The assembly may therefore require the 100K protein as a scaffolding protein. Once the hexon is formed, the 100K protein obviously dissociates from the trimeric unit. The assembly may involve the reactive serine within the nascent hexon polypeptide that can interact with diisopropylfluorophosphate (DFP), which only binds to active serines in proteases (DEVAUX and BOULANGER 1980). DFP obviously interacts with the denatured polypeptide in vitro and not with the trimeric unit. The assembly of the hexon thus requires the 100K protein, but the molecular detail of this process is still unclear. The efficiency of hexon formation is, however, remarkable. The hexon accumulates in the nuclei within 5 min of completion of the chain in the polysomes, and as much as 10% of the total protein content of the cell is hexon protein at the end of the infectious cycle.

9.2 Penton

The genes for the penton base and the fiber polypeptides are more than 10 kb apart on the viral genome, but both are located within the major late transcription unit in the L2 and L5 regions respectively (MILLER et al. 1980). Several *ts* mutants have been isolated which map in the fiber region of the genome, but no *ts* mutants have yet been found in the penton base region, with the possible exception of H2*ts*3 (D'HALLUIN et al. 1982).

The penton base and the fiber are separately synthesized on polyribosomes and within minutes the two subunits are accumulating in the nucleus of the infected cell (VELICER and GINSBERG 1970). Pulse-chase experiments suggest that the penton base is first synthesized and then assembled with the fiber to form complete pentons (HORWITZ et al. 1969; VELICER and GINSBERG 1970). During productive infection with wild-type virus an excess of fiber units is formed, and the formation of penton base is probably rate limiting in the assembly process, since only minute amounts of free base can be found (BOUDIN et al. 1979). Penton base accumulates, however, in cells infected with fiber defective *ts* mutants at the nonpermissive temperature (BOUDIN et al. 1979). An in vitro assembly system for pentons has recently been described (BOUDIN and BOULANGER 1982), using purified fiber from Ad2 wild-type infection and penton base accumulated in cells infected with a fiber defective *ts* mutant H2*ts*125. As revealed by two-dimensional immunoelectrophoresis assembly occurs over a broad pH range (5.5–9.0) and at ionic strength from 0.05 to 1.0. The association between the penton base and the fiber is a reversible reaction with a dissociation constant of 2×10^{-7} *M* determined on the basis of fiber molarity. The fiber appears to combine with the penton base via its C-terminal and the last 20 amino acids are obviously necessary for assembly. Recent structural studies (GREEN et al. 1983) suggest, however, that this region resides in the N-terminal end of the polypeptide. The formation of chimeric pentons containing units from different species and the formation of ts^+ recombinants in vivo establish that a genus specific area of the fiber is recognized by the penton base. The antigenicity of the fiber moiety is also changed during complex formation, possibly reflecting a three-dimensional structural change. Thus the penton capsomer can self-assemble, and no additional proteins appear to be necessary for formation of this capsid unit.

10 Formation of Empty Capsids

Adenovirus assembly was first analyzed by kinetic labeling (HORWITZ et al. 1969; SUNDQUIST et al. 1973b; ISHIBASHI and MAIZEL 1974b), but the large pools of structural proteins of which only a small fraction is incorporated in virus particles hampered pulse-chase kinetics experiments. In fact only 20% of the hexon trimers are incorporated into virions and only 10% of viral DNA enters virus particles (PHILIPSON and LINDBERG 1974). The excess pools in infected cells are therefore enormous, as around 100000 virus particles are produced per infected HeLa cell. In fact viral DNA and proteins must accumulate in amounts comparable to the total DNA and protein content of the host cell.

By instead following the labeling of single polypeptides (HORWITZ et al. 1969; SUNDQUIST et al. 1973b; ISHIBASHI and MAIZEL 1974b), it became clear that core proteins V and VII appear in intact virions within 2–5 min but hexons appear only 30–60 min later, suggesting a differential rate of insertion of the different proteins into virions.

An explanation was offered when the different subviral structures were studied in more detail. From 13 h after infection, cells infected by Ad2 or Ad3 contain both virions and empty capsids that can be readily separated from each other by CsCl gradients. These particles increase in concentration in parallel, and both are synthesized at maximum rates 24 h after infection. Analysis of the appearance of hexon in empty capsid and mature virus showed that the hexon trimers appeared without a lag in empty capsids, but 60–80 min elapsed before they entered mature virions (SUNDQUIST et al. 1973b).

Pulse-chase experiments with cells infected by Ad2 with a small pool of empty capsids suggest in fact a precursor-product relationship between empty and mature particles (SUNDQUIST et al. 1973b; ISHIBASHI and MAIZEL 1974b). The radioactivity in core proteins appears within minutes in complete virions (SUNDQUIST et al. 1973b), favoring the idea that mature virions are formed by the insertion of viral DNA and core proteins into preassembled empty capsids.

Analysis of the protein composition of empty capsids compared to full virions revealed that the former contain no core protein V or VII (MAIZEL et al. 1968b; PRAGE et al. 1972; ISHIBASHI and MAIZEL 1974b), and proteins VI, VIII, and X–XII are also missing. However, several other polypeptides not present in mature virions were identified in the low-density particles (PRAGE et al. 1972; SUNDQUIST et al. 1973b; WADELL et al. 1973; ISHIBASHI and MAIZEL 1974b; ROSENWIRTH et al. 1974; WINBERG and WADELL 1977; EDVARDSSON et al. 1976). Two of these correspond to the precursor polypeptides pVII and pVIII (EDVARDSSON et al. 1976), but others are only found in empty capsids and may be scaffolding proteins like the IVa_2, the 40K, and the 32K–33K proteins (EDVARDSSON et al. 1976; D'HALLUIN et al. 1978a). The latter are rapidly incorporated into empty capsids but do not persist in mature virions.

Combining the results from studies on kinetics of labeling and protein composition, it appears that when empty capsids are assembled they contain precursor polypeptides and possibly scaffolding proteins; viral DNA and core proteins are then inserted either together or separately, and the final step in morphogenesis appears to involve proteolytic cleavage of the precursor polypeptides in the virion.

We do not, however, understand in detail the mechanism of assembly of empty capsids. No in vitro systems are available, but a number of attempts to generate capsid-like structures with substructures of the virion have been successful. Hexons purified from infected cells do not all have the same charge (PETTERSSON 1971; BOULANGER 1975). Two classes can be separated by electrophoresis or ion exchange chromatography and they may be used to assemble nonamers of hexon associated with the faces of the icosahedron. The nonamers of hexon purified from virions disrupted with pyridine or sodium desoxycholate and heat can reaggregate at low pH. They can form dimers, rings of five, and in some cases icosahedral shells containing 20 nonamers similar to the perforated structures obtained after removal of the peripentonal regions (PEREIRA and WRIGLEY 1974). These results suggest that the nonamers of hexon might self-assemble to form a cage-like capsid. The role in this process of other hexon-associated proteins like VI and IX is unclear. An early step in virus assembly may therefore involve the formation of an icosahedral lattice through the interaction of nonamers of hexons, perhaps by a self-assembly reaction. Proteins pVI and pVIII might be necessary in vivo together with the recently identified presumptive scaffolding proteins (EDVARDSSON et al. 1976; D'HALLUIN et al. 1978a). Since the penton units and the IIIa protein are also associated with the empty capsids, the process in vivo involves several additional steps which cannot be reconstructed in vitro. It therefore seems unlikely that formation of empty capsids can be explained simply by a self-assembly mechanism. Formation of an empty capsid appears, however, to be a prerequisite for the insertion of the DNA in the virion.

11 Insertion of Viral DNA

Before discussing the mechanism of insertion of DNA into empty capsids, we must examine the properties of the defective and incomplete virus particles which contain limiting amounts of viral DNA. They are often isolated after sonication of cells by virtue of their buoyant density in CsCl gradients, which is intermediate between infectious virions and empty capsids. The infectivity associated with particles with small amounts of DNA is 3–4 orders of magnitude less than that of virions (PRAGE et al. 1972; WADELL et al. 1973; BURLINGHAM et al. 1974). These particles do not interfere with infection by virions or alter the yield of progeny (PRAGE et al. 1972; DANIELL 1976), and they have the same potential as intact virus to transform cells in culture (SCHALLER and YOHN 1974; IGARISHI et al. 1975), suggesting that the left-hand end of the adenovirus genome which is necessary for transformation (GRAHAM et al. 1974; GALLIMORE et al. 1974) is overrepresented.

A spectrum of defective particles ranging in density from empty capsids to intact virions has been observed among all adenoviruses, but viruses from subgenus B appear to accumulate empty capsids and low-density defective particles in larger amounts than those from subgenera A or C. The protein composition of the defective particles always mimics that of empty capsids, and instead of fully processed virion polypeptides, precursor and scaffolding polypeptides are encountered (SUNDQUIST et al. 1973a; ISHIBASHI and MAIZEL 1974b; EDVARDSSON et al. 1976).

Another class of defective particles accumulates in human Ad12 and bovine Ad3 injection after high multiplicity infection (MAK 1971; NIIYAMA et al. 1975; IGARISHI et al. 1975). They contain DNA with specific deletions and the protein composition mimics that of mature virions (MAK et al. 1979; NIIYAMA et al. 1975; VAN ROY et al. 1979).

Conflicting reports have appeared on whether only viral DNA is associated with defective particles. It has been claimed that they also contain cellular DNA (TIJA et al. 1977; HAMMARSKJÖLD et al. 1977; KHITTOO and WEBER 1981), but precautions to remove unspecifically adsorbed DNA by DNase digestion prior to purification of the particles have not always been taken. Cellular DNA has in one report been claimed to be covalently linked to adenovirus DNA as revealed by a two-step hybridization procedure (TIJA et al. 1977), but the functional significance of this observation is unclear. Since cellular DNA readily and unspecificially associates with capsids in vitro (TIBBETTS and GIAM 1979), caution is necessary in interpreting these results. Aberrant forms of viral DNA have in addition been detected in defective particles from subgenus B viruses, mainly molecules with long panhandles (DANIELL 1976). They are probably generated by defective DNA replication.

Since only about 10% of the total viral DNA is packaged into virions, recognition for packaging must be a discriminatory event. Quantitative analysis suggests that 95% of defective particles contain adenovirus DNA and only around 5% or less of the particles contain aberrant forms of viral DNA (TIBBETTS 1977; BROWN and WEBER 1982).

In human subgenus B viruses, each empty capsid interacts with only one viral DNA molecule, starting insertion at the molecular left-hand end of the conventional viral genomic map. Accordingly, different density classes of Ad7 incomplete particles were shown to contain a linear duplex sequence from the left-hand end of the Ad7 genome with a variable right terminal (TIBBETTS 1977). The control for this experiment was provided by DANIELL and MULLENBACH (1978), who demonstrated that both ends of the genome were equally represented in the pool of subgenomic DNA molecules in the cells.

This places the recognition signal for packaging of DNA in the left-hand end of the adenovirus genome. In an elegant study utilizing plaque-purified (PP) and multiple-passage (MP) strains of Ad16, HAMMARSKJÖLD and WINBERG (1980) showed that the PP virus inserted its DNA with preference for the left-hand end, but the MP strain containing a reduplication of the left-hand end at the right-hand end of the genome could insert either end in the capsid. A reduplication of only 390 bp was sufficient to allow packaging of DNA from the right-hand end. A variant of the MP virus containing a reduplication which was only 100 bp shorter was not encapsidated from the opposite end. The conclusion appears obvious; there is a region between 290 and 390 nucleotides from the left-hand end of the viral genome which is essential for the packaging of viral DNA. These results explain the failure to encapsidate cellular DNA into adenovirus empty capsids even if cellular DNA readily associates with empty capsids (TIBBETTS and GIAM 1979).

Placing the packaging signal internal to the inverted terminal repeat of 100 nucleotides explains how the left-hand end can be preferred. The pool of replicating DNA in infected cells contain both ends in equal concentration, confirming that the recognition sequence, which may span nucleotides 100–390, is really a packaging signal. This signal acts only in *cis*, suggesting that it does not code for a protein (HAMMARSKJÖLD and WINBERG 1980). It mimics in fact the recognition site in the lambda phage DNA where a specific sequence close to the left-hand end but outside the cohesive end sequence mediates the binding to proteins which are required to insert lambda DNA in the head structure (FEISS et al. 1979). In fact the cosmids used as cloning vectors in conjunction with the lambda phage in vitro packaging system also carry this piece of lambda DNA (COLLINS and HOHN 1978). This specific signal in adenovirus DNA must be included in adenovirus vectors designed for recombinant DNA in mammalian cells, and cloning of this region may help to identify the protein(s) specifically interacting with it.

12 Proteolytic Cleavage

Studies with one *ts* mutant, H2*ts*1, established a requirement for proteolytic cleavage in virion maturation (WEBER 1976). Cells infected with H2*ts*1 accumulate virions at the nonpermissive temperature, but the precursor polypeptides pVI, pVII, and pVIII are not processed. This mutation therefore blocks the appearance of the mature polypeptides VI, VII, and VIII but polypeptides X, XI, and XII, which normally accumulate in mature virions, are also missing. The proteolytic cleavages are inhibited by temperature shift-up, but occur after temperature shift-down. The mutant particles are not infectious, although they contain a full complement of DNA and the DNA retains infectivity. This *ts* mutant explained the separate step in the maturation pathway of the so-called young virions (ISHIBASHI and MAIZEL 1974b), which when exposed to an internal endopeptidase convert into infectious mature virions. The gene affected in H2*ts*1 has been identified by sequence analysis (KRUIJER et al. 1980; AKUSJÄRVI et al. 1981) following its localization by restriction analysis of interspecies ts^+ recombinants (HASSEL and WEBER 1978). It is located in the L3 region of the late transcription unit around coordinate 60 and beyond the coding sequence for the hexon. The endopeptidase is around 23K in molecular weight but the scarcity of the mRNA and the product has precluded its purification from infected cell. The enzymatic activity has, however, been detected in nuclei from infected cell and inside mature virions but not in empty capsids (BHATTI and WEBER 1979). The mRNA for the endopeptidase is present at a concentration 1000-fold lower than the hexon mRNA, which contains in a nontranslatable form at its 3′ terminal the encoded sequences for the endopeptidase. The Ad2 *ts*103 mutant may also be a protease mutant (D′HALLUIN et al. 1980a, 1982).

Recently it has been shown that several other virion proteins are processed during the final step in maturation. The terminal protein covalently linked to the DNA appears in vivo as an 80K species which during maturation of the virion is cleaved into the 55K species of infectious virions (CHALLBERG et al. 1980). In fact the in vitro translation product of this protein, derived from the E2B region, appears to be 87K, and in this form it accumulates in H2*ts*1 virions at the nonpermissive temperature (LEWIS and MATHEWS 1980; STILLMAN et al. 1981).

Polypeptide IIIa appears also to undergo proteolytic cleavage from 67K to 66K during the final steps of maturation of the virion (BOUDIN et al. 1980).

Directly from the DNA sequences it can be deduced that the recognition signal for the viral endopeptidase in the case of pVI involves a *gly-ala* bond (AKUSJÄRVI and PERSSON 1981) and the same peptide bond is probably cleaved in pVII and pVIII (TREMBLAY et al. 1983). However, biochemical evidence is still lacking for the latter two cleavage sites, as well as for those of pIIIa and the terminal protein.

13 Assembly Arrest

All *ts* mutants affecting proteins in the L2, L3, and L4 regions of the late transcription unit, with the one exception of the H2*ts*1 mutated in the protease (L3 region), fail to accumulate empty capsids and assembly intermediates (EDVARDSSON et al. 1978). Some mutants, e.g., H2*ts*3, which are defective in assembling the capsomers, do, however, process some precursor proteins (WEBER et al. 1977). On the other hand, *ts* mutants in proteins from the L1 and L5 regions form empty capsids but do not process precursors (KHITTOO and WEBER 1977; EDVARDSSON et al. 1978). The L1 and L5 regions encode the structural proteins IIIa and the fiber (IV) respectively (MILLER et al. 1980). Assembly of adenoviruses is thus impaired drastically in many *ts* mutants from several different areas of the genome.

Assembly is also inhibited if infected cells are maintained in medium lacking arginine (BONIFAS and SCHLESINGER 1959; ROUSE and SCHLESINGER 1967), or at high temperatures (LEIBOWITZ and HORWITZ 1974). In the absence of arginine all major structural proteins of the virion, including the arginine-rich core protein, are made (EVERITT et al. 1971; WINTERS and RUSSELL 1971; PRAGE and ROUSE 1976). DNA replication is also almost normal, but the yield of virus is three orders of magnitude lower than in the presence of arginine (ROUSE and SCHLESINGER 1967; RUSSELL and BECKER 1968). In the absence of another amino acid, e.g., isoleucine, a much higher yield of virus is obtained (HEILMAN and ROUSE 1980). The arginine-sensitive function is therefore probably not related to protein synthesis but affects a subtle reaction during maturation where arginine may be required in catalytic amounts.

At high temperature (42° C) the yield of infectious virus is decreased by two orders of magnitude but viral DNA and mRNA are synthesized at faster rates than at 37° C, even if the proportions of the polypeptides vary. Although the formation of hexon trimers is clearly temperature sensitive there also appears to be a block in assembly at the high temperature (OKUBO and RASKAS 1971; LEIBOWITZ and HORWITZ 1974).

All these results suggest that adenovirus assembly is a coordinated activity in which all structural units must participate in the correct proportions.

The chances of developing an in vitro system under these conditions are low, and more insight into the assembly pathway might only be achieved by identifying the events affected during defective assembly, for example during arginine deprivation, or identifying the specific defects in the *ts* polypeptides. The development of an in vitro system is, however, almost mandatory for comprehension of the details of the assembly pathway.

14 Intermediates in Assembly

The adenovirus assembly intermediates are fragile, and the DNA in the nucleoprotein is easily sheared by sonication or even deproteinized by salt during extraction from the cells or during isodensity centrifugation in CsCl gradients (EDVARDSSON et al. 1976). When the extruding nucleoprotein has been removed by shearing, the assembly intermediates are recovered at different density in CsCl gradients depending on the amount of DNA which has been encapsidated (TIBBETTS 1977). Reversible cross-linkers can be used to isolate the intermediates in CsCl gradients and to investigate their polypeptide composition after reversion of the cross-linkage (D'HALLUIN et al. 1978a). By sucrose gradient centrifugation, which preserves the intermediates more efficiently, two assembly intermediates sedimenting at 750S and 600S respectively have been isolated after pulse labeling from cells infected with several adenovirus types. The 750S species, which accumulates in a sharp peak, has the properties of complete virions with a density of 1.34 g/cm^3 (D'HALLUIN et al. 1978a). It contains intact viral DNA and a polypeptide complement which is similar to that identified in young virions, i.e., with all the precursor polypeptides in the uncleaved form. Both mature and young virions will sediment at this density, and they can therefore only be differentiated by SDS–polyacrylamide gel electrophoresis (ISHIBASHI and MAIZEL 1974b). The 600 S peak is heterogeneous, and after cross-linking it is resolved on CsCl gradients into two components with densities of 1.315 g/cm^3 and 1.37 g/cm^3. The lighter class is associated with DNA that sediments at 11S in alkaline gradients, and it contains precursor polypeptides pVI and pVIII but none of the core polypeptides V and pVII. The association of 11S instead of full-length DNA with the light intermediates may be caused by DNA shearing. These particles also contain 50K, 40K, and 32K–33K proteins believed to be scaffolding proteins necessary for forming the empty capsids (D'HALLUIN et al. 1978a; EDVARDSSON et al. 1976). The 50K protein is the IVa_2 adenovirus gene product (PERSSON et al. 1979), but the other two proteins have not yet been shown to be viral gene products. The heavier intermediates contain full-length viral DNA but no core polypeptides. The precursor polypeptides pVI, pVII, pVIII, and pIIIa and the DNA terminal protein are not processed in the intermediates and are still uncleaved in the young virions which accumulate in the nucleus of the infected cell (ISHIBASHI and MAIZEL 1974b; EDVARDSSON et al. 1976; WEBER 1976; BOUDIN et al. 1980; STILLMAN et al. 1981).

In kinetic labelling experiments, the light intermediate is detected before the heavy intermediate and the label can also be chased partially into the heavier form (D'HALLUIN et al. 1978b). These results suggest that DNA can associate with empty capsids before it associates with the core proteins. In a similar study, however, intermediates were formed containing different amounts of DNA, in all cases associated with the core proteins (EDVARDSSON et al. 1976). The only difference between these two experiments was that 0.3 $M(NH_4)_2SO_4$ was used to extract intermediates with naked DNA, while after sonication the core proteins were associated with the DNA. Whether DNA enters preformed empty capsids together with or before the core proteins is therefore still unclear. The salt resistance of the bonds between protein VII and DNA has recently been investigated (BLACK and CENTER 1979), the results suggesting that the initial interaction is salt sensitive but the complex resists 3 M NaCl. Similar studies on the interaction between the precursor pVII and DNA have not been performed but would help to establish whether an intermediate without core proteins is an artefact. The identification by electron microscopy of an intermediate in adenovirus assembly (MONCANY et al. 1980), comprising a viral particle directly associated with replicating DNA, may favor a direct insertion of DNA into the empty capsid. If so, it would explain the observation (D'HALLUIN et al. 1980b) that novobiocin, which inhibits initiation but not elongation of viral DNA replication, also blocks DNA encapsidation. On the other hand, since only a very small fraction of the newly replicated viral DNA in the cells is packaged into virions (GREEN 1962), this insertion mechanism is unlikely on stoichiometric grounds. The protein composition of the intermediates suggests that all the structural units are present, but it is difficult to determine quantitatively if the full complement of the vertex proteins has been inserted (EDVARDSSON et al. 1976).

Antigenic analysis of the intermediates during virus assembly suggests that several polypeptides in the vertex region (III, IIIa, and IV) are not recognized by antibodies to mature structural units until the final maturation involving proteolytic cleavage has occurred (D'HALLUIN et al. 1980a). Attempts to verify the tentative assembly pathway shown in Fig. 6 with *ts* mutants have been disappointing (KHITTOO and WEBER 1977; EDVARDSSON et al. 1978; D'HALLUIN et al. 1978b). Precise assignment of the protein affected by the *ts* mutation is lacking in most cases, and some of the *ts* mutants might be double mutants. Among several complementation groups of Ad5 *ts* mutants, only a few accumulate intermediates but little or no young or mature virions at the nonpermissive temperature (EDVARDSSON et al. 1978). *Ts* mutants from the fiber gene (WILLIAMS et al. 1974) accumulate intermediates and a small amount of young and mature virions. In

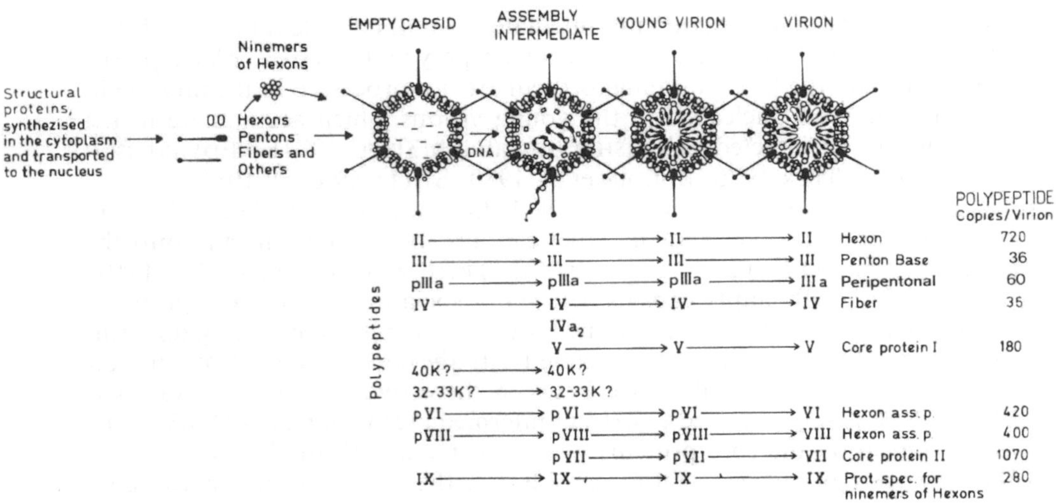

Fig. 6. A tentative assembly pathway for adenoviruses. The structural proteins are synthesized in the cytoplasm and rapidly transported to the nucleus, probably in the form of a trimer for the main capsid units, hexon, penton base, and fiber. From nonamers of hexon, empty capsids are formed, possibly by a process of self-assembly. The main assembly intermediates observed will contain varying portions of the DNA or varying amounts of core proteins depending on the procedure for extracting the particles from the cell, especially the sheer force and the salt concentration used to liberate the intermediates. It is not yet resolved whether the DNA is packaged before the core protein. The next step is the young virions where several polypeptides are present in a precursor form, followed by the final step involving proteolytic cleavage of at least five polypeptides in the virion structure. The four stages of virus assembly have been indicated at the *top* and the polypeptides present in each structure are indicated *below;* the number of copies of the polypeptides per virion is indicated at the *right*

addition, Ad5 *ts*19 and Ad5 *ts*58 accumulated intermediates but no young or mature virions. The Ad5 *ts*19 may map in the area for the pVIII polypeptide in the L4 region and Ad5 *ts*58 has been unequivocally located in the L1 region of the Ad5 genome (Frost and Williams 1978). Shift-down experiments revealed, however, that the intermediates formed at the nonpermissive temperature are thermolabile and only a fraction of them can be utilized for virus assembly after a temperature shift (Edvardsson et al. 1978). Experiments with the Ad2 mutant H2*ts*112 have, however, provided evidence that a fraction of the label incorporated in the light intermediates at the nonpermissive temperature could be chased into the heavy intermediates and then into mature virions under conditions where protein synthesis was inhibited by cycloheximide (D'Halluin et al. 1978b). The H2*ts*112 mutant has recently also been mapped in the L1 region of the genome by restriction enzyme analysis of the DNA of interspecies recombinants (D'Halluin et al. 1982). An arrest of assembly at the level of empty capsids has also been observed after infection at the nonpermissive temperature with the H2*ts*4 mutant, probably also located in the polypeptide IIIa gene region (KHITTOO and WEBER 1977). *Ts* mutants in the protein IIIa region are obviously stable enough to demonstrate partial assembly in shift-down experiments.

D'HALLUIN et al. (1980a) have also reported shift-down experiments with two fiber-defective mutants, the results of which also suggest that a fraction of the light intermediates can chase into mature virions. In this case, however, it appears that the mutation in the fiber created a deletion of the polypeptide which reduced its molecular weight by 2K. In spite of all these shortcomings, the arrest of the Ad2 *ts*1 mutant (WEBER 1976) at the stage of young virions and the possibility of chasing the young virions into mature virions after shift-down unequivocally identify one intermediate step in virus assembly.

In conclusion, because of the many polypeptides in the adenovirus virion and the heterogeneous effects of *ts* mutations on the expression of viral proteins, the *ts* mutants have not been as useful as the corresponding mutants of bacterial viruses in clarifying the assembly pathway. On the other hand, when several more of the available *ts* mutations have been mapped precisely on the viral genome, we will probably be able to reinterpret the accumulated results in a more comprehensible manner.

Acknowledgments. I would like to thank Dr. John Tooze for help with the editing and Mrs. Waltraud Ackermann and Ms. Nelly van der Jagt for expert secretarial assistance. I am also grateful to several colleagues who have contributed comments, preprints, and illustrations. Special gratitude goes to Dr. R.M. Burnett, who helped with advice and illustrations of the hexon model, and to Dr. W. Russell, who provided the figures of the whole virus and its disintegration pattern.

References

Akusjärvi G, Persson H (1981) Gene and mRNA for precursor polypeptide VI from adenovirus type 2. J Virol 38:469–482

Akusjärvi G, Philipson L, Pettersson U (1978) A protein kinase associated with adenovirus type 2. Virology 87:276–286

Akusjärvi G, Zabielski J, Perricaudet M, Pettersson U (1981) The sequence of the 3′ non-coding region of the hexon mRNA discloses a novel adenovirus gene. Nucleic Acids Res 9:1–17

Aleström P, Akusjärvi G, Perricaudet M, Mathews MB, Klessig DF, Pettersson U (1980) The gene for polypeptide IX of adenovirus type 2 and its unspliced messenger RNA. Cell 19:671–681

Anderson CW, Lewis JB (1980) Amino-terminal sequence of adenovirus type 2 proteins: hexon, fiber, component IX, early protein 1B–15K. Virology 104:27–41

Anderson CW, Baum PR, Gesteland RF (1973) Processing of adenovirus 2 induced proteins. J Virol 12:241–252

Arrand JR, Roberts RJ (1979) The nucleotide sequences at the termini of adenovirus 2 DNA. J Mol Biol 128:577–594

Bauer H, Wigand R (1963) Eigenschaften der Adenovirus-Hämagglutinine. Z Hyg Infekt-Kr 149:96–113

Berger J, Burnett RM, Franklin RM, Grütter M (1978) Small angle x-ray scattering studies on adenovirus type 2 hexon. Biochim Biophys Acta 535:233–240

Berns KI, Kelly TJ Jr (1974) Visualization of the inverted terminal repetition in adeno-associated virus DNA. J Mol Biol 82:267–271

Bhatti AR, Weber J (1979) Protease of adenovirus type 2. Subcellular localization. J Biol Chem 254:12265–12268
Black BC, Center MS (1979) DNA-binding properties of the major core protein of adenovirus 2. Nucleic Acids Res 6:2329–2353
Blair GE, Russell WC (1978) Identification of a protein kinase activity associated with human adenovirus. Virology 86:157–166
Bonifas V, Schlesinger RW (1959) Nutritional requirement for plaque production by adenovirus. Fed Proc 18:560
Bos JL, Polder L, Bernards R, Schrier P, van den Elsen P, van der Eb AJ, van Ormondt H (1981) The 2.2 kilobase E1B messenger RNA of human adenovirus type 12 and adenovirus type 5 codes for 2 tumor antigens starting at different AUG triplets. Cell 27:121–131
Boudin ML, Boulanger P (1981) Antibody-triggered dissociation of adenovirus penton capsomer. Virology 113:781–786
Boudin ML, Boulanger P (1982) Assembly of adenovirus penton base and fiber. Virology 116:589–604
Boudin ML, Moncany M, D'Halluin JC, Boulanger PA (1979) Isolation and characterization of adenovirus type 2 vertex capsomer (penton base). Virology 92:125–138
Boudin ML, D'Halluin JC, Cousin C, Boulanger P (1980) Human adenovirus type 2 protein IIIa. II. Maturation and encapsidation. Virology 101:144–156
Boulanger PA (1975) Adenovirus assembly: self-assembly of partially digested hexons. J Virol 16:1678–1682
Boulanger PA, Loucheux MH (1972) Conformational study of adenovirus type 2 hexon and fiber antigens. Biochem Biophys Res Commun 47:194–201
Boulanger PA, Puvion F (1973) Large-scale preparation of soluble adenovirus hexon, penton and fiber antigens in highly purified form. Eur J Biochem 39:37–42
Boulanger PA, Flamencourt P, Biserte G (1969) Isolation and comparative chemical study of structure proteins of the adenovirus 2 and 5: hexon and fiber antigens. Eur J Biochem 10:116–131
Boulanger P, Devaux C, Lemay P (1978) Isolation and characterization of a slow-migrating class of adenovirus type 2 hexons. Virology 84:456–468
Boulanger P, Lemay P, Blair GE, Russell WC (1979) Characterization of adenovirus protein IX. J Gen Virol 44:783–800
Boursnell MEG, Mautner V (1981) Recombination in adenovirus: crossover sites in intertypic recombinants are located in regions of homology. Virology 112:198–209
Brown M, Weber J (1980). Virion core-like organization of intranuclear adenovirus chromatin late in infection. Virology 107:306–310
Brown M, Weber J (1982) Discrete subgenomic DNA fragments in incomplete particles of adenovirus type 2. J Gen Virol 62:81–90
Brown DT, Westphal M, Burlingham BT, Winterhoff U, Doerfler W (1975) Structure and composition of the adenovirus type 2 core. J Virol 16:366–387
Bürki F, Hinaidy B, Böckmann J (1979) Bovine adenoviruses. III. Subclassification into two subgroups by means of immunodiffusion tests. Microbiologica 2:65–74
Burlingham BT, Doerfler W, Pettersson U, Philipson L (1971) Adenovirus endonuclease: association with the penton of adenovirus type 2. J Mol Biol 60:45–64
Burlingham BT, Brown DT, Doerfler W (1974) Incomplete particles of adenovirus: I. characteristics of the DNA associated with incomplete adenovirions of types 2 and 12. Virology 60:419–430
Burnett RM (1983) Structural investigations of hexon, the major coat protein of adenovirus. In: McPherson A, Jurnak FA (eds) The viruses. Wiley, New York, (in press) (Structural Biology, vol 2)
Burnett RM, Grütter MG, Markovic Z, White JL (1978) Determination of the molecular structure of a viral coat protein: adenovirus type 2 hexon. Acta Crystallogr A34:51
Cajean-Feroldi C, Chardonnet Y, Chantepie-Auray J (1977) Deoxyribonuclease activity associated with adenovirus 5 and 7. Eur J Biochem 74:457–462
Carusi EA (1977) Evidence for blocked 5′ termini in human adenovirus DNA. Virology 76:380–394

Caspar DLD (1966) Design principles in organized biological structures. In: Wolstenholme GEW, O'Connor M (eds) Principles of biomolecular organization. Churchill, London, pp 7–34
Cepko CL, Sharp PA (1982) Assembly of adenovirus major capsid protein is mediated by a non-virion protein. Cell 31:407–415
Cepko CL, Changelian PA, Sharp PA (1981) Immunoprecipitation with two-dimensional pools as a hybridoma screening technique: production and characterization of monoclonal antibodies against adenovirus 2 proteins. Virology 110:385–401
Challberg MD, Kelly TJ Jr (1979) Adenovirus DNA replication in vitro. Proc Natl Acad Sci USA 76:655–659
Challberg MD, Desiderio SV, Kelly TJ Jr (1980) Adenovirus DNA replication in vitro: characterization of a protein covalently linked to nascent DNA strands. Proc Natl Acad Sci USA 77:5105–5109
Cheng Chee-Sheung C, Ginsberg HS (1982) Characterization of a temperature-sensitive fiber mutant of type 5 adenovirus and effect of the mutation on virion assembly. J Virol 42:932–950
Colby WW, Shenk T (1981) Adenovirus type 5 virions can be assembled in vivo in the absence of detectable polypeptide IX. J Virol 39:977–980
Collins J, Hohn B (1978) Cosmids: a type of plasmid genecloning vector that is packageable in vitro in bacteriophage lambda heads. Proc Natl Acad Sci USA 75:4242–4246
Corden J, Engelking M, Pearson GD (1976) Chromatin-like organization of the adenovirus chromosome. Proc Natl Acad Sci USA 73:401–404
Cornick G, Sigler PB, Ginsberg HS (1971) Characterization of crystals of adenovirus type 5 hexon. J Mol Biol 57:397–401
Cornick G, Sigler PB, Ginsberg HS (1973) Mass of protein in the asymmetric unit of hexon crystals – a new method. J Mol Biol 73:533–537
Crowther RA, Franklin RM (1972) The structure of the groups of nine hexons from adenovirus. J Mol Biol 68:181–184
Daniell E (1976) Genome structure of incomplete particles of adenovirus. J Virol 19:685–708
Daniell E, Mullenbach T (1978) Synthesis of defective viral DNA in HeLa cells infected with adenovirus type 3. J Virol 26:61–70
Devaux C, Boulanger P (1980) Reactive serine in human adenovirus hexon polypeptide. Virology 102:94–106
Devaux C, Zulauf M, Boulanger P, Jacrot B (1982) Molecular weight of adenovirus serotype 2 capsomers. A new characterization. J Mol Biol 156:927–939
Devaux C, Berthé-Colominas C, Timmins PA, Jacrot B (1983) Crystal packing and stoichiometry of the fiber protein adenovirus type 2. J Mol Biol (in press)
D'Halluin JC, Martin GR, Torpier G, Boulanger PA (1978a) Adenovirus type 2 assembly analysed by reversible crosslinking of labile intermediates. J Virol 26:357–363
D'Halluin JC, Milleville M, Boulanger PA, Martin GR (1978b) Temperature-sensitive mutant of adenovirus type 2 blocked in virion assembly: accumulation of light intermediate particles. J Virol 26:344–356
D'Halluin JC, Milleville M, Martin GR, Boulanger P (1980a) Morphogenesis of human adenovirus type 2 studied with fiber- and fiber and penton base-defective temperature-sensitive mutants. J Virol 33:88–99
D'Halluin JC, Milleville M, Boulanger P (1980b) Effects of novobiocin on adenovirus DNA synthesis and encapsidation. Nucleic Acids Res 8:1625–1641
D'Halluin JC, Cousin C, Boulanger P (1982) Physical mapping of adenovirus type 2 temperature-sensitive mutations by restriction endonuclease analysis of interserotypic recombinants. J Virol 41:401–413
Dijkema R, Maat J, Dekker BMM, van Ormondt H, Boyer HW (1981) The gene for polypeptide IX of human adenovirus type 7. Gene 13:375–385
Doerfler W (1977) Animal virus-host genome interactions. In: Fraenkel-Conrat H, Wagner RR (eds) Comprehensive virology, vol 10. Plenum New York, pp 279–399
Doerfler W, Kleinschmidt A (1970) Denaturation pattern of the DNA of adenovirus type 2 as determined by electron microscopy. J Mol Biol 50:579–593

Doerfler W, Hellmann W, Kleinschmidt AK (1972) The DNA of adenovirus type 12 and its denaturation pattern. Virology 47:507–512
Dorsett PH, Ginsberg HS (1975) Characterization of an adenovirus type 5 fiber protein. J Virol 15:208–216
Edvardsson B, Everitt E, Jörnvall H, Prage L, Philipson L (1976) Intermediates in adenovirus assembly. J Virol 19:533–547
Edvardsson B, Ustacelebi S, Williams J, Philipson L (1978) Assembly intermediates among adenovirus type 5 temperature-sensitive mutants. J Virol 25:641–651
Ellens DJ, Sussenbach JS, Jansz HS (1974) Studies on the mechanism of replication of adenovirus DNA. III. Electron microscopy of replicating DNA. Virology 61:427–442
Engler JA (1981) The nucleotic sequence of the polypeptide IX gene of human adenovirus type 3. Gene 13:387–394
Epstein MA (1959) Observation on the fine structure of type 5 adenovirus. J Biophys Biochem Cytol 6:523–524
Everett SF, Ginsberg HS (1958) A toxin-like material separable from type 5 adenovirus particles. Virology 6:770–771
Everitt E, Philipson L (1974) Structural proteins of adenoviruses. XI. Purification of three low molecular weight virion proteins of adenovirus type 2 and their synthesis during productive infection. Virology 62:253–269
Everitt E, Sundquist B, Philipson L (1971) Mechanism of arginine requirement for adenovirus synthesis. I. Synthesis of structural proteins. J Virol 8:742–753
Everitt E, Sundquist B, Pettersson U, Philipson L (1973) Structural proteins of adenoviruses. X. Isolation and topography of low molecular weight antigens from the virion of adenovirus type 2. Virology 62:130–147
Everitt E, Lutter L, Philipson L (1975) Structural proteins of adenoviruses. XII. Location and neighbor relationship among proteins of adenovirion type 2 as revealed by enzymatic iodination, immuno-precipitation and chemical cross-linking. Virology 67:197–208
Feiss M, Fisher RA, Siegele DA, Nichols BP, Donelson JE (1979) Packaging of the bacteriophage lambda chromosome: a role for the base sequences outside cos. Virology 92:56–67
Fitzgerald DJP, Padmanabhan R, Pastan I, Willingham MC (1983) Adenovirus-induced release of epidermal growth factor and pseudomonas toxin into the cytosol of KB cells during receptor-mediated endocytosis. Cell 32:607–617
Flint SJ, Broker TR (1981) Lytic infection by adenoviruses, pp 443–546. In: Tooze J (ed) DNA Tumor viruses, 2nd ed. Molecular biology of tumor viruses, pt 2. Rev Cold Spring Harbor Laboratory, Cold Spring Harbor, pp 443–546
Franklin RM, Pettersson U, Akervall K, Strandberg B, Philipson L (1971) Structural proteins of adenoviruses. IV. On the size and structure of the adenovirus type 2 hexon. J Mol Biol 57:383–395
Frearson PM, Crawford LV (1972) Polyoma virus basic proteins. J Gen Virol 14:141–155
Frost E, Williams J (1978) Mapping temperature-sensitive and host-range mutations of adenovirus type 5 by marker rescue. Virology 91:39–50
Galibert F, Hérissé J, Courtois G (1979) Nucleotide sequence of the *Eco* RI F fragment of adenovirus 2 genome. Gene 6:1–22
Gallimore PH, Sharp PA, Sambrook J (1974) Viral DNA in transformed cells: II. a study of the sequences of adenovirus 2 DNA in nine lines of transformed rat cells using specific fragments of the viral genome. J Mol Biol 89:49–71
Garon CF, Berry KW, Rose JA (1972) A unique form of terminal redundancy in adenovirus DNA molecules. Proc Natl Acad Sci USA 69:2391–2395
Garon CF, Berry KW, Rose JA (1975) Arrangement of sequences in the inverted terminal repetition of adenovirus 18 DNA. Proc Natl Acad Sci USA 72:3039–3043
Gelderblom H, Maichle-Laupe I (1982) The fibers of fowl adenoviruses. Arch Virol 72:289–298
Ginsberg HS (1979) Adenovirus structural proteins. In: Fraenkel-Conrat H, Wagner RR (eds) Comprehensive virology, vol B. Plenum, New York, pp 409–457
Ginsberg HS, Pereira HG, Valentine RC, Wilcox WC (1966) A proposed terminology for the adenovirus antigens and virion morphological subunits. Virology 28:782–783
Ginsberg HS, Bello LJ, Levine AJ (1967) Control of biosynthesis of host macromolecules

in cells infected with adenovirus. In: Colter JS, Paranchych W (eds) The molecular biology of viruses. Academic, New York, pp 547–572
Goding CR, Russell WC (1983) Adenovirus cores can function as templates in in vitro DNA replication. The EMBO Journal 2:339–344
Graham FL, van der Eb AJ (1973) A new technique for the assay of the infectivity of human adenovirus 5 DNA. Virology 52:456–467
Graham FL, van der Eb AJ, Heijneker HL (1974) Size and location of the transforming region of human adenovirus type 5 DNA. Nature 251:687–691
Green M (1962) Studies on the biosynthesis of viral DNA. IV. Isolation, purification and chemical analysis of adenovirus. Cold Spring Harbor Symp Quant Biol 27:219–235
Green M, Piña M, Kimes R, Wensink PC, Machattie LA, Thomas CA (1967) Adenovirus DNA, I. Molecular weight and conformation. Proc Natl Acad Sci USA 57:1302–1309
Green NM, Wrigley NG, Russell WC, Martin SR, McLachlan AD (1983) Evidence for a repeating cross β-sheet structure in the adenovirus fibre. EMBO Journal 2:1357–1366
Grütter M, Franklin RM (1974) Studies on the molecular weight of the adenovirus type 2 hexon and its subunit. J Mol Biol 89:163–178
Haase AT, Pereira HG (1972) The purification of adenovirus neutralising activity: adenovirus type 5 hexon immunoadsorbent. J Immunol 108:633–636
Hammarskjöld ML, Winberg G (1980) Encapsidation of adenovirus 16 DNA is directed by a small DNA sequence at the left end of the genome. Cell 20:787–795
Hammarskjöld ML, Winberg G, Norrby E, Wadell G (1977) Isolation of incomplete adenovirus 16 particles containing viral and host cell DNA. Virology 82:449–461
Haruna J, Yaosi H, Kono R, Watanabe I (1961) Separation of adenovirus by chromatography on DEAE-cellulose. Virology 13:264–267
Hassell JA, Weber J (1978) Genetic analysis of adenovirus type 2: VIII. physical locations of temperature-sensitive mutations. J Virol 28:671–678
Heilman CA, Rouse H (1980) Adenovirus type 2 polypeptide synthesis in arginine-deprived cells. Virology 105:159–170
Hérissé J, Courtois G, Galibert F (1980) Nucleotide sequences of the *Eco* RI-D fragment of the adenovirus 2 genome. Nucleic Acids Res 8:2173–2191
Hérissé J, Rigolet M, Dupont de Dinechin S, Galibert F (1981) Nucleotide sequence of adenovirus 2 DNA fragment encoding for the carboxylic region of the fiber protein and the entire E4 region. Nucleic Acids Res 9:4023–4042
Hilleman RJ, Werner JH, Stewart MT (1955) Grouping and occurrence of RI (prototype RI-67) viruses. Proc Soc Exp Biol Med 90:555–562
Hollinshead AC, Bunnag B, Alford TC (1967) Relationship between the subunits and "T" antigens of adenovirus type 2. Nature 215:397–399
Horne RW, Brenner S, Waterson AP, Wildy P (1959) The icosahedral form of an adenovirus. J Mol Biol 1:84–86
Horwitz MS, Scharff MD, Maizel JV (1969) Synthesis and assembly of adenovirus 2. I. Polypeptide synthesis, assembly of capsomers and morphogenesis of the virion. Virology 39:682–694
Horwitz MS, Maizel JV, Scharff MD (1970) Molecular weight of adenovirus type 2 hexon polypeptide. J Virol 6:569–571
Hosakawa K, Sung MT (1976) Isolation and characterization of an extremely basic protein from adenovirus type 5. J Virol 17:924–934
Huebner RJ, Rowe WP, Ward TG, Parrot RH, Bell JA (1954) Adenoidal-pharyngal-conjunctival agents. A newly recognized group of common viruses of the respiratory system. N Engl J Med 251:1077–186
Igarishi K, Niiyama Y, Tsukamoto K, Kurokawa T, Sugino Y (1975) Biochemical studies on bovine adenovirus type 3. II. incomplete virus. J Virol 16:634–641
Ishibashi M, Maizel JV Jr (1974a) The polypeptides of adenovirus. VI. The early and late glycopeptides. Virology 58:345–361
Ishibashi M, Maizel JV Jr (1974b) The polypeptides of adenovirus. V. Young virions, structural intermediates between top components and aged virions. Virology 57:409–424
Jörnvall H, Philipson L (1980) Limited proteolysis and a reactive cysteine residue define accessi-

ble regions in the native conformation of the adenovirus hexon protein. Eur J Biochem 104:237–247
Jörnvall H, Pettersson U, Philipson L (1974a) Structural studies of adenovirus type 2 hexon protein. Eur J Biochem 48:179–192
Jörnvall H, Ohlsson H, Philipson L (1974b) An acetylated N-terminus of adenovirus 2 hexon protein. Biochem Biophys Res Commun 56:304–310
Jörnvall H, Akusjärvi G, Aleström P, von Bahr-Lindström H, Pettersson U, Apella E, Fowler A, Philipson L (1981) The adenovirus hexon protein: the primary structure of the polypeptide and its correlation with the hexon gene. J Biol Chem 256:6181–6204
Kasel JA, Huber M (1964) Relationship of the adenovirus erythrocyte-receptor modifying factor to the type specific complement fixing antigen. Proc Soc Exp Biol Med 116:16–18
Kasel JA, Rowe WP, Nemes JL (1961) Further characterization of the adenovirus erythrocyte receptor modifying factor. J Exp Med 114:717–728
Keegstra CS, van Wielink PS, Sussenbach JS (1977) The visualization of a circular DNA-protein complex from adenovirions. Virology 76:444–447
Kelly TJ Jr, Lechner RL (1979) The structure of replicating adenovirus DNA molecules: characterization of DNA-protein complexes from infected cells. Cold Spring Harbor Symp Quant Biol 43:721–728
Khittoo G, Weber JM (1981) The nature of the DNA associated with incomplete particles of adenovirus type 2. J Gen Virol 54:343–355
Kjellén L, Pereira HG (1968) Role of adenovirus antigens in the induction of virus neutralizing antibodies. J Gen Virol 2:177–185
Klemperer HG, Pereira HG (1959) Study of adenovirus antigens fractionated by chromatography on DEAE-cellulose. Virology 9:536–545
Köhler K (1965) Reinigung und Charakterisierung zweier Proteine des Adenovirus-type 2. Z Naturforsch 20b:747–752
Kozcot FJ, Carter BJ, Garon CF, Rose JA (1973) Selfcomplementarity of terminal sequences within plus and minus strands of adenovirus associated virus DNA. Proc Natl Acad Sci USA 70:215–219
Kruijer W, van Schaik FMA, Sussenbach JS (1980) Nucleotide sequence analysis of a region of adenovirus 5 DNA encoding a hitherto unidentified gene. Nucleic Acids Res 8:6033–6042
Kubinski H, Rose JA (1967) Regions containing repeating base-pairs in DNA from some oncogenic and non-oncogenic animal viruses. Proc Natl Acad Sci USA 57:1720–1725
Landgraf-Leurs M, Green M (1971) Adenovirus DNA. III. Separation of the complementary strands of adenovirus types 2, 7, and 12 DNA molecules. J Mol Biol 60:185–202
Laver WG (1970) Isolation of an arginine-rich protein from particles of adenovirus type 2. Virology 41:488–500
Laver WG, Pereira HG, Russell WC, Valentine R (1968) Isolation of an internal component from adenovirus type 5. J Mol Biol 37:379–386
Laver WG, Wrigley NG, Pereira HG (1969) Removal of pentons from particles of adenovirus type 2. Virology 39:599–605
Laver WG, Younghusband HB, Wrigley NG (1971) Purification and properties of chick-embryo-lethal orphan virus (an avian adenovirus). Virology 45:598–614
Leibowitz J, Horwitz MS (1974) Synthesis and assembly of adenovirus type 2 polypeptides. II. Reversible inhibition of hexon assembly at 42°. Virology 61:129–139
Leibowitz J, Horwitz MS (1975) Synthesis and assembly of adenovirus polypeptides. III. Reversible inhibition of hexon assembly in adenovirus type 5 temperature-sensitive mutants. Virology 66:10–24
Lemay P, Boulanger P (1980) Physiochemical characteristics of structural and non-structural proteins of human adenovirus 2. Ann Virol Inst Pasteur 131:259–277
Lemay P, Boudin ML, Milleville M, Boulanger P (1980) Human adenovirus type 2 protein IIIa. I. Purification and characterization. Virology 101:131–143
Levine AJ, Ginsberg HS (1967) Mechanism by which fiber antigen inhibits multiplication of type 5 adenovirus. J Virol 1:747–757
Lewis JB, Mathews M (1980) Control of adenovirus early gene expression: a class of immediate early products. Cell 21:303–313

Lichy JH, Field J, Horwitz MS, Hurwitz J (1982) Sepration of the adenovirus terminal protein precursor from its associated DNA polymerase: role of both proteins in the initiation of adenovirus DNA replication. Proc Natl Acad Sci USA 79:5225–5229
Lischwe MA, Sung MT (1977) A histone-like protein from adenovirus chromatin. Nature 267:552–554
Maat J, van Beveren CP, von Ormondt H (1980) The nucleotide sequence of adenovirus type 5 early region E1: the region between map positions 8.0 (*Hin*d III site) and 11.8 (SmaI site). Gene 10:27–38
Maizel JV Jr (1971) Polyacrylamide gel electrophoresis of viral proteins. In: Maramosch K, Koprowski H (eds) Methods in virology, vol 5. Academic, New York, pp 179–246
Maizel JV Jr, White DO, Scharff MD (1968a) The polypeptides of adenovirus. I. evidence for multiple protein components in the virion and a comparison of types 2, 7, and 12. Virology 36:115–125
Maizel JV Jr, White DO, Sharff MD (1968b) The polypeptides of adenovirus. II. Soluble proteins, cores, top components and the structure of the virion. Virology 36:126–136
Mak S (1971) Defective virions in human adenovirus type 12. J Virol 7:426–433
Mak I, Ezoe H, Mak S (1979) Structure and function of adenovirus type 12 defective virions. J Virol 32:240–250
Marusyk RG, Morgan AR, Wadell G (1975) Association of endonuclease activity with serotypes belonging to the three subgroups of human adenoviruses. J Virol 16:456–458
Mattern CFT (1969) Virus architecture as determined by x-ray diffraction and electron microscopy, 55–100. In: Levy HB (ed) Biochemistry of viruses. Dekker, New York, pp 55–100
Mautner V, Pereira HG (1971) Crystallization of a second adenovirus protein (the fiber). Nature 230:456–457
Miller JS, Ricciardi RP, Roberts BE, Patterson BM, Mathews MB (1980) Arrangement of messenger RNAs and protein coding sequences in the major late transcription unit of adenovirus 2. J Mol Biol 142:455–488
Moncany MLJ, Rėvet B, Girard M (1980) Characterization of a new adenovirus type 5 assembly intermediate. J Gen Virol 50:33–47
Murray RE, Green M (1973) Adenovirus DNA. IV. Topology of adenovirus genomes. J Mol Biol 74:735–738
Nagata K, Guggenheimer RA, Enomoto T, Lichy JH, Hurwitz J (1982) Adenovirus DNA replication in vitro: identification of a host factor that stimulates synthesis of the preterminal protein dCMP complex. Proc Natl Acad Sci USA 79:6438–6442
Nermut MV (1975) Fine structure of adenovirus type 5. Virology 65:480–495
Nermut MV (1978) Structural elements in adenovirus cores. Studies by means of freeze-fracturing and ultrathin sectioning. Arch Virol 57:323–337
Nermut MV (1979) Structural elements in adenovirus cores. Evidence for a "Core Shell" and linear structures in "Relaxed" cores. Arch Virol 62:101–116
Nermut MV (1980) The architecture of adenoviruses: recent views and problems. Brief review. Arch Virol 64:175–196
Nermut MV, Perkins WJ (1979) Consideration of the three-dimensional structure of the adenovirus hexon from electron microscopy and computer modelling. Micron 10:247–266
Neurath AR, Rubin BA, Stasny JT (1968) Cleavage by formamide or intercapsomer bonds in adenovirus type 4 and 7 virions and hemagglutinins. J Virol 2:1086–1095
Neurath AR, Hartzell RW, Rubin BA (1969) Solubilization and some properties of erythrocyte receptors for adenovirus type 7 hemagglutinin. Nature 221:1069–1071
Nevins JR, Chen-Kiang S (1981) Processing of adenovirus nuclear RNA to mRNA. Adv Virus Res 26:1–35
Nicolson MO, MacAllister RM (1972) Infectivity of human adenovirus. I. DNA. Virology 48:14–21
Niiyama Y, Igarishi K, Tsukamoto K, Kurokawa T, Sugino Y (1975) Biochemical studies on bovine adenovirus type 3: I. Purification and properties. J Virol 16:621–633
Norrby E (1966a) The relationship between the soluble antigens and the virion of adenovirus type 3. I. Morphological characteristics. Virology 28:236–248
Norrby E (1966b) The relationship between the soluble antigens and the virion of adenovirus

type 3. II. Identification and characterization of an incomplete hemagglutinin. Virology 30:608–617
Norrby E (1968) Biological significance of structural adenovirus components. Curr Top Microbiol Immunol 43:1–43
Norrby E (1969a) The relationship between the soluble antigens and the virion of adenovirus type 3. IV. Immunological characteristics. Virology 37:565–576
Norrby E (1969b) The structural and functional diversity of adenovirus capsid components. J Gen Virol 5:221–236
Norrby E, Skaaret P (1967) The relationship between soluble antigens and the virion of adenovirus type 3. III. Immunological identification of fiber antigen and isolated vertex capsomer antigen. Virology 32:489–502
Norrby E, Wadell G (1969) Immunological relationship between hexons of certain human adenoviruses. J Virol 4:663–670
Norrby E, Marusyk HL, Hammarskjöld M-L (1969a) The relationship between the soluble antigens and the virion of adenovirus type 3. V. identification of antigen specificities available at the surface of virions. Virology 38:477–482
Norrby E, Wadell G, Marusyk H (1969b) Fiber-associated incomplete and complete hemagglutinins of adenovirus type 6. Arch Gesamte Virusforsch 28:239–244
Öberg B, Saborio J, Persson T, Everitt E, Philipson L (1975) Identification of the in vitro translation products of adenovirus mRNA by immunoprecipitation. J Virol 15:199–207
Okubo CK, Raskas, HJ (1971) Thermo-sensitive events in the replication of adenovirus type 2 at 42°. Virology 46:175–182
Oosterom-Dragon EA, Ginsberg HS (1981) Characterization of two temperature-sensitive mutants of type 5 adenovirus with mutations in the 100,000-dalton protein gene. J Virol 40:491–500
Patch CT, Lewis AM, Levine AS (1972) Evidence for a transcriptional control region of SV40 in the adenovirus 2-SV40 hybrid, $Ad2^{+}ND1$. Proc Natl Acad Sci USA 69:3375–3379
Pereira HG (1958) A protein factor responsible for the early cytopathic effect of adenoviruses. Virology 6:601–611
Pereira HG, de Figueiredo MVT (1962) Mechanism of hemagglutination by adenovirus types 1, 2, 4, 5, and 6. Virology 18:1–8
Pereira HG, Kelly B (1957) Dose response curves of toxic and infective actions of adenovirus in HeLa cell cultures. J Gen Microbiol 17:517–564
Pereira HG, Laver WG (1970) Comparison of adenovirus types 2 and 5 hexons by immunological and biochemical techniques. J Gen Virol 9:163–167
Pereira HG, Wrigley NG (1974) In vitro reconstitution, hexon bonding and handedness of incomplete adenovirus capsid. J Mol Biol 85:617–631
Pereira HG, Allison AC, Farthing C (1959) Study of adenovirus antigens by immuno-electrophoresis. Nature 183:895–896
Pereira HG, Valentine RC, Russell WC (1968) Crystallization of an adenovirus protein. Nature 219:946–947
Persson H, Öberg B, Philipson L (1977) In vitro translation with adenovirus polyribosomes. J Virol 21:187–198
Persson H, Pettersson U, Mathews MB (1978) Synthesis of a structural adenovirus polypeptide in the absence of viral DNA replication. Virology 90:67–79
Persson H, Mathisen B, Philipson L, Pettersson U (1979) A maturation protein in adenovirus morphogenesis. Virology 93:198–208
Pettersson U (1971) Structural proteins of adenoviruses. VI. On the antigenic determinants of the hexon. Virology 43:123–136
Pettersson U, Höglund S (1969) Structural proteins of adenoviruses: III. Purification and characterization of the adenovirus type 2 penton antigen. Virology 39:90–106
Pettersson U, Philipson L, Höglund S (1967) Structural proteins of adenoviruses. I. Purification and characterization of adenovirus type 2 hexon antigen. Virology 33:575–590
Pettersson U, Philipson L, Höglund S (1968) Structural proteins of adenoviruses. II. Purification and characterization of adenovirus type 2 fiber antigen. Virology 35:204–215

Philipson L (1960) Separation on DEAE-cellulose of components associated with adenovirus reproduction. Virology 10:459–465

Philipson L (1979) Adenovirus proteins and their messenger RNAs. Adv Virus Res 25:357–405

Philipson L, Pettersson U (1973) Structure and function of virion proteins of adenoviruses. Prog Exp Tumor Res 18:1–55

Philipson L, Lindberg U (1974) Reproduction of adenoviruses. In: Fraenkel-Conrat H, Wagner RR (eds) Comprehensive virology, vol 3. Plenum, New York, pp 143–227

Philipson L, Lonberg-Holm K, Pettersson U (1968) Virus receptor interaction in an adenovirus system. J Virol 2:1064–1075

Philipson L, Pettersson U, Lindberg U (1975) Molecular biology of adenoviruses, In: Hallauer C, Gard S (eds) Virology monographs, vol 14. Springer, Vienna New York, pp 1–115

Philipson L, Everitt E, Lonberg-Holm K (1976) Molecular aspects of virus-receptor interaction in the adenovirus system. In: Beers F, Bassett EG (eds) Cell membrane receptor for viruses antigens and antibodies. Ninth Miles int symp, Raven Press, New York, pp 203–216

Prage L, Pettersson U (1971) Structural proteins of adenoviruses. VII. Purification and properties of an arginine-rich core protein from adenovirus type 2 and type 3. Virology 45:364–373

Prage L, Rouse HC (1976) Effect of arginine starvation on macromolecular synthesis in infection with type 2 adenovirus. III. Immunofluorescence studies of the synthesis of the hexon and the major core antigen (AAP). Virology 69:352–356

Prage L, Pettersson U, Philipson L (1968) Internal basic proteins in adenovirus. Virology 36:508–511

Prage L, Pettersson U, Höglund S, Lonberg-Holm K, Philipson L (1970) Structural proteins of adenoviruses. IV. Sequential degradation of the adenovirus type 2 virion. Virology 42:341–358

Prage L, Höglund S, Philipson L (1972) Structural proteins of adenoviruses. III. Characterization of incomplete particles of adenovirus type 3. Virology 49:745–757

Reif UM, Winterhoff U, Lundholm U, Philipson L, Doerfler W (1977a) Purification of an endonuclease from adenovirus-infected KB cells. Eur J Biochem 73:313–325

Reif UM, Winterhoff U, Doerfler W (1977b) Characterization of the pH 4.0 endonuclease from adenovirus-type 2-infected cells. Eur J Biochem 73:327–333

Rekosh DMK, Russell WC, Bellett AJD, Robinson AJ (1977) Identification of a protein linked to the ends of adenovirus DNA. Cell 11:283–295

Robinson AJ, Bellett AJD (1975) Complementary strands of CELO virus DNA. J Virol 15:458–465

Robinson AJ, Younghusband HB, Bellett AJD (1973) A circular DNA-protein complex from adenoviruses. Virology 56:54–69

Roblin R, Härle E, Dulbecco R (1971) Polyoma Virus Proteins: I. multiple virion components. Virology 45:555–566

Rosen L (1958) Hemagglutination of adenoviruses. Virology 5:574–577

Rosen L (1960) A hemagglutination-inhibition technique for typing adenoviruses. Am J Hyg 71:120–128

Rosen L, Hovis JF, Bell JA (1962) Further observation on typing adenovirus and a description of two possible additional serotypes. Proc Soc Exp Biol Med 110:710–713

Rosenwirth B, Tija S, Westphal M, Doerfler W (1974) Incomplete particles of adenovirus. II. Kinetics of formation and polypeptide composition of adenovirus type 2. Virology 60:431–437

Rouse HC, Schlesinger RW (1967) An arginine-dependent step in the maturation of type 1 adenovirus. Virology 33:513–522

Rowe WP, Hartley JW, Roizmann B, Levey HB (1958) Characterization of a factor formed in the course of adenovirus infection of tissue cultures causing detachment of cells from glass. J Exp Med 108:713–729

Russell WC, Becker Y (1968) A maturation factor for adenovirus. Virology 35:18–27

Russell WC, Precious B (1982) Nucleic acid binding properties of adenovirus structural polypeptides. J Gen Virol 63:69–79

Russell WC, Valentine RC, Pereira HG (1967) The effect of heat on the anatomy of the adenovirus. J Gen Virol 1:509–522

Russell WC, McIntosh K, Skehel JJ (1971) The preparation and properties of adenovirus cores. J Gen Virol 11:35–46
Russell WC, Patel G, Precious B, Sharp I, Gardner PS (1981) Monoclonal antibodies against adenovirus type 5: preparation and preliminary characterization. J Gen Virol 56:393–408
Sato K, Hosokawa K (1981) The structure of adenovirion chromatin revealed by ultraviolet light induced crosslinking. Biochem Biophys Res Commun 101:1318–1323
Schaller JP, Yohn DS (1974) Transformation potentials of the non-infectious (defective) component in pools of adenoviruses type 12 and simian adenovirus 7. J Virol 14:392–401
Schmidt NJ, King CJ, Lennette EH (1965) Hemagglutination and hemagglutination-inhibition with adenovirus type 12. Proc Soc Exp Biol Med 118:208–211
Sharp PA, Gallimore PH, Flint SJ (1975) Mapping of adenovirus 2 RNA sequences in lytically infected cells and transformed cell lines. Cold Spring Harbor Symp Quant Biol 39:457–474
Sharp PA, Moore C, Haverty J (1976) The infectivity of adenovirus 5-DNA protein complex. Virology 75:442–456
Shortridge KF, Biddle F (1970) The proteins of adenovirus type 5. Arch Gesamte Virusforsch 29:1–24
Smith KO, Gehle WD, Trousdale MD (1965) Architecture of the adenovirus capsid. J Bacteriol 90:254–261
Stansy JT, Neurath AR, Rubin BA (1968) Effect of formamide on the capsid morphology of adenovirus types 4 and 7. J Virol 2:1429–1442
Steenbergh PH, Maat J, van Ormondt H, Sussenbach JS (1977) The nucleotide sequence at the termini of adenovirus type 5 DNA. Nucleic Acids Res 4:4371–4389
Stillman BW, Lewis JB, Chow LT, Mathews MB, Smart JE (1981) Identification of the gene and mRNA for the adenovirus terminal protein precursor. Cell 23:497–508
Stinski MF, Ginsberg HS (1974) Antibody to the type 5 adenovirus hexon polypeptide: detection of nascent polypeptides in cytoplasm of infected KB cells. Intervirology 4:226–236
Stinski MF, Ginsberg HS (1975) Hexon polypeptides of type 2, 3, and 5 adenoviruses and their relationship to hexon structure. J Virol 15:898–905
Sundquist B, Pettersson U, Thelander L, Philipson L (1973a) Structural proteins of adenoviruses. IX. Molecular weight and subunit composition of adenovirus type 2 fiber. Virology 51:252–256
Sundquist B, Everitt E, Philipson L, Höglund S (1973b) Assembly of adenoviruses. J Virol 11:449–459
Sung MT, Lischwe MA, Richards JC, Hosokawa K (1977) Adenovirus chromatin. I. Isolation and characterization of the major core protein VII and precursor PRO-VII. J Biol Chem 252:4981–4987
Sung MT, Cao TM, Coleman RT, Budelier KA (1983) Gene and protein sequences of adenovirus protein VII, a hybrid basic chromosomal protein. Proc Nat Acad Sci USA 80:2902–2906
Sussenbach JS (1967) Early events in the infection process of adenovirus type 5 in HeLa cells. Virology 33:567–574
Sussenbach JS, Kuijk MG (1977) Studies on the mechanism of replication of adenovirus DNA. V. The location of termini of replication. Virology 77:149–162
Sussenbach JS, Ellens DJ, Jansz HS (1973) Studies on the mechanism of replication of adenovirus DNA. I. The nature of single-stranded DNA in replicative intermediates. J Virol 12:1131–1138
Tate V, Philipson L (1979) Parental adenovirus DNA accumulates in nucleosome-like structures in infected cells. Nucleic Acids Res 6:2769–2785
Tejg-Jensen B, Furugren B, Lindquist I, Philipson L (1972) A small angle X-ray study of adenovirus type 2 hexon. Monatsh Chem 103:1730–1736
Tibbetts C (1977) Viral DNA sequences from incomplete particles of human adenovirus type 7. Cell 12:243–249
Tibbetts C, Giam C-Z (1979) In vitro association of empty adenovirus capsids with double-stranded DNA. J Virol 32:995–1005
Tibbetts C, Pettersson U, Johansson K, Philipson L (1974) Transcription of the adenovirus type 2 genome. I Relationship of cytoplasmic RNA to the complementary strands of viral DNA. J Virol 13:370–377

Tija S, Fanning E, Schick J, Doerfler W (1977) Incomplete particles of adenovirus type 2. III. Viral and cellular DNA sequences in incomplete particles. Virology 76:365–379

Tremblay ML, Déry CV, Talbot BC, Weber J (1983) In vitro cleavage specificity of the adenovirus type 2 protease. Biochim Biophys Acta 743:239–245

Valentine RC, Pereira HG (1965) Antigens and structure of the adenovirus. J Mol Biol 13:13–20

Van der Eb AJ, van Kesteren LW (1966) Structure and molecular weight of the DNA of adenovirus type 5. Biocheim Biophys Acta 129:441–444

Van der Eb AJ, van Kesteren LW, van Bruggen EFJ (1969) Structural properties of adenovirus DNAs. Biochim Biophys Acta 182:530–541

Van Roy F, Engler G, Fiers W (1979) Isolation and characterization of a specific deletion mutant of human adenovirus type 2. Virology 96:486–502

Vayda ME, Rogers AE, Flint SJ (1983) The structure of nucleoprotein cores released from adenovirions. Nucleic Acids Res 11:441–460

Velicer LF, Ginsberg HS (1970) Synthesis, transport and morphogenesis of type 5 adenovirus capsid proteins. J Virol 5:338–352

Von Bahr-Lindström H, Jörnvall H, Althin S, Philipson L (1982) Structural differences between hexons from adenovirus types 2 and 5: correlation with differences in size and immunological properties. Virology 118:353–362

Wadell G (1969) Hemagglutination with adenovirus serotypes belonging to Rosen's subgroup II and III. Proc Soc Exp Biol Med 132:413–421

Wadell G (1970) Structural and biological properties of capsid components of human adenoviruses. Thesis, Karolinska Institut, Stockholm

Wadell G (1972) Sensitization and neutralization of adenovirus by specific sera against capsid subunits. J Immunol 108:622–632

Wadell G, Norrby E (1969a) The soluble hemagglutinins of adenoviruses belonging to Rosen's subgroup III. II. The slowly sedimenting hemagglutinin. Arch Gesamte Virusforsch 26:53–62

Wadell G, Norrby E (1969b) Immunological and other biological characteristics of pentons of human adenoviruses. J Virol 4:671–680

Wadell G, Norrby E, Skaaret P (1969) The soluble hemagglutinins of adenovirus belonging to Rosen's subgroup III. I. The rapidly sedimenting hemagglutinin. Arch Gesamte Virusforsch 26:33–52

Wadell G, Hammerskjöld ML, Varsanyi T (1973) Incomplete particles of adenovirus type 16. J Gen Virol 20:287–302

Wadell G, Hammarskjöld M-L, Winberg G, Varsanyi TW, Sundell G (1980) Genetic variability of adenoviruses. Ann NY Acad Sci 354:16–42

Wasmuth EH, Tytell AA (1966) Physical studies with adenovirus hexon antigens. Life Sci 6:1063–1068

Weber J (1976) Genetic analysis of adenovirus type 2. III. Temperature-sensitivity of processing of viral proteins. J Virol 17:462–471

Weber J, Begin M, Carstens EB (1977) Genetic analysis of adenovirus type 2. IV. Coordinate regulation of polypeptides 80k, IIIa and V. Virology 76:709–724

White DO, Scharff MD, Maizel JV Jr (1969) The polypeptides of adenoviruses. III. Synthesis in infected cells. Virology 38:395–406

Wigand R, Bartha A, Dreizin RS, Esche H, Ginsberg HS, Green M, Hierholzer JC, Kalter SS, McFerran JB, Pettersson U, Russell WC, Wadell G (1982) Adenoviridae, second report. Intervirology 18:169–176

Wilcox WC, Ginsberg HS (1961) Purification and immunological characterization of types 4 and 5 adenovirus soluble antigens. Proc Natl Acad Sci USA 47:512–526

Wilcox WC, Ginsberg HS (1963a) Structure of type 5 adenovirus. I. Antigenic relationship of virus structural proteins to virus specific soluble antigens from infected cells. J Exp Med 118:295–306

Wilcox WC, Ginsberg HS (1963b) Production of specific neutralizing antibody with soluble antigens of type 5 adenovirus. Proc Soc Exp Biol Med 114:37–42

Wilcox WC, Ginsberg HS, Anderson TF (1963) Structure of type 5 adenovirus. II. Fine struc-

ture of virus subunits, morphologic relationship of structural subunits to virus-specific soluble antigens from infected cells. J Exp Med 118:307–314
Wilcox N, Mautner V (1976) Antigenic determinants of adenovirus capsids. I. Measurement of antibody cross-reactivity. J Immunol 116:19–24
Williams JF, Young CS, Austin PE (1974) Genetic analysis of human adenovirus type 5 in permissive and non-permissive cells. Cold Spring Harbor Symp Quant Biol 39:427–437
Winberg G, Wadell G (1977) Structural polypeptides of adenovirus type 16 incomplete particles. J Virol 22:389–401
Winters WD, Russell WC (1971) Studies on the assembly of adenovirus in vitro. J Gen Virol 10:181–194
Winters WD, Brownstone A, Pereira HG (1970) Separation of adenovirus penton base antigens by preparative gel electrophoresis. J Gen Virol 9:105–110
Wolfson J, Dressler D (1972) Adenovirus 2 DNA contains an inverted terminal repetition. Proc Natl Acad Sci USA 69:3054–3057
Younghusband HB, Bellett AJD (1971) Mature form of DNA from chicken-embryo-lethal orphan virus. J Virol 8:265–274

The Mechanism of Adenovirus DNA Replication and the Characterization of Replication Proteins

John S. Sussenbach and Peter C. van der Vliet

1 Introduction

The replication of the genome of human adenoviruses in permissive host cells is a highly efficient process. During the infection cycle a total amount of viral DNA is synthesized comparable to the entire chromosomal DNA content of the infected cell. In order to unravel the mechanism of adenovirus DNA replication this process has been studied in a number of experimental systems.

Biochemical and electron microscopical analysis of adenovirus type 5 (Ad5) replicative intermediates isolated from infected nuclei (Sussenbach et al. 1972; Sussenbach and Kuijk 1977, 1978) or infected cells (Lechner and Kelly 1977) has revealed that Ad5 DNA replication proceeds via a displacement mechanism. Replication may start at either end of the linear double-stranded genome and proceeds by displacement of a parental strand. After completion of the displacement synthesis, the displaced parental strand is converted into a normal daughter molecule by complementary strand synthesis.

Laboratory for Physiological Chemistry, State University of Utrecht, Utrecht, The Netherlands

Current Topics in Microbiology and Immunology, Vol. 109

More insight in the processes of initiation of DNA replication and chain elongation was obtained from the analysis of an in vitro DNA replication system employing nuclear extracts (CHALLBERG and KELLY 1979a). With this system it has been demonstrated that the preferential template in Ad5 DNA replication is a DNA-protein complex consisting of a linear double-stranded DNA molecule with a MW of 23×10^6 and a protein with a MW of 55000 (REKOSH et al. 1977). Each terminus of the DNA molecule is covalently linked to one copy of this protein (terminal protein; *TP*) via a phosphodiester bond between a serine residue in TP and the 5′-terminal deoxycytidylic acid residues (DESIDERIO and KELLY 1981).

Replication of viral DNA starts by linkage of the first nucleotide of each daughter strand to an 80000-dalton precursor of the terminal protein (pTP). This reaction is catalyzed by an adenovirus-coded DNA polymerase which can also perform chain elongation (LICHY et al. 1982; STILLMAN et al. 1982b). Further, viral DNA replication requires the presence of a functional virus-coded DNA-binding protein (DBP) with a MW of 72000 (VAN DER VLIET and LEVINE 1973; VAN DER VLIET et al. 1975; KAPLAN et al. 1979). This protein plays a role in initiation as well as chain elongation (VAN DER VLIET and SUSSENBACH 1975; VAN DER VLIET et al. 1977). In addition to these virus-coded proteins, cellular proteins are able to stimulate adenovirus DNA replication in the in vitro system (NAGATA et al. 1982). This report deals with both the mechanism and the enzymology of Ad5 DNA replication. Data will be presented on the structure of the origin of DNA replication using fragments devoid of TP. In addition, the characterization of the virus-coded proteins required for initiation and elongation will be discussed.

2 Adenovirus DNA Replication In Vitro with DNA-TP and with Protein-Free DNA Fragments

After the initial experiments described by CHALLBERG and KELLY (1979a), it soon became clear that nuclear extracts from cells, infected in the presence of 3–10 m*M* hydroxyurea, contain all the essential components to replicate a DNA-TP template at a rate which almost equals the in vivo rate. So far, the observed polymerization reactions are limited to displacement synthesis. Both parental DNA-TP and progeny DNA-pTP can be used effectively as template, as indicated by the appearance of labeled single-stranded DNA after prolonged incubation (HORWITZ and ARIGA 1981; VAN BERGEN and VAN DER VLIET 1983).

We have studied the replication in nuclear extracts of Ad5 DNA-TP that has been predigested with *Xba*I. Although preferential labeling of the TP-containing fragments occurs, internal fragments also become labeled aspecifically (Fig. 1, lanes 1 and 2). This repair-like reaction, also observed

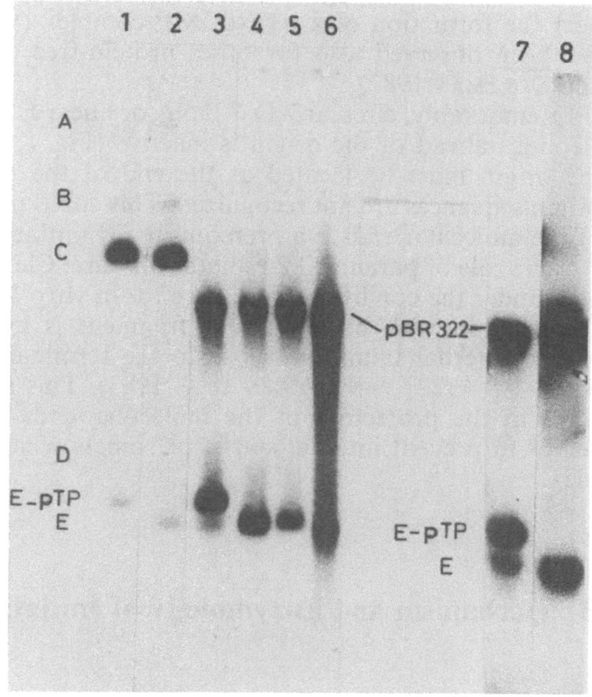

Fig. 1. Replication of origin-containing Ad5 DNA fragment that does not carry the TP. Nuclear extracts from Ad5-infected (*lanes 1–5, 7, 8*) or uninfected (*lane 6*) HeLa cells were prepared as described (CHALLBERG and KELLY 1979a) and incubated with *Xba*I-digested DNA-TP (lanes 1, 2), *Eco*RI digested XD-7 DNA (*lanes 3, 4, 6, 7*), XD-7 DNA digested with *Xba*I and *Hin*dIII (*lane 5*), or undigested XD-7 DNA (*lane 8*). The samples in lanes 2 and 4 were treated with pronase before agarose-SDS gel electrophoresis. The undigested XD-7 DNA (*lane 8*) was digested with *Eco*RI after incubation in the nuclear extract. Details of the incubation conditions are given elsewhere (VAN BERGEN et al. 1983). The positions of the protein-free *Xba*I Ad5 DNA fragments (*A–E*) are indicated

when deproteinized DNA-TP is used as template (CHALLBERG and KELLY 1979b), can be suppressed in part by preincubation of the extracts and also by addition of aphidicolin at low concentration (10–30 μ*M*). This suggests that DNA polymerase α is involved in this reaction.

The difference between the repair reaction and true replication of the terminal fragments C and E can be observed after gel electrophoresis in the presence of sodium dodecyl sulphate (SDS Fig. 1, lanes 1 and 2). Newly synthesized fragments C and E contain the 82000-dalton pTP while the aspecifically labeled C and E fragments contain the 55000-dalton TP. This leads to a mobility difference which can be detected most easily for the smaller E fragment. We have used this criterium to study the replication of protein-free E fragment obtained after *Eco*RI digestion of the plasmid XD-7. Such a fragment is completely devoid of the peptides which remain after proteolytic digestion of DNA-TP and which may be inhibitory for initiation. As shown in Fig. 1, lane 3, the newly synthesized E fragment comigrates with the pTP-containing E fragment. A number of observations indicate that the E fragment indeed contains pTP and has replicated by the same protein-primed mechanism as DNA-TP. These observations are (a): The position of the E fragment is shifted after proteolysis (Fig. 1, lane 4), (b) Protein-containing E fragments are only synthesized in extracts from infected cells (Fig. 1, lane 6). (c) The protein-free E fragment can support the formation of a pTP-dCMP complex (van Bergen et al. 1983) as has been observed also for other protein-free origin fragments (Tamanoi and Stillman 1982).

Remarkably, circular XD-7 DNA or linear E fragment containing 29 nucleotides ahead of the origin is inactive (Fig. 1, lane 5). This suggests that the origin must be located at the end of the molecule and that internal origin sequences are not recognized. This could mean that partial unwinding of the molecular ends is a prerequisite for initiation.

The role of parental TP remains obscure. Clearly, this protein is dispensable under the conditions used here for in vitro DNA replication. However, the reaction with protein-free E fragment is less efficient and also gives rise to internal initiations close to the terminal nucleotide (Tamanoi and Stillman 1982; van Bergen et al. 1983). This may indicate that TP functions in the protection of the molecular ends against nucleolytic attack, serves to prevent internal starts, or somehow stabilizes the initiation complex.

3 Mechanism and Enzymology of Initiation

3.1 Role of pTP in Initiation

Early studies of the properties of replicative intermediates from infected cells have shown that newly synthesized adenovirus DNA contains covalently bound protein (Coombs et al. 1978; Kelly and Lechner 1978; Stillman and Bellett 1979; van Wielink et al. 1979). Although a similarity with the TP was suspected, these studies did not reveal the identity of the protein. In vitro studies made it clear that the protein in question is an 80000- to 87000-dalton precursor of the TP encoded by the E2B region (Challberg et al. 1980; Lichy et al. 1981; Stillman et al. 1981). These results were confirmed for replicative intermediates from infected cells (Challberg and Kelly 1981). pTP becomes bound to the newly synthesized DNA as part of the initiation reaction. pTP remains bound to the replicating molecule and is only processed to TP late in infection (Challberg and Kelly 1981).

The initiation reaction can be studied easily by incubation of nuclear extracts in the presence of [α-^{32}P]dCTP and 2′-3′-dideoxy ATP to block any further elongation. A labeled pTP-dCMP complex is formed in the presence of DNA-TP or protein-free DNA, but not with DNA-TP deproteinized by proteolytic enzymes, presumably because the remaining peptides are inhibitory. The reaction requires Mg^{++} and ATP, which can be substituted by dATP or GTP but not efficiently by pyrimidine triphosphates (de Jong et al. 1983).

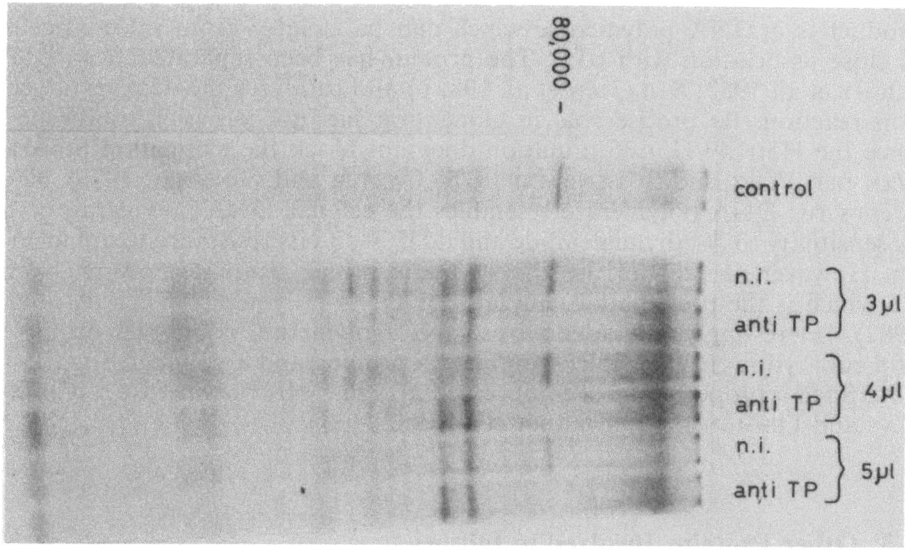

Fig. 2. In vitro initiation is inhibited by anti-TP IgG. In vitro initiation was performed with intact DNA-TP as cofactor in the presence of various amounts of anti-TP IgG or nonimmune (*n.i.*) IgG. Products were analyzed on a 10% polyacrylamide gel. The position of the 80000-dalton pTP-dCMP complex is indicated

The role of TP and pTP in Ad5 DNA replication was studied immunologically employing an antiserum raised against TP and pTP. This antiserum was obtained by immunization of a guinea pig with viral TP isolated from purified DNA-TP by cleavage with nuclease S_1 (RIJNDERS et al. 1983). The IgG fraction of this serum was tested in the in vitro DNA replication system employing the terminal *Xba*I E fragment obtained by digestion of the plasmid XD-7 by Eco RI.

Direct analysis of the initiation reaction shows that anti-TP IgG inhibits the formation of the pTP-dCMP complex (Fig. 2) (RIJNDERS et al., to be published). Even limited amounts of anti-TP IgG are sufficient to block this reaction.

3.2 Functions of the Adenovirus DNA Polymerase

An important finding was the temperature-sensitive initiation reaction in extracts from cells infected with H5*ts*36 or H5*ts*149 (STILLMAN et al. 1982b; OSTROVE et al. 1983; VAN BERGEN and VAN DER VLIET 1983). These mutants map in early region E2B outside the pTP, suggesting that another E2B protein is required for pTP-dCMP formation. Complementation studies (STILLMAN et al. 1982b; OSTROVE et al. 1983) have shown that the H5*ts*36/H5*ts*149 product is a DNA polymerase which can be isolated from infected cells in close association with pTP. The protein has been separated from pTP (LICHY et al. 1982; STILLMAN et al. 1982b) and catalyzes the deoxycytidylation reaction. Its precise role in elongation has not yet been established since the H5*ts*149/H5*ts*36 mutation does not block the elongation process (VAN DER VLIET and SUSSENBACH 1975; CARTER and GINSBERG 1976). The adenovirus DNA polymerase resembles the cellular DNA polymerase γ in its sensitivity to *N*-ethylmaleimide and ddTTP and its resistance to aphidicolin. However, it can not use the synthetic primer-template poly rA-oligo dT which is the preferential template for DNA polymerase γ (LICHY et al. 1982). Interestingly, the adenovirus DNA polymerase is another example of a eukaryotic enzyme which combines a priming and a polymerizing function. Such a property has been observed also in DNA polymerase α (CONAWAY and LEHMAN 1982; YAGURA et al. 1982).

3.3 Other Proteins Involved in Initiation

In addition to pTP and DNA polymerase, at least two other proteins may be required for initiation. One is a cellular protein of 47000 daltons (factor I) which stimulates the reaction (NAGATA et al. 1982). Another is the 72000-dalton DBP which stimulates the reaction in the presence but inhibits in the absence of factor I (NAGATA et al. 1982). A role of DBP in initiation was inferred from earlier in vivo experiments on the phenotype of H5*ts*125 (VAN DER VLIET and SUSSENBACH 1975). The results obtained with the purified proteins seem to contradict the observation that in crude H5*ts*125 nuclear extracts the inititation reaction is not temperature sensitive (CHALLBERG et al. 1982; VAN BERGEN and VAN DER VLIET 1983; FRIEFELD et al. 1983). A possible explanation is that a protein from uninfected cells can substitute for DBP during initiation.

3.4 Nature of the Origin of Replication

The deoxycytidylation of pTP is a template-dependent process. The reaction occurs preferentially with DNA-TP from the homologous serotype, but DNA-TP from five different serological subgroups can support initiation in Ad2 nuclear extracts (STILLMAN 1981; STILLMAN et al. 1982a). This suggests that a conserved sequence, especially the region between base pairs 9 and 22, is involved in initiation. A pTP-dCMP complex also forms in the presence of protein-free templates (TAMANOI and STILLMAN 1982; VAN BERGEN et al. 1983), indicating that nucleic acid–protein interactions occur during initiation and that an interaction between the parental TP and initiation

Table 1. Replication in vitro of various protein-free template DNAs

Template	3′ Sequence			Replication
Ad5 – TP	GTAGTAGT ▲	TATTATATGGAAT	A – –	+
XD-7	GTAGTAGT ▲ ▲	TATTATATGGAAT	A – –	+
pAd12 RICl	GTACGATAGAT	TATTATATGGAAT	A – –	+
pAd12 RIC3	GTAC	TATTATATGGAAT	A – –	+/–
pAd12 RIC7	GTACTAGAT	TATTATATGGAAT	A – –	–
pAsc2	GGGATAGAT	TATTATATGGAAT	A – –	+
pAd12 RIA		GTACTGGAAT	A – –	–
pLA1	GGGTAGTAGT ▲ ▲	TATTATATGGAAT	A – –	+

The plasmid DNA was linearized with *Eco*RI to obtain the 3′ sequence depicted. Except for the Ad5-TP, the 5′ ends (not shown) contain, in addition to the complementary bases, a 5′-AATT protruding end. *Dashed vertical lines* border the conserved sequence retained in most serotypes. *Triangles* indicate the position where new strands can be initiated. The results of pLA1 are from TAMANOI and STILLMAN (1982). pAsc 2 was obtained from P.H. GALLIMORE (BYRD et al. 1982), the other Ad12 plasmids from J.C. BOS and R. BERNARDS (BOS et al. 1981), and XD-7 from J. CORDEN

proteins, if present, is not sufficient. The activity of protein-free DNA opens the possibility to study the nucleotide requirements in more detail using recombinant DNA technology and site-directed mutagenesis of the origin region. We have studied a number of Ad12 terminal fragments containing various deletions and mutations in the terminal nucleotides (BOS et al. 1981; BYRD et al. 1982). A deletion extending into the conserved region 9–22 was inactive as template, in agreement with data obtained from heterologous systems. Some deletions in the first eight nucleotides are permitted while others are not. A summary of the results is given in Table I. It appears that the nucleotide sequence surrounding the exact starting point can vary considerably. The results can be explained by the following model: The pTP-DNA polymerase complex recognizes the internal conserved sequence 9–22. Next, a G-residue in the template strand, preceding this region, is used to basepair with the dCMP, which becomes covalently bound to pTP. The distance between the G-residue and the conserved sequence, rather than the sequence surrounding the guanine, determines whether initiation can occur. Based on the results with the small number of deletion mutants studied so far, a distance of four to eight nucleotides seems permitted, although the efficiency differs depending upon the position of the G-residue. Direct binding studies will be required to establish whether, and in what manner, the pTP-DNA polymerase complex binds to DNA.

4 DNA Chain Elongation

4.1 ATP Dependency

The aspect of adenovirus DNA replication that has been studied first in vitro is the elongation of endogenous replicative intermediates that have initiated DNA replication in intact cells. Systems capable of elongation and deficient for initiation are either intact nuclei (VAN DER VLIET and SUSSENBACH 1972; WINNACKER 1975) or replication complexes isolated from these nuclei (WILHELM et al. 1976; YAMASHITA et al. 1977; KAPLAN et al. 1977).

A remarkable feature of adenovirus DNA chain elongation is its relatively low ATP dependency. Reduction of the ATP concentration in isolated nuclei from the optimal (8 m*M*) to the endogenous value (0.16 μ*M*) reduced adenovirus DNA replication only by 55%–70% (CHEN et al. 1980; DE JONG et al. 1983), while host cell DNA replication was almost tenfold reduced. Also, ATPγ-S which inhibits cellular DNA replication competitively with ATP only slightly affects viral DNA replication (DE JONG et al. 1983). Similar cases of ATP-independent DNA replication have been described for herpes simplex virus and vaccinia virus DNA replication (HIRSCHHORN and ABRAMS 1978; BERGER et al. 1978). The reason for the reduced need for ATP during elongation, which contrasts to its requirement in initiation, remains unclear. Possibly, dNTP hydrolysis occuring during elongation or the binding of DBP provides enough energy for the elongation process.

4.2 Sensitivity to 2′-3′-Dideoxynucleoside Triphosphates and Aphidicolin

Adenovirus DNA chain elongation is extremely sensitive to 2′-3′-dideoxynucleoside triphosphates (ddNTPs) and differs in this aspect also from host cell DNA replication. This result was originally taken as evidence for the participation of the cellular DNA polymerase γ in adenovirus DNA replication (VAN DER VLIET and KWANT 1978; KROKAN et al. 1979; CHEN et al. 1980; SHAW et al. 1980). However, since the discovery that the E2B region codes for a DNA polymerase (LICHY et al. 1982; STILLMAN et al. 1982b) it seems less likely that DNA polymerase γ is involved. The adenovirus DNA polymerase displays a sensitivity to ddTTP similar to that of DNA polymerase γ (LICHY et al. 1982).

An as yet unexplained phenomenon is the sensitivity of adenovirus DNA synthesis to high concentrations of aphidicolin. This tetracyclic terpenoid inhibits cellular DNA replication at low concentrations, most likely by inhibition of DNA polymerase α (SPADARI et al. 1982). Adenovirus DNA repli-

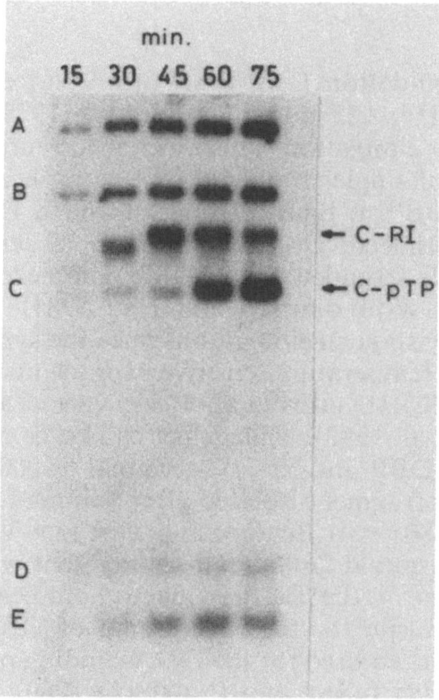

a

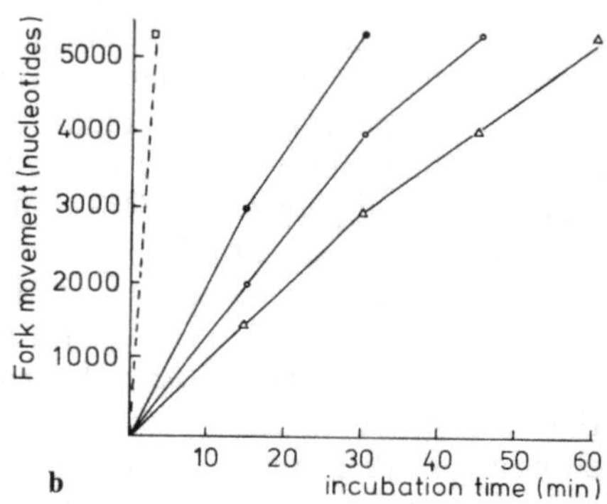

b

Fig. 3a, b. Individual rate of fork movement is inhibited by aphidicolin. **a** Ad5 DNA-TP was digested with *Xba*I and incubated for various periods in nuclear extracts as indicated. Aphidicolin was present at 300 μ*M*. The positions of the *Xba*I fragments are indicated (*A–E*). C and E are the terminal fragments. *C-RI* is the replicative intermediate consisting of newly synthesized C fragment containing a full-length displaced single strand. Molecules migrating between C-pTP and C-RI are in various stages of replication. **b** The rate of fork movement was calculated from the position of the replicative intermediate in the gel, assuming a semilogarithmic relation between the mobility and the length of the displaced strand. Aphidicolin concentrations were: ●, 100 μ*M*; ○, 300 μ*M*; △, 600 μ*M*. In the absence of aphidicolin, C-RI was first observed after only 2.5 min (□)

cation is only inhibited by 300- to 400-fold higher concentrations of the drug (LONGIARU et al. 1979; KROKAN et al. 1979; KWANT and VAN DER VLIET 1980; PINCUS et al. 1981), but even under these conditions the formation of a pTP-dCMP complex is uninhibited. This indicates that the drug acts on the level of DNA chain elongation. Direct evidence of this has been obtained in vitro (PINCUS et al. 1981; VAN DAM and VAN DER VLIET, unpublished) (see also Fig. 3). The adenovirus DNA polymerase is completely resistant to aphidicolin at concentrations which inhibit DNA chain elongation (LICHY et al. 1982). Thus, either aphidicolin must have another target within the replication fork, or the adenovirus DNA polymerase must become sensitive to the drug by interaction with other replication proteins. At present, no adequate explanation for the intriguing aphidicolin sensitivity can be given.

4.3 Role of DBP

Using an antiserum against the 72000-dalton DBP (VAN DER VLIET et al. 1975; KEDINGER et al. 1978) or the H5*ts*125 mutant (HORWITZ 1978), it was shown that the DBP is required for elongation. This was not too surprising in view of the abundant amount of single-stranded DNA in replicative intermediates and the ability of the DBP to bind to single-stranded DNA (VAN DER VLIET and LEVINE 1973). Indeed, electron microscopic studies have confirmed the presence of single-stranded DNA-DBP complexes in infected cells (KEDINGER et al. 1978; WOHLGEMUTH and HSU 1981). The DBP appears to be required at an early stage in elongation, since the synthesis of the first 26 nucleotides in vitro is temperature sensitive in crude nuclear extracts from H5*ts*125-infected cells (CHALLBERG et al. 1982; VAN BERGEN and VAN DER VLIET 1983; FRIEFELD et al. 1983). This defect can be complemented both by intact 72000-dalton DBP and by a C-terminal 34000- or 45000- to 48000-dalton DNA-binding fragment isolated after a limited chymotrypsin treatment (ARIGA et.al. 1980; VAN BERGEN and VAN DER VLIET 1983; FRIEFELD et al. 1983). The N-terminal 26000-dalton fragment which carries most of the phosphate groups of the DBP is inactive, indicating that phosphorylation is not a prerequisite for the functioning of DBP in elongation. Phosphorylation is also not required for the DNA-binding properties of the protein (KLEIN et al. 1979; LINNE and PHILIPSON 1980). The DBP may function primarily in the protection of single-stranded DNA, or possibly in the formation of a DNA-DBP structure which is suitable for elongation.

4.4 Role of TP and pTP in Elongation

As described in a previous section, a specific antiserum against TP and pTP is able to inhibit in vitro DNA replication and block the initiation reaction. A possible effect of the anti-TP serum on elongation was studied in isolated nuclei of infected cells and in the in vitro DNA replication system. Isolated nuclei from infected and uninfected HeLa cells were preincubated with anti-TP IgG, while as a control incubations were also performed with anti-DBP IgG. After preincubation of the nuclei with IgG for 1 h at 0 °C, DNA synthesis was measured by the incorporation of [α-^{32}P]dCTP in TCA-insoluble material. The results of these incubations are shown in Fig. 4. It appears that without additions the infected nuclei incorporate considerably more radioactivity than uninfected nuclei. Addition of anti-DBP IgG inhibits DNA replication in infected nuclei by about 50%, in agreement with previous observations (VAN DER VLIET et al. 1977). Interestingly, addition of anti-TP IgG reduced DNA synthesis in infected nuclei to the level of DNA synthesis in uninfected nuclei. Thus, anti-TP IgG has a strong effect on the elongation of replicating adenovirus DNA, suggesting that TP and/or pTP are involved not only in initiation, but also in chain elongation. In order to detect whether anti-TP IgG has a

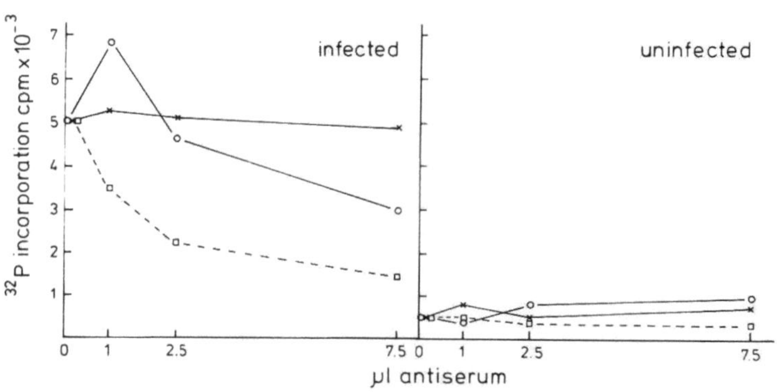

Fig. 4. DNA replication in isolated nuclei. DNA replication was measured in nuclei from Ad5-infected HeLa cells or uninfected HeLa cells in the presence of [α-^{32}P]dCTP according to VAN DER VLIET and SUSSENBACH (1972). Nuclei were preincubated with nonimmune IgG (×——×), anti-DBP IgG (○——○), or anti-TP IgG (□——□)

similar effect on chain elongation in nuclear extracts, undigested DNA-TP was incubated in the presence of [α-^{32}P]dCTP for 10 min at 37 °C allowing initiation and partial chain elongation. Then anti-TP IgG or nonimmune IgG was added and the incubation continued for an additional 50 min. Fig. 5 shows that after 10 min incubation only the two terminal *Xba*I fragments E and C are labeled (lane 1), whereas after 60 min labeling in the absence of IgG the internal fragments also contain radioactivity (lane 3). However, in the presence of anti-TP IgG, labeling of internal fragments is strongly inhibited (lane 5). A quantitative analysis of the labeled bands indicates that during the initial 10 min, replication proceeds only 15% inwards of the adenovirus genome. After 60 min incubation in the absence of IgG's, radioactivity is found in all fragments, including the internal fragments A and D. This internal labeling is slightly reduced by addition of nonimmune IgG, but is strongly inhibited by addition of anti-TP IgG.

These results indicate that addition of anti-TP IgG almost completely blocks chain elongation both in nuclear extracts and in isolated nuclei, suggesting that TP and/or pTP are involved not only in initiation, but also in chain elongation. Unfortunately, we cannot discriminate between the possibilities: whether TP, pTP, or both are involved in elongation. We assume that these proteins comigrate in the replication fork with the DNA polymerase (Fig. 6). From the alternatives indicated in Fig. 6, the first is the most attractive. According to this model, parental TP is associated with the DNA polymerase. An attractive consequence of this model is that the two ends of the displaced parental single-strand are brought close to each other at the end of the displacement synthesis, thus facilitating the formation of a panhandle structure. This structure arises due to the inverted terminal repetition (STEENBERGH et al. 1977) and has been proposed to be an intermediate in complementary strand synthesis (SAMBROOK, cited in DANIELL 1976; LECHNER and KELLY 1977). For steric reasons, alternatives 2 and 3 are less likely.

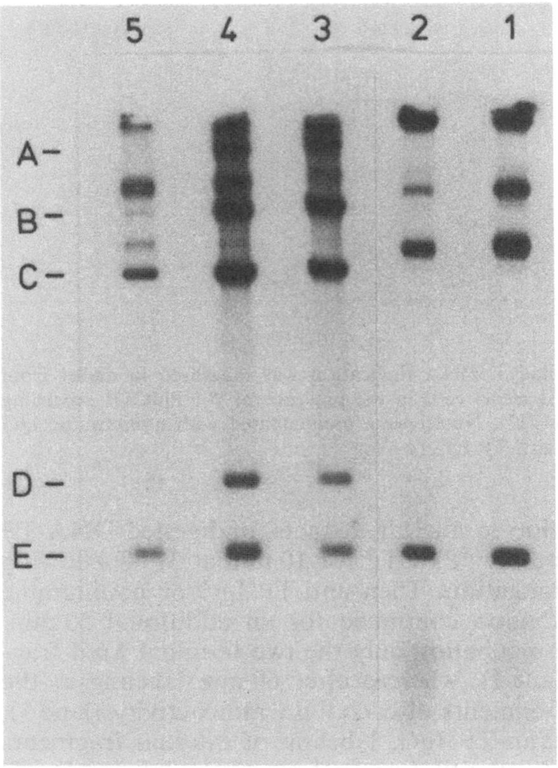

Fig. 5. Inhibition of the elongation reaction using nuclear extracts. Undigested Ad5 DNA-TP was used as template. After the incubations the DNA was purified from the replication mixtures, digested with *XbaI*, and analyzed on a 1% agarose gel. *Lane 1*: 10 min incubation in the absence of IgGs. *Lane 2*: 10 min incubation in the presence of 4 µl nonimmune IgG. *Lane 3*: 60 min incubation in the absence of IgGs. *Lane 4*: 10 min incubation in the absence of IgGs followed by 50 min incubation in the presence of 4 µl nonimmune IgG. *Lane 5*: 10 min incubation in the absence of IgGs followed by 50 min incubation in the presence of 4 µl anti-TP IgG. Lanes 1 and 2 were exposed for 60 h, lanes 3, 4, and 5 for 16 h

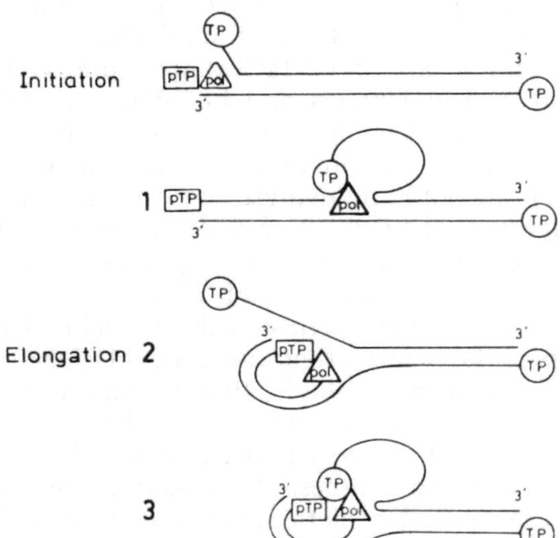

Fig. 6. Models for the involvement of TP or pTP in the elongation reaction. After initiation elongation might proceed in one of three ways. *1* The parental terminal protein ((TP)) is associated with the DNA polymerase (pol) during displacement synthesis. *2* The precursor of the terminal protein (pTP) comigrates with the DNA polymerase. *3* TP and pTP are both associated with the DNA polymerase

5 Relation Between Structure and Function of DBP

Phenotypic analysis of the mutants H5*ts*125 and H5*ts*107 has revealed that DBP is a multifunctional protein involved in initiation and chain elongation of viral DNA synthesis, in early gene expression, and in transformation (GINSBERG et al. 1974; VAN DER VLIET and SUSSENBACH 1975; VAN DER VLIET et al. 1977; HORWITZ 1978; CARTER and BLANTON 1978; NEVINS and JENSEN-WINKLER 1980; BABICH and NEVINS 1981). Characterization of the mutant H5*hr*404 has shown that DBP is also involved in late gene expression (KLESSIG and GRODZICKER 1979; KLESSIG and CHOW 1980). In order to unravel the relation between the structure and the functions of this protein, we have studied in detail the genes of DBPs from wild-type Ad5 and some mutants and the primary structures of the DBPs as derived from the nucleotide sequences of the genes (KRUIJER et al. 1981, 1983a). These studies have revealed that wild-type Ad5 DBP is a protein of 529 amino acid residues and that the mutants H5*ts*125 and H5*ts*107 are located in a different part of the DBP molecule than H5*hr*404 (KRUIJER et al. 1981, 1983a). The H5*ts*125 and H5*ts*107 mutations are located in the C-terminal 45000- to 48000-dalton fragment of DBP that binds to single-stranded DNA, while the H5*hr*404 mutation has been mapped in the N-terminal 26000-dalton fragment that is phosphorylated and does not bind to single-stranded DNA. These results suggest that DBP is composed of at least two domains: a C-terminal domain involved in DNA replication and early gene expression and an N-terminal domain involved in late gene expression.

A major role of the C-terminal fragment in DNA replication is supported by the results of the characterization of revertants of the mutants H5*ts*125 and H5*ts*107 (KRUIJER et al. 1983a). Both mutants appear to carry a proline → serine substitution at position 413 in the C-terminal region of DBP, which leads to a temperature-sensitive DNA replication and a thermolabile DBP molecule as demonstrated by the increased sensitivity of mutant DBP to proteolysis.

Several revertants of H5*ts*125 and H5*ts*107 with a temperature-independent DNA replication in HeLa cells carry, in addition to the original H5*ts*125/H5*ts*107 mutation, second-site mutations which are all located within the C-terminal fragment (KRUIJER et al. 1983a). This emphasizes an important role of this domain in viral DNA replication. It is noteworthy that revertant DBP has not regained wild-type thermostability in all cases, indicating that the tertiary structure of revertant DBP is not always identical to the wild-type structure.

Further information on the relation between structure and function was derived from a comparison of the primary structure of Ad5 and Ad12 DBP as deduced from the nucleotide sequence of the corresponding genes (KRUIJER et al. 1983b). This analysis shows that Ad12 and Ad5 DBP contain 484 and 529 amino acids respectively. In order to align the Ad5 and Ad12 DBP amino acid sequences with the highest degree of homology, gaps have to be introduced at various positions in both sequences (Fig. 7).

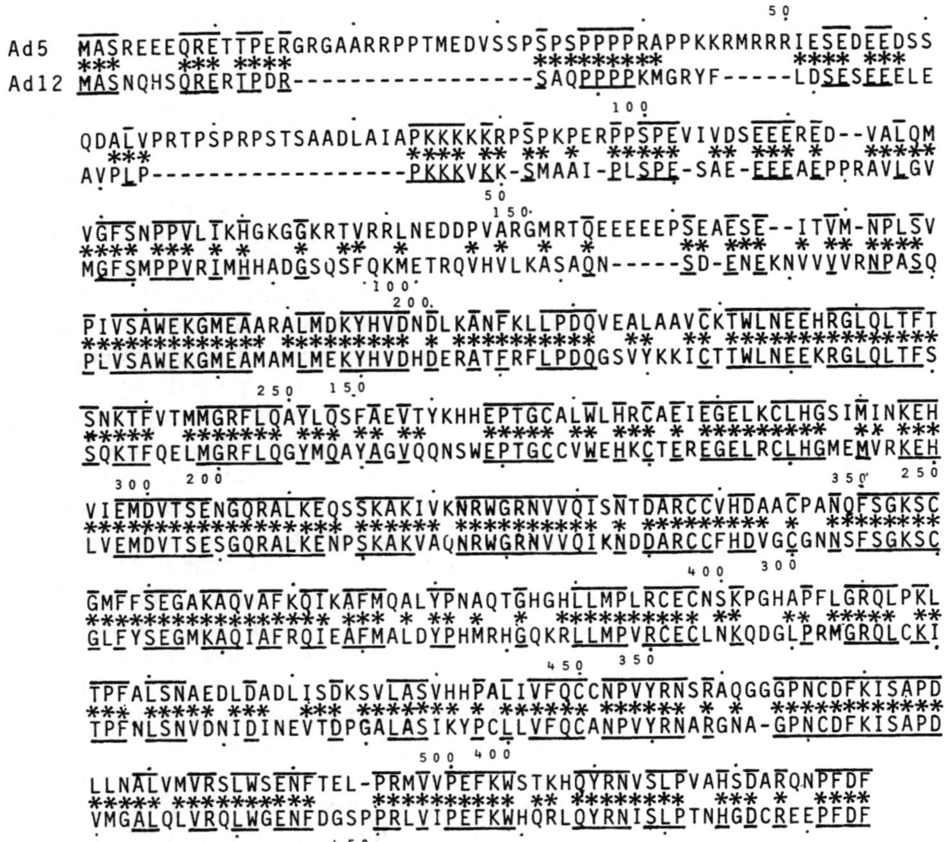

Fig. 7. Comparison of the Ad12 (*lower* sequence) and Ad5 (*upper* sequence) amino acid sequences. Positions with similar amino acids are indicated by *asterisks*, those with identical amino acids by *lines* above and below the sequences

The location of the gaps indicates that mainly stretches of amino acids from the N-terminal 170 amino acids of Ad5 DBP are absent in Ad12 DBP. In contrast, no extensive deletions or insertions are found in the C-terminal 359 amino acids of either DBPs. Only amino acid residue 449 in the Ad12 sequence is absent in the Ad5 sequence, while amino acid residue 463 in the Ad5 sequence is not found in the Ad12 sequence.

Comparison of the amino acid sequences of Ad5 and Ad12 DBP provides further evidence that DBP consists of at least two distinct domains. It appears that the C-terminal regions are homologous to a high degree (80%), while the N-terminal regions show only a limited degree of homology (45%). Obviously, in the N-terminal domain, which is involved in late gene expression, extensive structural alterations are tolerated without loss of function. On the other hand, the high degree of homology of the C-terminal region suggests that for optimal functioning of this domain only limited amino acid changes are allowed (Kruijer et al. 1983b).

Unfortunately, it is not possible to substantiate these conclusions on the relation between structure and function further without detailed information on the conformation of the DBP molecule. Since a three-dimensional model of DBP is not available, we have employed cumputerized versions of the methods of CHOU and FASMAN and LIM to predict the secondary structures of Ad5 and Ad12 DBP (LIM 1974; LENSTRA et al. 1977; CHOU and FASMAN 1978). The hydrophobicity of various regions in Ad5 and Ad12 DBP was calculated employing a five-point moving average (ROSE 1978). CHOU and FASMAN's method predicts 30.5% α-helix, 14.9% β-sheet, and 16.4% turn structure for Ad5 DBP and 34.4% α-helix, 22.9% β-sheet, and 13.4% turn structure for Ad12 DBP. LIM's procedure predicts 19.7% α-helix and 18.0% β-sheet structure for Ad5 DBP and 19.2% α-helix and 18.2% β-sheet structure for Ad12 DBP. At the positions of hydrophobic minima α-helices and turns are frequently found (ROSE 1978). Although there are many regions of homology in the hydrophobicity patterns, the predicted positions of α-helices and β-sheets in Ad5 and Ad12 DBP according to CHOU and FASMAN and LIM show only limited similarity (Fig. 8). This is rather surprising for the C-terminal fragment, since Ad5 and Ad12 are 80% homologous in this region. However, it should be noted that these predictions are not valid for regions which are involved in interactions with other components. Since this is almost certainly the case with DBP, the secondary structure predictions should be interpreted with caution.

As for the role of DBP in early and late gene expression, this is almost completely obscure. It has been shown that DBP can form complexes with RNA, and UV cross-linking studies have revealed that the binding site for RNA is located in the C-terminal fragment of DBP (BABICH and KLESSIG, personal communication). We assume that binding to RNA is essential for the effect of DBP on the stability of early mRNA, as detected in H5*ts*125 (CARTER and BLANTON 1978; BABICH and NEVINS 1981).

Phenotypic analysis of H5*hr*404 and H5r(*ts*125)13 has shown that the effect of DBP on late viral transcripts and autoregulation of DBP levels is host cell specific (KLESSIG and GRODZICKER 1979; NICOLAS et al. 1982). This suggests that DBP functions in these processes by interactions with different host cell components.

6 Concluding Remarks

With the help of an in vitro DNA replication system, many details of the mechanism of adenovirus DNA replication have been elucidated. It is evident that three virus-coded proteins play a major role in this proces. Initiation of DNA replication requires the presence of pTP and the adenovirus DNA polymerase, while the DNA-TP complex from virions is the preferred template. There are indications that with purified replication proteins, the virus-coded DBP stimulates initiation. A stimulatory effect has also been shown for a host cell protein (NAGATA et al. 1982). For DNA chain elonga-

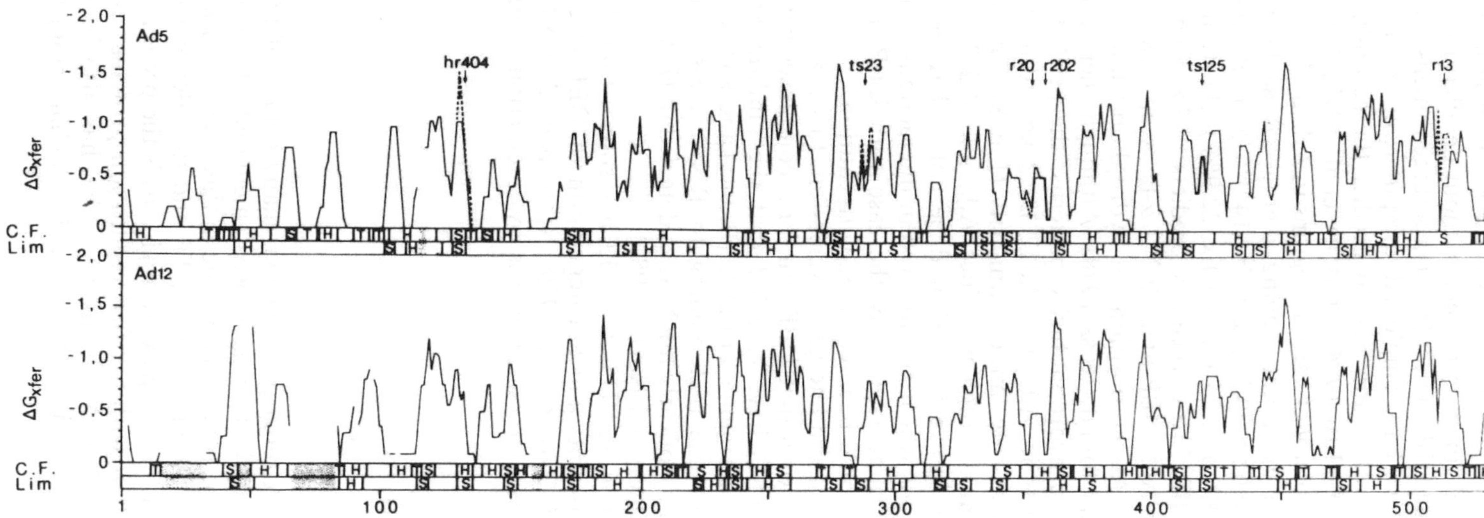

Fig. 8. Hydrophobicity profiles and predicted secondary structure of Ad5 and Ad12 DBP. The hydrophobicity profiles according to ROSE (1978) and the predicted positions of α-helices (*H*), β-sheets (*S*) and turn structures (*T*) (CHOU and FASMAN (*C.F.*) 1978; LIM 1974) were determined for Ad5 and Ad12 DBP amino acid sequences. Hydrophobicity is expressed as $\Delta G \times$fer, the free energy of transfer in kcal/mole of amino acid residues from water to organic solvents. The Ad5 and Ad12 DBP amino acid sequences were aligned to obtain the highest degree of homology between them with the N-terminal amino acids at position 1. For maximal homology, gaps (*shaded areas*) were introduced in the Ad5 sequence at residues 113–114, 165–166, 170–171, and 494–495, and in the Ad12 sequence at residues 15–16, 28–29, 33–34, 50–51, 55–56, 60–61, 63–64, 113–114, 115–116, and 417–418. *Numbers* on the horizontal axis indicate positions in the aligned sequences. The sites of mutations and second-site reversions in mutant DBPs are indicated by *arrows*

tion, it is evident that DNA polymerase and DBP are required. We have shown that pTP is probably also involved in this reaction, while a role for host proteins in chain elongation is likely.

The enzymology of the displacement reaction has thus been characterized in some detail, but our knowledge of the complementary strand synthesis remains very limited, due to the lack of a suitable in vitro system and difficulties in separating displacement synthesis and complementary strand synthesis in vivo. It has been proposed that the displaced single-stranded DNA can form a panhandle structure which provides identical molecular ends, as present in double-stranded DNA. Thus initiation of the complementary strand synthesis may use a mechanism similar to that in displacement synthesis. However, kinetic studies (Bodnar and Pearson 1980) have indicated that the rate of initiation of complementary strand synthesis is 50 times greater than the initiation rate on double-stranded molecules, suggesting a different initiation mechanism. Indirect evidence that panhandle formation occurs in vivo comes from studies of Ad5 variants containing deletions in the left part of the inverted repetition (Stow 1982). Viable deletions were in all cases shorter than the 102-bp-long repetition, and the progeny consisted of normal wild-type virus DNA, suggesting that repair of the deletion could occur guided by the intact right-hand end. From kinetic observations it appears that the rate of complementary strand elongation is 2.6 times slower than that of displacement synthesis. This may reflect the difference in template and the difficulty in removing DBP from the single-stranded DNA.

The in vitro system for displacement replication has been developed in such a way that it has become possible to unravel in further detail the specific interactions between the different replication proteins and the DNA-TP template. The nature of the origin of DNA replication can be studied employing site-directed mutagenesis of the inverted terminal repetition. In general, a combined genetic and biochemical approach will be very fruitful for thorough understanding of adenovirus DNA replication.

Acknowledgments. The authors gratefully acknowledge gifts of XD-7 by J. Corden, of Ad12 plasmids from J.C. Bos, R. Bernards, and P.H. Gallimore and of aphidicolin from A.H. Todd (ICI). They thank B.G.M. van Bergen, W. Kruijer, A.W.M. Rijnders, P.J. de Jong, M.M. Kwant, M.G. Kuijk, P. van der Ley, W. van Driel, F.M.A. van Schaik and R. Schiphof for collaboration and assistance. Experiments performed in the laboratory of the authors was supported in part by the Netherlands Organization for Chemical Research (SON), with financial aid from the Netherlands Organization for the Advancement of Pure Research (ZWO).

References

Ariga H, Klein H, Levine AJ, Horwitz MS (1980) A cleavage product of the adenoviral DNA binding protein is active in DNA replication in vitro. Virology 101:307–310

Babich A, Nevins JR (1981) The stability of early adenovirus mRNA is controlled by the viral 72-kD binding protein. Cell 26:371–379

Berger NA, Kauff RA, Sikorsky GW (1978) ATP-independent DNA synthesis in vaccinia-infected L cells. Biochim Biophys Acta 520:531–538

Bodnar JW, Pearson GD (1980) Kinetics of adenovirus DNA replication. II Initiation of adenovirus DNA replication. Virology 105:357–370

Bos JC, Polder LJ, Bernards R, Schrier PI, van den Elsen PJ, van der Eb AJ, van Ormondt H (1981) The 2.2 kB Elb mRNA of human Ad12 and Ad5 codes for two tumor antigens starting at different AVG triplets. Cell 27:121–131

Byrd PJ, Chia W, Rigby PWJ, Gallimore PH (1982) Cloning of DNA fragments from the left end of the adenovirus type 12 genome: transformation by cloned early region I. J Gen Virol 60:279–293

Carter TH, Blanton RA (1978) Possible role of the 72000 dalton DNA binding protein in regulation of adenovirus type 5 early gene expression. J Virol 25:664–674

Carter TH, Ginsberg HS (1976) Viral transcription in KB cells infected by temperature-sensitive early mutants of adenovirus type 5. J Virol 18:156–166

Challberg MD, Kelly TJ (1979a) Adenovirus DNA replication in vitro. Proc Natl Acad Sci USA 76:655–659

Challberg MD, Kelly TJ (1979b) Adenovirus DNA replication in vitro: origin and direction of daughter strand synthesis. J Mol Biol 135:999–1012

Challberg MD, Kelly TJ (1981) Processing of the adenovirus terminal protein. J Virol 38:272–277

Challberg MD, Desiderio SV, Kelly TJ (1980) Adenovirus DNA replication in vitro: characterization of a protein covalently linked to nascent DNA strands. Proc Natl Acad Sci USA 77:5105–5109

Challberg MD, Ostrove JM, Kelly TJ (1982) Initiation of adenovirus DNA replication: detection of covalent complexes between nucleotide and the 80 kD terminal protein. J Virol 41:265–270

Chen S, Zubay G, Ginsberg HS (1980) The replication pattern of adenovirus DNA in vivo reproduced in vitro. Eur J Biochem 104:587–594

Chou PY, Fasman GD (1978) Prediction of the secondary structure of proteins from their amino acid sequence. Adv Enzymol 47:45–148

Conaway RC, Lehman IR (1982) Synthesis by the DNA primase of Drosophila Melanogaster of a primer with a unique chain length. Proc Natl Acad Sci USA 79:4584–4588

Coombs DH, Robinson AJ, Bodnar JW, James CJ, Pearson GD (1978) Detection of DNA-protein complexes: the adenovirus DNA-terminal protein and HeLa DNA protein complexes. Cold Spring Harbor Symp Quant Biol 43:741–753

Daniell E (1976) Genome structure of incomplete particles of adenovirus. J Virol 19:685–708

De Jong PJ, Kwant MM, van Driel W, Jansz HS, van der Vliet (1983) The ATP requirements of adenovirus type 5 DNA replication and cellular DNA replication. Virology 124:45–58

Desiderio SV, Kelly TJ (1981) Structure of the linkage between adenovirus DNA and the 55000 molecular weight terminal protein. J Mol Biol 145:319–337

Friedfeld BR, Krevolin MD, Horwitz MS (1983) Effects of the adenovirus H5*ts*125 and H5*ts*107 DNA binding proteins on DNA replication in vitro. Virology 124:380–389

Ginsberg HS, Ensinger MJ, Kauffman RS, Mayer AJ, Lundholm U (1974) Cell transformation: a study of regulation with types 5 and 12 adenovirus temperature-sensitive mutants. Cold Spring Harbor Symp Quant Biol 39:419–426

Hirschhorn RR, Abrams R (1978) Synthesis of herpes simplex virus DNA in soluble nuclear extracts. Biochim Biophys Res Commun 84:1129–1135

Horwitz MS (1978) Temperature-sensitive replication of H5*ts*125 adenovirus DNA in vitro. Proc Natl Acad Sci USA 75:4291–4295

Horwitz MS, Ariga M (1981) Multiple rounds of adenovirus synthesis in vitro. Proc Natl Acad Sci USA 78:1476–1480

Kaplan LM, Kleinman RE, Horwitz MS (1977) Replication of adenovirus type 2 DNA in vitro. Proc Natl Acad Sci USA 74:4425–4429

Kaplan LM, Ariga H, Hurwitz J, Horwitz MS (1979) Complementation of the temperature-sensitive defect in H5*ts*125 adenovirus DNA replication in vitro. Proc Natl Acad Sci USA 76:5534–5538

Kedinger C, Brison O, Perrin F, Wilhelm J (1978) Structural analysis of replicative intermediates from adenovirus type 2 infected HeLa cell nuclei. J Virol 26:364–379
Kelly TJ, Lechner RL (1978) The structure of replicating adenovirus DNA molecules: characterization of DNA-protein complexes from infected cells. Cold Spring Harbor Symp Quant Biol 43:721–728
Klein H, Maltzman W, Levine AJ (1979) Structure function relationships of the adenovirus DNA binding protein. J Biol Chem 254:11051–11060
Klessig DF, Chow LT (1980) Incomplete splicing and deficient accumulation of the fiber mRNA in monkey cells infected by human adenovirus type 2. J Mol Biol 139:221–242
Klessig DF, Grodzicker T (1979) Mutations that allow human Ad2 and Ad5 to express late genes in monkey cells map in the viral gene encoding the 72 kD DNA binding protein. Cell 17:957–966
Krokan H, Schaffer P, de Pamphilis ML (1979) Involvement of eucaryotic DNA polymerase α and γ in the replication of cellular and viral DNA. Biochemistry 18:4431–4443
Kruijer W, van Schaik FMA, Sussenbach JS (1981) Structure and organization of the gene coding for the DNA binding protein of adenovirus type 5. Nucleic Acids Res 9:4439–4457
Kruijer W, Nicolas JC, van Schaik FMA, Sussenbach JS (1983a) Structure and function of DNA binding proteins from revertants of adenovirus type 5 mutants with a temperature-sensitive DNA replication. Virology 124:425–433
Kruijer W, van Schaik FMA, Speijer JG, Sussenbach JS (1983b) Structure and function of adenovirus DNA binding protein. Comparison of the amino acid sequences of Ad5 and Ad12 proteins derived from the nucleotide sequence of the corresponding genes. Virology 120:140–153
Kwant MM, van der Vliet (1980) Differential effect of aphidicolin on adenovirus DNA synthesis and cellular DNA synthesis. Nucleic Acids Res 8:3993–4007
Lechner RL, Kelly TJ (1977) The structure of replicating adenovirus 2 DNA molecules. Cell 12:1007–1020
Lenstra JA, Hofsteenge J, Beintema JJ (1977) Invariant features of the structure of pancreatic ribonuclease. J Mol Biol 109:185–193
Lichy JH, Horwitz MS, Hurwitz J (1981) Formation of a covalent complex between the 80000 dalton adenovirus terminal protein and 5′-dCMP in vitro. Proc Natl Acad Sci USA 78:2678–2682
Lichy JH, Field J, Horwitz MS, Hurwitz J (1982) Separation of the adenovirus terminal protein precursor from its associated DNA polymerase: role of both proteins in the initiation of adenovirus DNA replication. Proc Natl Acad Sci USA 79:5225–5229
Lim VI (1974) Structural principles of the globular organization of protein chains. A stereochemical theory of globular protein secondary structure. J Mol Biol 88:857–872
Linne T, Philipson L (1980) Further characterization of the phosphate moiety of the adenovirus type 2 DNA-binding protein. Eur J Biochem 103:259–270
Longiaru M, Ikeda J, Jarkovsky Z, Horwitz SB, Horwitz MS (1979) The effect of aphidicolin on adenovirus DNA synthesis. Nucleic Acids Res 6:3369–3386
Nagata K, Guggenheimer RA, Enomoto T, Lichy JH, Hurwitz J (1982) Adenovirus DNA replication in vitro: identification of a host factor that stimulates synthesis of the preterminal protein-dCMP complex. Proc Natl Acad Sci USA 79:6438–6442
Nevins JR, Jensen-Winkler J (1980) Regulation of early adenovirus transcription: a protein product from early region 2 specifically represses region 4 transcription. Proc Natl Acad Sci USA 77:1893–1987
Nicolas JC, Ingrand D, Sarnow P, Levine AJ (1982) A mutation in the adenovirus type 5 DNA binding protein that fails to autoregulate the production of the DNA binding protein. Virology 122:481–485
Ostrove JM, Rosenfeld P, Williams JR, Kelly TJ (1983) In vitro complementation as an assay for purification of adenovirus DNA replication proteins. Proc Natl Acad Sci USA 80:935–939
Pincus S, Robertson W, Rekosh D (1981) Characterization of the effect of aphidicolin on adenovirus DNA replication: evidence in support of a protein primer model of initiation. Nucleic Acids Res 9:4919–4338

Rekosh DMK, Russell WC, Bellett AJD, Robinson AJ (1977) Identification of a protein linked to the ends of adenovirus DNA. Cell 11:283–295

Rijnders AWM, van Maarschalkerweerd MW, Visser L, Reemst AMCB, Sussenbach JS, Rozijn TH (1983) Expression of integrated viral DNA sequences outside the transforming region of eight adenovirus transformed cell lines. Biochim Biophys Acta 739:48–56

Rijnders AWM, van Bergen BGM, van der Vliet PC, Sussenbach JS (to be published) Immunological characterization of the role of adenovirus terminalprotein in viral DNA replication. Virology

Rose GD (1978) Prediction of chain turns in globular proteins on a hydrophobic base. Nature 272:586–590

Shaw CH, Rekosh DM, Russell WC (1980) Adenovirus DNA synthesis in vitro is catalyzed by DNA polymerase γ. J Gen Virol 48:231–236

Spadari S, Sala F, Pedrali-Noy G (1982) Aphidicolin: a specific inhibitor of nuclear DNA replication in eukaryotes. Trends Biochem Sci 7:29–31

Steenbergh PH, Maat J, van Ormondt H, Sussenbach JS (1977) The nucleotide sequence at the termini of adenovirus 5 DNA. Nucleic Acids Res 4:4371–4390

Stillman BW (1981) Adenovirus DNA replication in vitro: a protein linked to the 5′-end of nascent DNA strands. J Virol 37:139–147

Stillman BW, Bellett AJD (1979) An adenovirus protein associated with the ends of replicating DNA molecules. Virology 93:69–79

Stillman BW, Lewis JB, Chow LT, Mathews MB, Smart JE (1981) Identification of the gene and mRNA for the adenovirus terminal protein precursor. Cell 23:497–508

Stillman BW, Topp WC, Engler JA (1982a) Conserved sequences at the origin of adenovirus DNA replication. J Virol 44:530–537

Stillman BW, Tamanoi F, Mathews MB (1982b) Purification of an adenovirus-coded DNA polymerase that is required for initiation of DNA replication. Cell 31:613–623

Stow ND (1982) The infectivity of adenovirus genomes lacking DNA sequences from their left-hand termini. Nucleic Acids Res 10:5105–5119

Sussenbach JS, Kuijk MG (1977) Studies on the mechanism of replication of adenovirus DNA. V The location of termini of replication. Virology 77:149–157

Sussenbach JS, Kuijk MG (1978) The mechanism of replication of adenovirus DNA. VI Localization of the origins of the displacement synthesis. Virology 84:509–517

Sussenbach JS, van der Vliet PC, Ellens DJ, Jansz HS (1972) Linear intermediates in the replication of adenovirus DNA. Nature 239:47–49

Tamanoi F, Stillman BW (1982) Function of adenovirus terminal protein in the initiation of DNA replication. Proc Natl Acad Sci USA 79:2221–2225

Van Bergen BGM, van der Vliet PC (1983) Temperature-sensitive initiation and elongation of adenovirus DNA replication in vitro with nuclear extracts from H5*ts*36, H5*ts*149 and H5*ts*125 infected HeLa cells. J Virol 42:642–648

Van Bergen BGM, van der Ley PA, van Driel W, van Mansfeld ADM, van der Vliet PC (1983) Replication of origin containing adenovirus DNA fragments that do not carry the terminal protein. Nucleic Acids Res 11:1975–1990

Van der Vliet PC, Kwant MM (1978) Role of DNA polymerase γ in adenovirus DNA replication. Nature 276:532–534

Van der Vliet PC, Levine AJ (1973) DNA binding proteins specific for cells infected by adenoviruses. Nature 246:1709–174

Van der Vliet PC, Sussenbach JS (1972) The mechanism of adenovirus DNA synthesis in isolated nuclei. Eur J Biochem 30:584–592

Van der Vliet PC, Sussenbach JS (1975) An adenovirus type 5 gene function required for initiation of viral DNA replication. Virology 67:415–426

Van der Vliet, Levine AJ, Ensinger M, Ginsbergs HS (1975) Thermolabile DNA binding proteins from cells infected with a temperature-sensitive mutant of adenovirus defective in viral DNA synthesis. J Virol 15:348–354

Van der Vliet PC, Zandberg J, Jansz HS (1977) Evidence for a function of the adenovirus DNA binding protein in initiation of DNA synthesis as well as in elongation on nascent DNA chains. Virology 80:98–110

Van Wielink PS, Naaktgeboren N, Sussenbach JS (1979) Presence of protein at the termini of intracellular adenovirus type 5 DNA. Biochim Biophys Acta 563:89–99
Wilhelm J, Brison O, Kedinger C, Chambon P (1976) Characterization of adenovirus type 2 transcriptional complexes isolated from infected HeLa cell nuclei. J Virol 15:744–758
Winnacker EL (1975) Adenovirus type 2 DNA replication. I Evidence for discontinuous replication. J Virol 15:744–758
Wohlgemuth DJ, Hsu M-T (1981) Visualization of nascent RNA transcripts and simultaneous transcription and replication in viral nucleoprotein complexes from adenovirus 2-infected HeLa cells. J Mol Biol 147:247–268
Yagura T, Kozu T, Seno T (1982) Mouse DNA polymerase accompanied by a novel RNA polymerase activity: purification and partial characterization. J Biochem 91:607–618
Yamashita T, Arens M, Green M (1977) Adenovirus DNA replication. Isolation of a soluble replication system and analysis of the in vitro DNA product. J Biol Chem 252:7940–7954

The Origin of Adenovirus DNA Replication

FUYUHIKO TAMANOI and BRUCE W. STILLMAN

1 Introduction

Current efforts to understand regulation of replication and cell growth often focus on initiation of DNA replication at an origin sequence. It is most likely, although not yet proven in eukaryotes, that the ultimate step in the regulation of the G to S transition in the cell cycle is the priming of DNA synthesis at an origin. The study of this process in eukaryotic cells has progressed slowly, and most of the attention has centered upon the replication of virus DNAs in infected cells. Although this provides practical advantages, it may be limited in its generality since viral DNAs replicate many times throughout the course of an infection, disregarding the constraints normally placed upon chromosome replication. Nevertheless, an understanding of the mechanism of initiation in eukaryotes is necessary, and the availability of virus origins dictates that they be studied first.

In the past decade, replication origins from many organisms have been identified and the DNA sequences contained in them determined (reviewed by KORNBERG 1980, 1982). Such studies have uncovered the presence of specific DNA regions which direct the formation of a complex of proteins that is capable of initiating DNA synthesis at the origin. This has most clearly been shown in prokaryotic systems, where suitable assays for origin function (both in vivo and in vitro) have been developed. These origins have been defined by creating deletions, insertions, and base substitutions within the DNA sequence. Initially, studies with the origins from the single-strand DNA phages from ΦX174, M13, and f1 and the double-strand DNA phages lambda, T7, and T4 (KORNBERG 1980, 1982) provided information

Cold Spring Harbor Laboratory, P.O. Box 100, Cold Spring Harbor, NY 11724, USA

Current Topics in Microbiology and Immunology, Vol. 109

on the nature of an origin, as well as the identification of a vast array of replication proteins that interact with the origin. In some cases, such as phage T7 (TABOR et al. 1981), the presence of a specific, primary origin region was obscured by the presence of secondary origin sequences which functioned when the primary site was deleted. Such studies have provided the groundwork for current investigations on the function of the *Escherichia coli oriC* region, both in vivo and in vitro (HIROTA et al. 1981; ZYSKIND et al. 1981; FULLER et al. 1981). This 245-bp sequence was identified by its ability to support autonomous replication of plasmid DNA in vivo and recently by its ability to act as a template for initiation of replication in vitro (FULLER et al. 1981).

Using a similar approach, sequences in some eukaryotic virus DNAs have been identified as origin sequences, particularly the origins of the papovaviruses SV40 and polyoma virus. These sequences have been defined by their ability to support the replication of extrachromosomal (plasmid-like) DNA in eukaryotic cells that provide, in trans, all the known replication proteins (reviewed by DEPAMPHILIS and WASSARMAN 1982). Whether similar unique origin sequences exist in the cell chromosome is still not clear, although the recent isolation of an apparent origin sequence from the amplified dihydrofalate reductase (DHFR) gene region (HEINTZ et al. 1983), strongly supports the conclusion that the chromosome does contain unique origin sequences. Autonomously replicating sequences (ars) have been detected in yeast DNA (BEACH et al. 1980; CHAN and TYE 1980) and may represent unique origins in the chromosome.

When origin sequences from closely related organisms are compared, certain common features are found, and in some cases appear to correspond to the recognition sequences for enzymes involved in initiation. For example, the origins of lambdoid phages lambda, ϕ80, and ϕ82 contain a highly conserved sequence that is repeated four times, in tandem (MOORE et al. 1979; HOBOM et al. 1979). Recently it has been shown that the lambda O protein, which is an essential enzyme for the initiation of lambda DNA replication, binds to these repeated sequences (TSURIMOTO and MATSUBURA 1981). The papovaviruses, SV40, polyoma virus, and BK virus all contain closely related sequences at their origins, and in the case of SV40, three sites that overlap with the origin of replication serve as binding sites for large T antigen, which is known to be required for the initiation of DNA synthesis.

2 The Adenovirus System

As indicated in this volume, studies with adenoviruses have pioneered modern molecular biology of eukaryotes. This has also been true with the replication of adenovirus DNA; a system that has yielded and continues to yield surprising and novel results. Discoveries such as a "protein-priming" mechanism for the initiation of DNA replication and novel replication enzymes such as the terminal protein precursor (pTP) and the adenovirus DNA polymerase have made adenovirus an exceptional system. Now we are just beginning to probe the regulation of the onset of DNA synthesis in a lytic infection, and initial observations suggest that this may occur at the stage of initiation of DNA synthesis.

The general mechanism of adenovirus DNA replication has been reviewed recently by CHALLBERG and KELLY 1982 and KELLY 1982, as well as in other sections of this volume. Here we have focused our attention on the latest efforts to define the nature of the origin of adenovirus DNA replication and the interaction between the origin and proteins required for initiation.

The origins of adenovirus DNA replication lie at each end of the linear genome, within an inverted, terminally repeated sequence (ARRAND and ROBERTS 1979; SCHILLING et al. 1975; TOLUN and PETTERSSON 1975; HORWITZ 1976; WEINGÄRTNER et al. 1976; ARIGA and SHIMOJO 1977; SUSSENBACH and KUIJK 1977; ARENS and YAMASHITA 1978). DNA replication is initiated at either end of the DNA, and chain elongation proceeds via a strand-displacement mechanism, displacing the nontemplate strand as a single-strand DNA. All intracellular DNAs, including nascent DNA strands, contain protein covalently attached to the 5′ termini (STILLMAN and BELLETT 1979a, b; VAN WIELINK et al. 1979; COOMBS et al. 1979; KELLY and LECHNER 1979; CHALLBERG and KELLY 1981). The protein is now known to be the virus-coded pTP, and this 80000-dalton precursor protein is cleaved to the 55000-dalton virion protein during morphogenesis of the virus particle (STILLMAN et al. 1981; CHALLBERG and KELLY 1981). Thus virion DNA contains the 55000-dalton terminal protein covalently linked to each 5′ end (ROBINSON et al. 1973; ROBINSON and BELLETT 1974; REKOSH et al. 1977). These studies in vivo were consistent with a model for initiation of adenovirus DNA replication, first proposed by Bellett and his co-workers (REKOSH et al. 1977), in which the terminal protein (precursor) covalently bound dCMP and this complex was the primer for subsequent chain elongation. The model, originally proposed as an explanation of how a primer for DNA polymerase might be synthesized at the very end of the linear adenovirus DNA molecule, has withstood the test of time.

Undoubtedly the single most important contribution to the elucidation of the mechanism of adenovirus DNA replication was the development of a cell-free system that could initiate adenovirus DNA synthesis on exogenous template DNA (CHALLBERG and KELLY 1979a, b). The use of this system has enabled detailed studies in vitro on the proteins required for initiation of DNA replication and the interaction with specific origin sequences. Three proteins have so far been identified as necessary for initiation of DNA replication in vitro, which has been defined as the formation of the covalent complex between the pTP and 5′-dCMP (LICHY et al. 1981; PINCUS et al. 1981; CHALLBERG et al. 1982; TAMANOI and STILLMAN 1982).

The first protein, the pTP itself, becomes covalently attached to nascent DNA strands synthesized in vitro (CHALLBERG et al. 1980; STILLMAN 1981;

IKEDA et al. 1982). Non-DNA-bound protein is present in infected cells (GREEN et al. 1981) and has been purified in a functional form (ENOMOTO et al. 1981) as a single polypeptide of 80000 daltons. This protein is encoded by early region E2B of the virus genome (STILLMAN et al. 1981; BINGER et al. 1982; SMART and STILLMAN 1982).

A second virus-coded protein required for initiation at the origin of replication is the recently discovered adenovirus DNA polymerase. This protein is also encoded by early region E2B (STILLMAN et al. 1982a) and was first identified as a 140000-dalton protein which copurified with pTP (ENOMOTO et al. 1981). Subsequently, this protein was purified to homogeneity and shown to complement the defective nuclear extracts prepared from temperature-sensitive mutants Ad5*ts*149 and Ad5*ts*36 (STILLMAN et al. 1982a; OSTROVE et al. 1983; FRIEFELD et al. 1983; VAN BERGEN and VAN DER VLIET 1983). The purified 140000-dalton protein contains a DNA polymerase activity that is different from the host cell polymerases (LICHY et al. 1982) and is required for formation of the covalent complex between pTP and dCMP.

The third protein that has been identified as being required for initiation at the origin of replication is a host cell protein of apparent molecular weight of 47000 (NAGATA et al. 1982). This protein stimulates initiation of replication when double-strand origin DNA is used as the template, but can be discarded when single-strand DNA is used. The exact role of this and possibly other host factors in the initiation reaction has not been determined.

In addition to these proteins, specific DNA sequences are required for initiation of DNA synthesis on double-strand DNA (TAMANOI and STILLMAN 1982). This study also demonstrated that, at least for pTP-dCMP complex formation, the terminal protein on the template DNA is dispensable. It is the nature of these sequences at the origin of replication that is discussed below.

3 Comparison of DNA Sequences Contained in the Inverted Terminal Repetition

Since initiation begins at the termini of the DNA, the expectation was that DNA sequencing of the inverted terminal repetition would reveal the presence of any structural features that might be required for initiation at the origin. DNA sequences within this region from human adenoviruses Ad2 and Ad5 (group C viruses: ARRAND and ROBERTS 1979; SHINAGAWA and PADMANABHAN 1979; STEENBURGH et al. 1977), Ad12, Ad18, and Ad31 (group A viruses: SHINAGAWA and PADMANABHAN 1980; SUGISAKI et al. 1980; TOLUN et al. 1979; GARON et al. 1982; STILLMAN et al. 1982b) Ad3 and Ad7 (group B viruses: TOLUN et al. 1979; DIJKEMA and DEKKER 1979), Ad9 and Ad10 (group D viruses: STILLMAN et al. 1982b), and Ad4 (group E

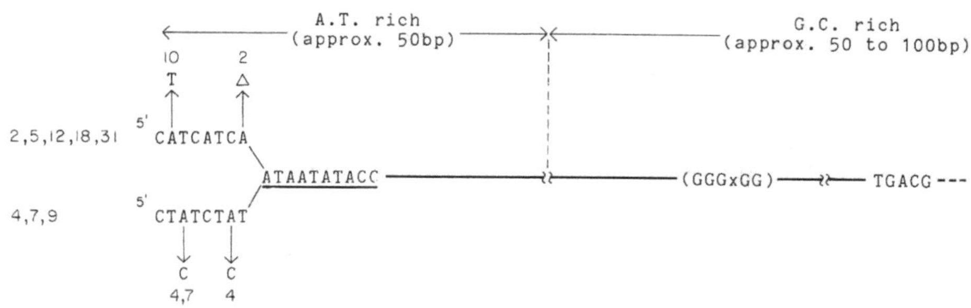

Fig. 1. Structure of the inverted terminal repetition in human adenovirus DNA. The DNA sequence of the inverted terminal repetition from all known human adenoviruses can be divided into A·T-rich and G·C-rich regions. The conserved 18-bp region at the end of all human DNAs can be derived from two prototype sequences as shown: the *numbers* refer to the adenovirus serotype and the *arrows* point to single base changes in some serotype DNAs. The perfectly conserved 10-bp region is underlined. Within the G·C-rich region, the sequence GGG × GG, shown in parentheses, is repeated many times in all human DNAs and the sequence TGACG occurs at or near the end of each terminally repeated sequence. Only the 5′-strand sequence is shown. For references, see text

virus: STILLMAN et al. 1982b; TOKUNAGA et al. 1982) have been determined. The DNA sequences of inverted terminal repetitions of simian adenovirus SA7 (TOLUN et al. 1979), murine adenovirus FL (TEMPLE et al. 1981), canine (ICHV) and equine (T-1) adenoviruses (SHINAGAWA et al. 1983), and the avian adenovirus CELO (ALESTRÖM et al. 1982; SHINAGAWA et al. 1983) have also been determined.

Comparison of these DNA sequences has revealed some common structural features. First, the inverted terminal repetition for human adenovirus can be divided into two regions; a A·T-rich region of 50–52 bp at the end of the viral DNA and a G·C-rich region of 50–110 bp which is adjacent to the A·T-rich region. DNA sequences within the A·T-rich region are highy conserved among all human adenoviruses and are also partially conserved in the simian, murine, canine, equine, and avian adenoviruses. On the other hand, the G·C-rich region is not highly conserved and is also variable in size. Such asymmetry of base composition also occurs in many other origin sequences. However, unlike other origins, particularly prokaryote origins, there is a lack of secondary structure such as hairpins and repeated segments within the adenovirus origin.

The A·T-rich region contains an 18-nucleotide sequence at its very end which is highly conserved in all human adenovirus DNAs and is partially conserved in other nonhuman adenoviruses (Fig. 1). Three features are noteworthy. First, all DNAs contain a 5′-terminal dC residue, which is consistent with the priming of DNA synthesis by pTP-dCMP. Second, there is a perfectly conserved 10-bp region in all human DNAs located from 8 or 9 bp to 17 or 18 bp from the DNA terminus (some Ad2 strains carry a deletion of dA from base 8, relative to the closely related Ad5 sequence; Fig. 1). Third, between the 5′-terminal dC residue and this 10-bp perfectly conserved region are 6–7 base pairs containing overlapping repeat sequences of CATCA (Ad5 is the prototype) or CTAT (Ad7 is the prototype). Within this region, all human adenovirus sequences can be derived from one of these two prototypes by single base changes (Fig. 1).

Other conserved blocks appear outside the A·T-rich region. One is a sequence GGGCGG which is present in multiple copies within the G·C-rich region of all human adenoviruses and also in the G·C-rich region of the simian and equine adenoviruses. This sequence is similar to the sequence GGGXGGAG which is present in multiple copies at the origins of the papovaviruses BK virus, SV40, and polyoma virus (reviewed by SEIF et al. 1979). The region containing these repeated sequences in SV40 has been shown to enhance DNA replication in vivo (BERGSMA et al. 1982), and these repeats may play a similar role for adenovirus. The other conserved sequence is TGACG, which occurs at or near the end of all human adenovirus inverted terminal repetition. A similar sequence occurs in murine and avian adenoviruses; in the case of murine adenovirus FL it is TCACG (right end) or TGACG (left end), and in the avian adenovirus CELO, the sequence appears to be replaced by the sequence TGTCG. However, the function of this sequence is not known.

As a first attempt to assess the role of the conserved sequences in adenovirus DNA replication, the template activity of DNA-protein complexes from several serotypes of adenovirus have been compared (STILLMAN 1981; STILLMAN et al. 1982b). DNA-protein complexes were isolated from Ad2, Ad4, Ad7, Ad9, and Ad31, and their ability to direct adenovirus DNA replication was examined using nuclear extracts from Ad2-infected cells. All these template DNA-protein complexes supported initiation as well as elongation of replication, although the efficiency varied considerably. These results were consistent with the idea that the conserved sequence plays some role in initiation of DNA replication, but the interpretation of the results was complicated by the presence of the heterologous terminal protein on each template DNA. To circumvent this problem, initiation of DNA replication has been pursued using cloned adenovirus DNA sequences in the absence of the terminal protein on the template DNA.

4 Plasmid DNAs Containing the Origin of Adenovirus DNA Replication

Plasmid DNAs containing the origin of adenovirus DNA replication have proved invaluable for studies on the DNA sequence requirements for the initiation of DNA replication. Fig. 2 shows three different plasmid DNAs that have been used for such studies. Plasmid *8 was constructed by N. Stow (personal communication) by using a dG-dC tailing procedure to circumvent problems associated with the presence of residual amino acids of the terminal protein that remain linked to the 5′ terminus of each strand

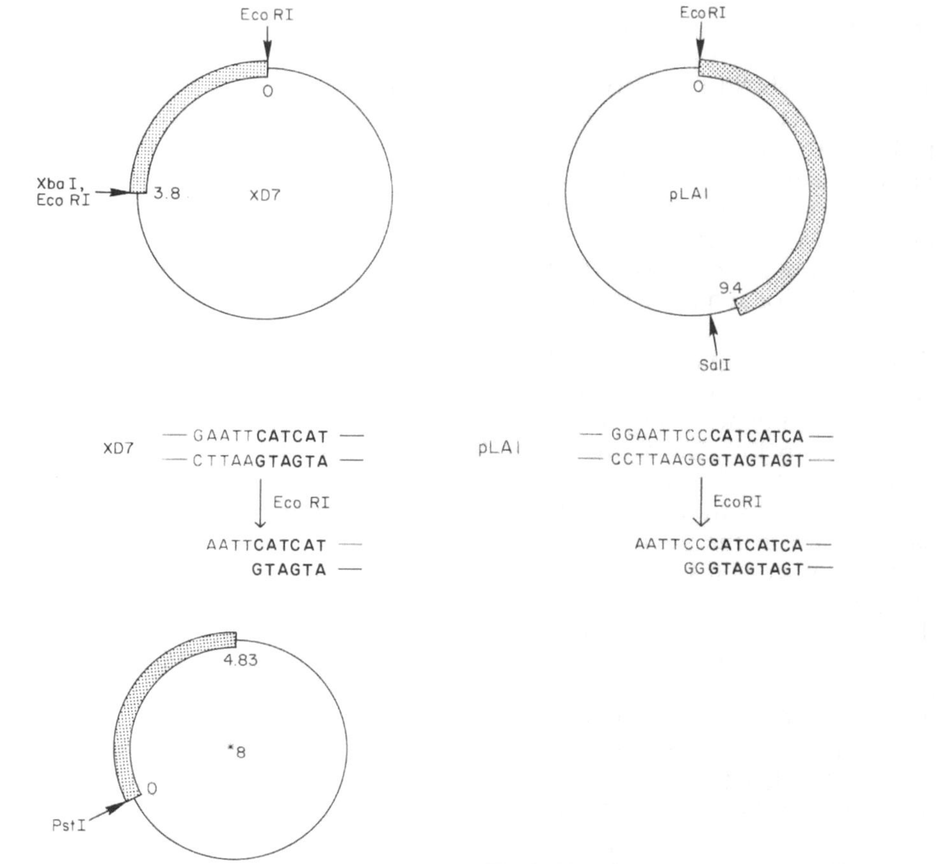

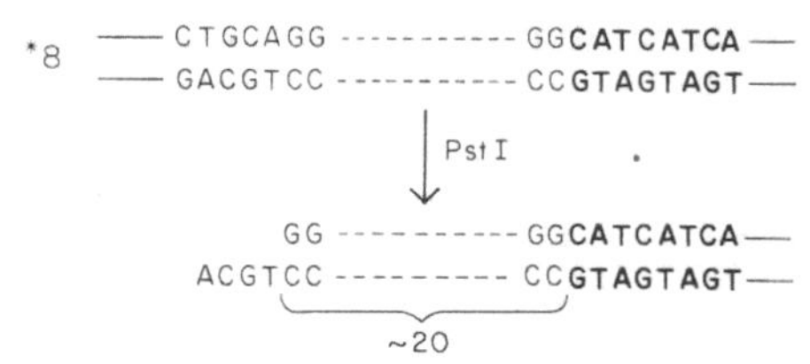

Fig. 2. Plasmids containing cloned sequences of the adenovirus origin. *Top:* the plasmids XD-7 (PEARSON et al. 1981), pLA1, and *8 (TAMANOI and STILLMAN 1982) are shown with the adenovirus DNA as the *shaded area.* The *numbers* refer to the map coordinates (in percent) on the adenovirus genome. *Bottom:* the DNA sequences of the linearized plasmid DNAs obtained by digestion with the appropriate enzyme which would place the adenovirus DNA terminus close to the end of the linear molecule. The adenovirus DNA sequences are shown in *bold letters*

after protease treatment of the viral DNA. Piperidine or alkali can also be used to remove the terminal protein (TAMANOI and STILLMAN 1982). After renaturation of the piperidine-treated DNA, the blunt ends can be ligated to DNA linkers, facilitating cloning of the terminal fragments of viral DNA into restriction sites in pBR322 DNA. pLA1 DNA was constructed using the piperidine method. Also shown in this figure is plasmid XD-7, which was constructed by direct blunt-end ligation of the repaired adenovirus DNA to the repaired plasmid DNA (Pearson et al. 1981).

The pLA1 DNA supported the initiation of adenovirus DNA replication in vitro, provided the DNA was linearized with *Eco*RI so that the adenovirus terminal sequence was located at the end of the molecule. This was demonstrated by its ability to support the formation of a covalent pTP-dCMP complex using an enzyme fraction obtained by chromatography of nuclear extracts from adenovirus-infected cells on a denatured DNA cellulose column (Tamanoi and Stillman 1982). A limited amount of DNA chain elongation also occurred on the pLA1 DNA. Van Bergen et al. (1983) have also shown that plasmid XD-7, which has a similar structure to that of pLA1, replicates in nuclear extracts from adenovirus-infected cells.

The requirement for the adenovirus terminal sequence to be located very close to the end of the linear DNA molecule is rather strict. Fig. 2 shows the DNA sequence at the end of the plasmid DNAs after they have been linearized with appropriate enzymes. The presence of a few extra base pairs added to the adenovirus terminal sequence, due to the presence of a DNA linker, did not affect initiation (pLA1, XD-7). However, addition of approximately 20 bp of dG-dC destroyed the template activity, since the *8 DNA linearized with *Pst*I did not support initiation.

Exact positioning of the initiation sites on the pLA1 DNA is possible by carrying out elongation in the presence of dATP, dCTP, TTP, and ddGTP. ddGTP blocks elongation at the first dG residue, which occurs at the 26th nucleotide from the adenovirus terminus. Therefore determination of the length of DNA contained in the covalent complex between the pTP and the nascent DNA exactly positions the initiation site. Results of such experiments demonstrated that the initiation of DNA synthesis on *Eco*RI digested pLA1 DNA takes place on the dC at the very terminus of the adenovirus DNA, as well as the next dC four nucleotides inside the inverted terminal repeat.

The observation that a plasmid DNA containing the origin of adenovirus DNA replication functioned as a template for initiation in vitro suggested that the presence of terminal protein on the template DNA was not an absolute requirement for template activity. However, it was shown that pronase-treated adenovirus DNA-protein complex did not support initiation in vitro (Challberg and Kelly 1979b). This apparent contradiction was due to the inhibition of initiation by residual amino acids remaining on the DNA after the pronase treatment. Similarly, terminal restriction fragments of adenovirus DNA-protein complex functioned as a template but pronase-treated terminal fragments did not. However, when these fragments were treated with pronase, followed by incubation with piperidine and then reannealed, they supported initiation (Tamanoi and Stillman 1982). Therefore a DNA sequence on the adenovirus DNA-protein complex is primarily responsible for the template activity and the parental terminal protein plays a subordinate role.

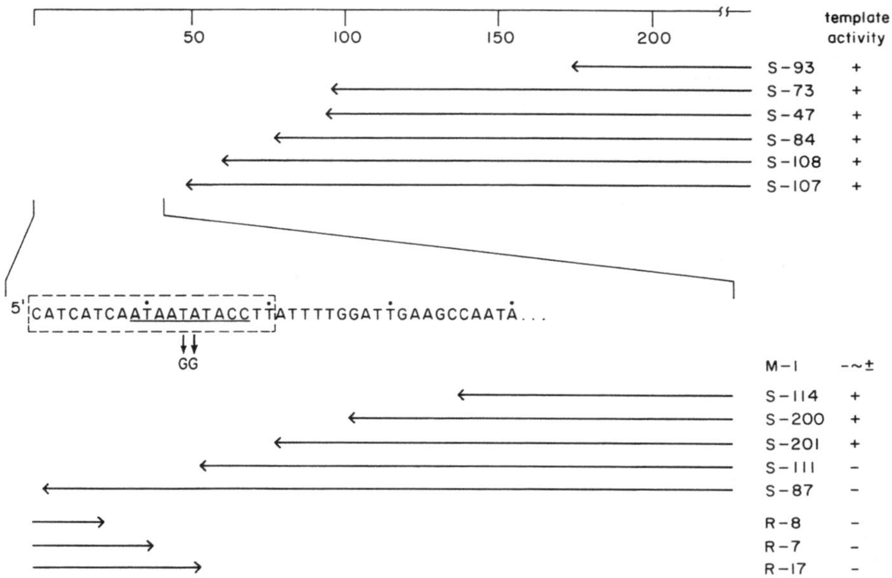

Fig. 3. Effect of deletions and mutations introduced into plasmid pLA1 DNA. Deletions (indicated by the *arrows*) and mutation (*TA → GG*) in the adenovirus origin were constructed in plasmid pLA1 DNA (TAMANOI and STILLMAN, to be published). The *top line* shows the number of nucleotides in adenovirus DNA from the *Eco*RI site within the plasmid. The *bottom* half of the figure is an expanded region comprising the first 40 bases from the 5′-strand sequence. Deletions were constructed from an internal restriction site or from the termini (*Eco*RI site). The base mutations at positions 13 and 14 were constructed with synthetic oligonucleotides. The template activity of these mutant plasmids, measured by formation of the pTP-dCMP complex in vitro, is shown on the *right*

5 Mutagenesis of a Cloned Origin

Use of plasmid DNAs containing the origin of DNA replication provided a system with which to investigate DNA sequence requirements for initiation of replication in vitro. Deletions have been constructed in the plasmid pLA1 DNA, and their template activities for formation of the pTP-dCMP complex examined (TAMANOI and STILLMAN, to be published). Fig. 3 shows the deletions that were constructed. Deletions beginning from an internal restriction site and extending toward the adenovirus terminus did not affect initiation if more than 20 base pairs of adenovirus DNA from the terminus were retained. However, when the deletions left only 14 base pairs or less of adenovirus terminal sequence, the template activity was lost. Deletions which removed sequences from the very end of the adenovirus DNA has a dramatic effect on the template activity. No pTP-dCMP complex was formed when 6, 10, or 14 base pairs from the terminus were removed. Thus the terminal 20 base pairs, boxed in by a broken line in Fig. 3, constitute a minimum size of adenovirus DNA required for the initiation of DNA synthesis in vitro. The 20-bp sequence was also sufficient to support a limited amount of elongation.

Not surprisingly, the functional origin thus defined corresponds to the highly conserved region described above and shown in Fig. 1. To confirm the importance of the highly conserved region in the origin sequence, site-directed mutants have been constructed using synthetic oligonucleotides, (Tamanoi and Stillman, to be published). One such mutant changes nucleotides 13 and 14 in the 5′-strand sequence from TA to GG (Fig. 3). This mutation, when inserted into plasmid DNA and examined for template function, is defective in its ability to support pTP-dCMP complex formation.

6 Summary

The sequences within the adenovirus genome that are required for initiation of DNA replication in vitro have been defined. These sequences constitute one-fifth of the inverted terminal repetition and are highly conserved among all human adenoviruses. It is likely that these conserved sequences act as a recognition signal for one or more of the virus proteins required for initiation of DNA replication, and further physical studies should determine whether this is correct. Furthermore, these sequences may also play some role in the regulation of initiation throughout the lytic cycle. Bodnar and Pearson (1980) have suggested that initiation of DNA synthesis throughout the replication cycle may be regulated since there is a slow accumulation of progeny DNA molecules. Regulation of initiation may also occur at the level of synthesis of the initiation enzymes pTP and DNA polymerase (Stillman et al. 1982). Further analysis of these processes should prove fruitful.

A relatively small region of the inverted terminal repetition is required for initiation of replication in vitro. However, we must be careful not to overinterpret these results and certainly should not conclude that other sequences within the terminal repetition are unnecessary for replication in vivo. Two points should be noted. First, even though use of the cell-free replication system has generated many interesting results, replication in vitro is limited to only one round of DNA synthesis (type I replication), and initiation of the second, displaced strand does not occur. Sequences within the inverted terminal repetition may be required for this process. Second, if we are to learn anything from RNA transcription studies, both in vivo and in vitro, we must consider that an origin of DNA replication in vivo contains not only a recognition signal for initiation enzymes, but also might contain enhancer or auxillary sequences. Such sequences are seen in the SV40 origin region (see Bergsma et al. 1982) and may lie in the G·C-rich region of the adenovirus terminally repeated sequence. Enhancer sequences can be overlooked by limiting analysis to cell-free systems. Further analysis of the adenovirus origin should resolve these questions. Whether chromosome origins of replication are similarly composed of a number of functional elements will be central to future studies of eukaryotic DNA replication and its regulation.

Acknowledgments. The authors' research was supported by a Cancer Center Grant from the National Cancer Institute. BWS is a Rita Allen Foundation Scholar.

References

Aleström P, Stenlund A, Li P, Pettersson U (1982) A common sequence in the inverted terminal repetitions of human and avian adenoviruses. Gene 18:193–197

Arens M, Yamashita T (1978) In vitro termination of adenovirus DNA synthesis by a soluble replication complex. J Virol 25:698–702

Ariga H, Shimojo H (1977) Initiation and termination sites of adenovirus type 12 DNA replication. Virology 78:415–424

Arrand JR, Roberts RJ (1979) The nucleotide sequence of the termini of adenovirus-2 DNA. J Mol Biol 128:577–594

Beach D, Piper M, Shall S (1980) Isolation of chromosomal origins of replication in yeast. Nature 284:185–187

Bergma DJ, Olive DM, Hartzell SW, Subramanian KN (1982) Territorial limits and functional anatomy of the simian virus 40 replication origin. Proc Natl Acad Sci USA 79:381–385

Binger MH, Flint SJ, Rekosh DM (1982) Expression of the gene encoding the adenovirus DNA terminal protein in productively infected and transformed cells. J Virol 42:488–501

Bodnar JW, Pearson GD (1980) Kinetics of adenovirus DNA replication. II. Initiation of adenovirus DNA replication. Virology 105:357–370

Challberg MD, Kelly TJ Jr (1979a) Adenovirus DNA replication in vitro. Proc Natl Acad Sci USA 76:655–659

Challberg MD, Kelly TJ Jr (1979b) Adenovirus DNA replication in vitro: origin and direction of daughter strand synthesis. J Mol Biol 135:999–1012

Challberg MD, Kelly TJ Jr (1981) Processing of the adenovirus terminal protein. J Virol 38:272–277

Challberg MD, Kelly TJ Jr (1982) Eukaryotic DNA replication: viral and plasmid model systems. Annu Rev Biochem 51:901–934

Challberg MD, Desiderio SV, Kelly TJ Jr (1980) Adenovirus DNA replication in vitro: characterization of a protein covalently linked to nascent DNA strands. Proc Natl Acad Sci USA 77:5105–5109

Challberg MD, Ostrove JM, Kelly TJ Jr (1982) Initiation of adenovirus DNA replication: detection of covalent complexes between nucleotide and the 80Kd terminal protein. J Virol 41:265–270

Chan CM, Tye BK (1980) Autonomously replicating sequences in *Saccharomyces cerevisiae.* Proc Natl Acad Sci USA 77:6329–6333

Coombs DH, Robinson AJ, Bodnar JW, Jones CJ, Pearson GD (1979) Detection of DNA-protein complexes: the adenovirus DNA-terminal protein and HeLa DNA-protein complexes. Cold Spring Harbor Symp Quant Biol 43:741–753

De Pamphilis ML, Wassarman PM (1982) Organization and replication of papovavirus DNA. In: Kaplan AS (ed) Organization and replication of viral DNA. CRC, Florida, pp 37–114

Dijkema R, Dekker BMM (1979) The inverted terminal repetition of the DNA of weakly oncogenic adenovirus type 7. Gene 8:7–15

Enomoto T, Lichy JH, Ikeda JE, Hurwitz J (1981) Adenovirus DNA replication in vitro: purification of the terminal protein in a functional form. Proc Natl Acad Sci USA 78:6779–6783

Friefeld BR, Lichy JH, Hurwitz J, Horwitz MS (1983) Evidence for an altered adenovirus DNA polymerase in cells infected with the mutant H5*ts*149. Proc Natl Acad Sci USA 80:1589–1593

Fuller RS, Kaguni TM, Kornberg A (1981) Enzymatic replication of the origin of the *Escherichia coli* chromosome. Proc Natl Acad Sci USA 78:7370–7374

Garon CF, Parr RP, Padmanabhan RK, Roninson I, Garrison JW, Rose JA (1982) Structural characterization of the adenovirus 18 inverted terminal repetition. Virology 121:230–239

Green M, Symington J, Brackmann KH, Cartas MA, Thorton H, Young L (1981) Immunological and chemical identification of intracellular forms of adenovirus type 2 terminal protein. J Virol 40:541–550

Heintz NH, Milbrandt JD, Greisen KS, Hamlin JL (1983) Cloning of the initiation region of a mammalian chromosomal replicon. Nature 302:439–441

Hirota Y, Oka A, Sugimoto K, Asada K, Sasaki H, Takanami M (1981) *Escherichia coli* origin of replication: structural organization of the region essential for autonomous replication and the recognition frame model. ICN-UCLA Symposia of Molecular and Cellular Biology XXII:1–2

Hobom G, Grosschedl R, Lusky M, Scherer G, Schwarz E, Kössel H (1979) Functional analysis of the replicator structure of lambdoid bacteriophage DNAs. Cold Spring Harbor Symp Quant Biol 43:165–178

Horwitz MS (1976) Bidirectional replication of adenovirus type 2 DNA. J Virol 18:307–315

Ikeda J-E, Enomoto T, Hurwitz J (1982) Adenoviral protein-primed DNA replication in vitro. Proc Natl Acad Sci 79:2442–2446

Kelly TJ Jr (1982) Organization and replication of adenovirus DNA. In: Kaplan AS (ed) Organization and replication of viral DNA. CRC, Florida, pp 115–146

Kelly TJ Jr and Lechner RL (1979) The structure of replicating adenovirus DNA molecules: characterization of DNA-protein complexes from infected cells. Cold Spring Harbor Symp Quant Biol 43:721–728

Kornberg A (1980) DNA replication. Freeman, San Francisco

Kornberg A (1982) Supplement to DNA replication. Freeman, San Francisco

Lichy JH, Horwitz MS, Hurwitz J (1981) Formation of a covalent complex between the 80000 dalton adenovirus terminal protein and 5′-dCMP in vitro. Proc Natl Acad Sci USA 78:2678–2682

Lichy JH, Field J, Horwitz MS, Hurwitz J (1982) Separation of the adenovirus terminal protein precursor from its associated DNA polymerase: role of both proteins in the initiation of adenovirus DNA replication. Proc Natl Acad Sci USA 79:5225–5229

Moore DD, Denniston-Thompson K, Kruger KE, Furth ME, Williams BG, Daniels DL, Blattner FR (1979) Dissection and comparative anatomy of the origins of replication of lambdoid phages. Cold Spring Harbor Symp Quant Biol 43:155–163

Nagata K, Guggenheimer RA, Enomoto T, Lichy JH, Hurwitz J (1982) Adenovirus DNA replication in vitro: identification of a host factor that stimulates synthesis of the preterminal protein-dCMP complex. Proc Natl Acad Sci USA 79:6438–6442

Ostrove JM, Rosenfeld P, Williams J, Kelly TJ Jr (1983) In vitro complementation as an assay for the purification of adenovirus DNA replication proteins. Proc Natl Acad Sci USA 80:935–939

Pearson GD, Chow K-C, Corden JL, Harpst JA (1981) Replication directed by a cloned adenovirus origin. ICN-UCLA Symposia on Molecular and Cellular Biology XXII:581–595

Pincus S, Robertson W, Rekosh DMK (1981) Characterization of the effect of aphidicolin on adenovirus DNA replication: evidence in support of a protein primer model of initiation. Nucleic Acids Res 9:4919–4938

Rekosh DMK, Russell WC, Bellett AJD, Robinson AJ (1977) Identification of a protein linked to the ends of adenovirus DNA. Cell 11:283–295

Robinson AJ, Bellett AJD (1974) A circular DNA-protein complex from adenoviruses and its possible role in DNA replication. Cold Spring Harbor Symp Quant Biol 39:523–531

Robinson AJ, Younghusband HB, Bellett AJD (1973) A circular DNA-protein complex from adenoviruses. Virology 56:54–69

Schilling R, Weingärtner B, Winnacker EL (1975) Adenovirus type 2 DNA replication. II Termini of DNA replication. J Virol 16:767–774

Seif I, Khoury G, Dhar R (1979) The genome of human papovavirus BKV. Cell 18:963–977

Shinagawa M, Padmanabhan R (1979) Nucleotide sequences at the inverted terminal repetition of Ad2 DNA. Biochem Biophys Res Commun 87:671–678

Shinagawa M, Padmanabhan R (1980) Comparative sequence analysis of the inverted terminal repetitions from different adenoviruses. Proc Natl Acad Sci USA 77:3831–3835

Shinagawa M, Ishiyama T, Padmanabhan R, Fujinaga K, Kamada M, Sato G (1983) Comparitive sequence analysis of the inverted terminal repetition in the genomes of animal and avian adenoviruses. Virology 125:491–495

Smart JE, Stillman BW (1982) Adenovirus terminal protein precursor: partial amino acid sequence and site of covalent linkage to virus DNA. J Biol Chem 257:13499–13506

Steenbergh PH, Maat J, van Ormondt H, Sussenbach JS (1977) The nucleotide sequence at the termini of adenovirus type 5 DNA. Nucleic Acids Res 4:4371–4389

Stillman BW (1981) Adenovirus DNA replication in vitro: a protein linked to the 5′ end of nascent DNA strands. J Virol 37:139–147

Stillman BW, Bellett AJD (1979a) Replication of DNA in adenovirus infected cells. Cold Spring Harbor Symp Quant Biol 43:729–739

Stillman BW, Bellett AJD (1979b) An adenovirus protein associated with the ends of replicating DNA molecules. Virology 93:69–79

Stillman BW, Lewis JB, Chow LT, Mathews MB, Smart JE (1981) Identification of the gene and mRNA for the adenovirus terminal protein precursor. Cell 23:497–508

Stillman BW, Tamanoi F, Mathews MB (1982a) Purification of an adenovirus-coded DNA polymerase that is required for initiation of DNA replication. Cell 31:613–623

Stillman BW, Topp WC, Engler JA (1982b) Conserved sequences at the origin of adenovirus DNA replication. J Virol 44:530–537

Sugisaki H, Sugimoto K, Takanami M, Shiroki K, Saito I, Shimojo Y, Sawada Y, Uemizu Y, Uesugi S-I, Fujinaga K (1980) Structure and gene organization of the transforming *Hind*III-G fragment of Ad12. Cell 20:777–786

Sussenbach JS, Kuijk MG (1977) Studies on the mechanism of replication of adenovirus DNA V. The location of the termini of replication. Virology 77:149–157

Tabor S, Engler MJ, Fuller CW, Lechner RL, Matson SW, Romano LJ, Saito H, Tamanoi F, Richardson CC (1981) Initiation of bacteriophage T7 DNA replication ICN-UCLA Symposia of Molecular and Cellular Biology XXII:387–408

Tamanoi F, Stillman BW (1982) Function of the adenovirus terminal protein in the initiation of DNA replication. Proc Natl Acad Sci USA 79:2221–2225

Tamanoi F, Stillman BW (to be published) Initiation of adenovirus DNA replication in vitro requires a specific DNA sequence. Proc Natl Acad Sci USA

Temple M, Antoine G, Delius H, Stahl S, Winnacker E-L (1981) Replication of mouse adenovirus strain FL DNA. Virology 109:1–12

Tokunaga O, Shinagawa M, Padmanabhan R (1982) Physical mapping of the genome and sequence analysis at the inverted terminal repetition of adenovirus type 4 DNA. Gene 18:329–334

Tolun A, Petterson U (1975) Termination sites for adenovirus type 2 DNA replication. J Virol 16:759–766

Tolun A, Aleström P, Pettersson U (1979) Sequence of inverted terminal repetitions from different adenoviruses: demonstration of conserved sequences and homology between SA 7 termini and SV40 DNA Cell 17:705–713

Tsurimoto T, Matsubura K (1981) Purified bacteriophage lambda O protein binds to four repeating sequences at the lambda replication origin. Nucleic Acids Res 9:1789–1799

Van Bergen BGM, van der Vliet PC (1983) Temperature sensitive initiation and elongation of adenovirus DNA replication in vitro with nuclear extracts from H5*ts*36, H5*ts*149 and H5*ts*125 infected HeLa cells. J Virol 46:642–648

Van Bergen BGM, van der Ley PA, van Driel W, van Mansfeld ADM, van der Vliet PC (1983) Replication of origin containing adenovirus DNA fragments that do not carry the terminal protein. Nucleic Acids Res 11:1975–1989

Van Wielink PS, Naaktgeboren N, Sussenbach JS (1979) Presence of protein at the termini of intracellular adenovirus type 5 DNA. Biochim Biophys Acta 563:89–99

Weingärtner B, Winnacker EL, Tolun A, Petterson U (1976) Two complementary strand specific termination sites for adenovirus DNA replication. Cell 9:259–268

Zyskind JW, Harding NE, Takeda Y, Cleary JM, Smith DW (1981) The DNA replication origin region of the Enterobacteriaceae. ICN-UCLA Symposia of Molecular and Cellular Biology XXII:13–25

A New Mechanism for the Initiation of Replication of Φ29 and Adenovirus DNA: Priming by the Terminal Protein

MARGARITA SALAS

1 Introduction

Bacteriophage Φ29 from *Bacillus subtilis* contains a double-stranded DNA of molecular weight 11.8×10^6 (SOGO et al. 1979) that can be isolated from the viral particles as a circular molecule closed noncovalently by protein (ORTÍN et al. 1971). In this review I will describe our present knowledge of the protein linked to Φ29 DNA. I will present first the structural aspects of the problem: characterization of the protein and the nature of the protein-DNA linkage and sequence at the terminals of Φ29 DNA and of the gene coding for the terminal protein. Next I will describe the mechanism of

Centro de Biología Molecular (CSIC-UAM), Universidad Autónoma, Canto Blanco, E-Madrid-34

Current Topics in Microbiology and Immunology, Vol. 109

Φ29 DNA replication and how the terminal protein functions as a "primer" in the initiation of replication. Other requirements for the initiation reaction will be also described, as well as the production of specific mutations at the carboxyl end of the terminal protein. Finally, I will compare the Φ29 and adenovirus DNA-protein complexes and the novel mechanism used for the initiation of replication in the two systems.

2 Characterization of the Protein Covalently Linked to the 5′ Terminals of Φ29 DNA

The DNA-protein complex was isolated from phage Φ29 by different methods which included treatments that dissociate proteins noncovalently bound to the DNA. The protein was shown to be very strongly attached to the DNA; it did not migrate into an SDS-polyacrylamide gel unless the sample was treated with nuclease. The protein which migrated into the gel after the nuclease treatment had the same electrophoretic mobility as p3 (SALAS et al. 1978; HARDING et al. 1978; YEHLE 1978), an early viral protein essential for Φ29 DNA replication (CARRASCOSA et al. 1976; HAGEN et al. 1976; YANOFSKY et al. 1976). Tryptic peptide analysis demonstrated that the protein attached to the DNA was indeed p3 (SALAS et al. 1978). Evidence that the protein is linked to the 5′ ends of the DNA was obtained from the fact that Φ29 DNA could neither be phosphorylated by polynucleotide kinase and [γ-^{32}P]ATP nor be degraded by λ exonuclease, which degrades DNA from the 5′ ends, while it was susceptible to degradation by exonuclease III, which degrades DNA from the 3′ ends (SALAS et al. 1978; ITO 1978; YEHLE 1978).

2.1 Circularization of Φ29 DNA by Protein-Protein Interaction

The molecules of protein p3 at the 5′ ends of Φ29 DNA interact with each other, giving rise to circular molecules or concatemeres (ORTÍN et al. 1971; SALAS et al. 1978) which, upon treatment with proteolytic enzymes, are converted into unit-length linear DNA. Fig. 1 A shows a linear Φ29 DNA molecule with the protein at the ends and Fig. 1 B shows a circular molecule, formed by protein-protein interaction, in which the protein can be visualized. Adenovirus DNA, which also has a protein covalently linked at the 5′ ends (REKOSH et al. 1977; CARUSI 1977), gives rise, similarly, to circular molecules and concatemeres (ROBINSON et al. 1973; BROWN et al. 1975). Whether or not this type of circularization of Φ29 and adenovirus DNAs has biological significance is an open question. Very recently, structures

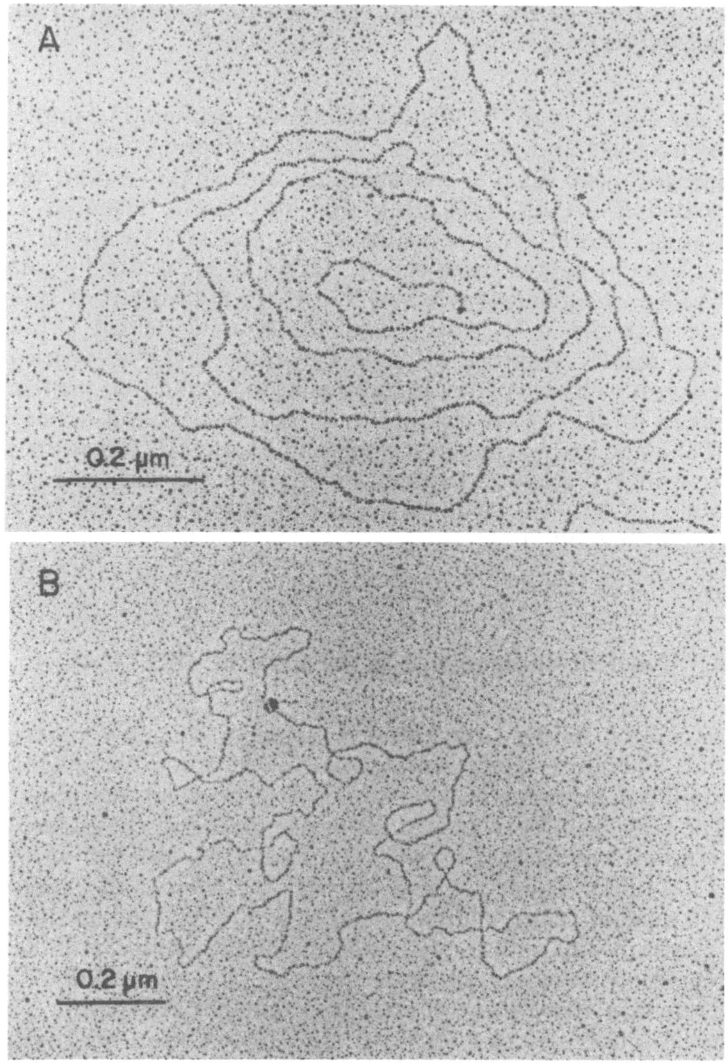

Fig. 1 A, B. Electron micrographs of Φ29 DNA–protein p3 complex. **A** Linear Φ29 DNA-protein p3. **B** Circular Φ29 DNA-protein p3. The ethidium bromide protein-free spreading technique was used as described by (Salas et al. 1978). The electron micrographs were taken by J.M. Sogo

resulting from covalent head-to-tail joining of viral DNA molecules have been found in Ad5-infected cells and these structures are due, at least in part, to the formation of covalently closed circles (Ruben et al. 1983). The possible role of these structures in replication or transformation remains to be elucidated.

2.2 Nature of the Linkage Between the Terminal Protein p3 and *Φ*29 DNA

After digestion of ^{32}P-labeled *Φ*29 DNA-protein p3 complex with micrococcal nuclease and pronase, a nucleotidyl peptide was obtained which, when digested with alkaline phosphatase and snake venom phosphodiesterase, yielded 5′-dAMP, indicating that this is the terminal nucleotide at the 5′ ends. The DNA-protein linkage is sensitive to alkali. Treatment of the nucleotidyl peptide with 0.1 *M* NaOH at 37 °C for 3 h after digestion with alkaline phosphatase released 5′-dAMP. Hydrolysis of the nucleotidyl peptide with 5.8 *N* HCl at 110 °C for 90 min yielded *O*-phosphoserine. All these results, together with the fact that the DNA-protein p3 linkage was resistant to hydroxylamine treatment, indicated that protein p3 is covalently linked to *Φ*29 DNA through a phosphoester bond between the OH group of a serine residue in the protein and 5′-dAMP (Hermoso and Salas 1980). A similar type of linkage exists between the adenovirus terminal protein and 5′-dCMP, the terminal nucleotide at the 5′ ends of the DNA (Desiderio and Kelly 1981).

2.3 Nucleotide Sequence at the Ends of *Φ*29 DNA

As will be shown later, the replication of *Φ*29 DNA starts at either DNA end. Therefore it was important to determine the nucleotide sequence at the DNA terminals which should correspond to the origin(s) of replication. The ends of *Φ*29 DNA are flush and there is a six-nucleotide-long inverted terminal repetition (AAAGTA) (Escarmís and Salas 1981; Yoshikawa et al. 1981). There are also internal repetitions, the homology in the first 150 nucleotides at the two ends of *Φ*29 DNA being 40%. The biological significance of both the terminal inverted repetition and the internal inverted repetitions remains an open question, but it is likely that they function in protein recognition for the initiation of replication. Adenovirus DNA has a longer inverted terminal repetition, about 100 nucleotides, and there is a region of 14 nucleotides which has been conserved in different types of adenovirus DNA (Tolun et al. 1979). There is no homology among the sequences at the ends of *Φ*29 DNA and those of adenovirus DNAs. Nevertheless, there is one important common feature of the inverted terminal repetition of *Φ*29 DNA and the end proximal 14 base pairs common to different adenovirus DNAs; they are very rich in A·T pairs. A·T-rich regions are also present in several replication origins sequenced, and it seems likely that such regions are important in DNA sites where a local melting of DNA is needed.

The sequences at the ends of other *B. subtilis* *Φ*29-related phages like *Φ*15, M2Y, Nf and GA-1, which also have protein linked at the 5′ ends of the DNA, have been determined (Yoshikawa and Ito 1981). The se-

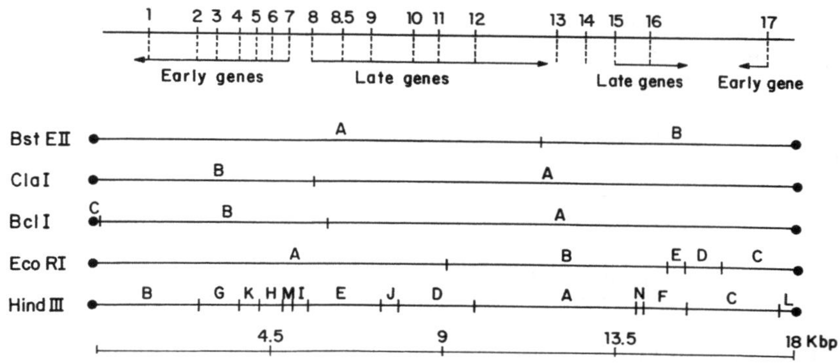

Fig. 2. Genetic and physical map of bacteriophage Φ29 DNA. The genetic map was drawn relative to the *Hin*dIII map as described by ESCARMÍS and SALAS (1982). The *Eco*RI cuts were taken from SOGO et al. (1979) and the *Bst*EII, *Cla*I, and *Bcl*I cuts from YOSHIKAWA and ITO (1982)

quence of the inverted terminal repetition at the ends of Φ15 DNA is AAAGTA, identical to that of Φ29 DNA, in the case of M2Y and Nf DNAs is AAAGTAAG, and for GA-1 is AAATAGA, in all cases very rich in A·T pairs.

It is also interesting to note that the *Streptococcus pneumoniae* phage Cp1, which also has protein covalently linked to the 5′ ends of the DNA (GARCÍA et al. 198b), has a sequence at the ends (AAAGCA) very similar to that of Φ29 DNA, although they differ in the length of the inverted terminal repetition, which in the case of Cp1 DNA is 236 nucleotides long (ESCARMÍS et al., submitted), even longer than in adenovirus DNA.

2.4 Nucleotide Sequence of Gene 3, Coding for the Terminal Protein

The sequence of the Φ29 DNA *Hin*dIII G fragment (1150 nucleotides) and 286 nucleotides from the right end of the *Hin*dIII B fragment (see Fig. 2), shown to contain genes 3 and 4 (MELLADO and SALAS 1982), has been determined (Fig. 3) (ESCARMÍS and SALAS 1982). There are two open reading frames, partially overlapping, one coding for a protein of 125 amino acids and the other for a protein of 266 amino acids, in agreement with the expected molecular weight of proteins p4 (MELLADO and SALAS 1982) and p3 (SALAS et al. 1978) respectively. This was confirmed by sequencing the same DNA region of mutants *sus*3(91) and *sus*4(56). For gene 3, the wild-type and *sus*3 sequences differed in only one nucleotide, at position 737, where a CAA triplet in the wild-type sequence had been converted into a UAA nonsense triplet in the mutant. For gene 4, the same type of change of a CAA to a UAA triplet occurs at position 225. These results unambiguously assign the open reading frame coding for a protein of 125 amino

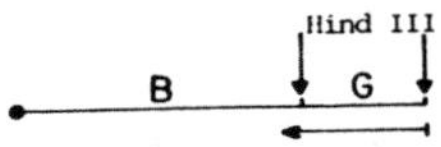


```
                                                                                                   90
5' AGCTTGAAACGTTTAAGGTTAAAGTGGTTCAAGGAACATCTAGTAAAGGTAACGTATTCTTTAGCTTACAACTATCCCTATAAACAGGAG
 Hind III
                                                                                                  180
GTAAAATATAGATGCCTAAAACACAAAGAGGTATCTATCATAACTTGAAGGAATCTGAATACGTGGCATCTAACACCGATGTCACGTTTT
           Met Pro Lys Thr Gln Arg Gly Ile Tyr His Asn Leu Lys Glu Ser Glu Tyr Val Ala Ser Asn Thr Asp Val Thr Phe

                                      sus4(56)→T                                                  270
TCTTTTCAAGTGAATTGTATTTGAACAAGTTTCTCGATGGATACCAAGAATACAGGAAGAAATTTAATAAGAAGATAGAACGGGTCGCTG
Phe Phe Ser Ser Glu Leu Tyr Leu Asn Lys Phe Leu Asp Gly Tyr Gln Glu Tyr Arg Lys Lys Phe Asn Lys Lys Ile Glu Arg Val Ala

                                                                                                  360
TTACACCGTGGAATATGGATATGCTCGCAGACATCACGTTCTATTCAGAAGTTGAAAAGCGTGGTTTCCATGCTTGGTTGAAAGGAGATA
Val Thr Pro Trp Asn Met Asp Met Leu Ala Asp Ile Thr Phe Tyr Ser Glu Val Glu Lys Arg Gly Phe His Ala Trp Leu Lys Gly Asp

                                                                                                  450
ACGCAACATGGCGAGAAGTCCACGTATACGCATTAAGGATAATGACAAAGCCGAATACGCTCGATTGGTCAAGAATACAAAAGCCAAGAT
Asn Ala Thr Trp Arg Glu Val His Val Tyr Ala Leu Arg Ile Met Thr Lys Pro Asn Thr Leu Asp Trp Ser Arg Ile Gln Lys Pro Arg
          Met Ala Arg Ser Pro Arg Ile Arg Ile Lys Asp Asn Asp Lys Ala Glu Tyr Ala Arg Leu Val Lys Asn Thr Lys Ala Lys Ile
                                                                                                  540
TGCGAGAACGAAGAAAAAGTATGGTGTAGACCTTACCGCTGAAATTGATATACCTGACCTTGATTCATTTGAAACACGGGCGCAGTTCAA
Leu Arg Glu Arg Arg Lys Ser Met Val END
 Ala Arg Thr Lys Lys Lys Tyr Gly Val Asp Leu Thr Ala Glu Ile Asp Ile Pro Asp Leu Asp Ser Phe Glu Thr Arg Ala Gln Phe Asn
                                                                                                  630
TAAGTGGAAGGAACAAGCGTCCTCTTTCACTAACCGTGCTAATATGCGTTATCAGTTCGAAAAGAATGCATACGGTGTGGTGGCTAGTAA

 Lys Trp Lys Glu Gln Ala Ser Ser Phe Thr Asn Arg Ala Asn Met Arg Tyr Gln Phe Glu Lys Asn Ala Tyr Gly Val Val Ala Ser Lys
                                                                                                  720
AGCTAAGATAGCTGAGATTGAACGTAACACAAAAGAGGTTCAGCGGTTAGTAGATGAGAAAATCAAGGCTATGAAAGACAAAGAATACTA

 Ala Lys Ile Ala Glu Ile Glu Arg Asn Thr Lys Glu Val Gln Arg Leu Val Asp Glu Lys Ile Lys Ala Met Lys Asp Lys Glu Tyr Tyr
          sus3(91)→T                                                                              810
TGCAGGCGGTAAGCCGCAAGGGACAATTGAACAACGGATAGCTATGACAAGTCCTGCACACGTTACAGGAATTAATAGACCCCATGATTT

 Ala Gly Gly Lys Pro Gln Gly Thr Ile Glu Gln Arg Ile Ala Met Thr Ser Pro Ala His Val Thr Gly Ile Asn Arg Pro His Asp Phe
                                                                                                  900
TGACTTTAGCAAGGTGCGAAGCTATAGCCGTTTGCGAACCCTAGAAGAAAGCATGGAGATGAGAACAGACCCTCAGTATTATGAAAAGAA

 Asp Phe Ser Lys Val Arg Ser Tyr Ser Arg Leu Arg Thr Leu Glu Glu Ser Met Glu Met Arg Thr Asp Pro Gln Tyr Tyr Glu Lys Lys
                                                                                                  990
AATGATACAGTTACAGTTAAACTTTATTAAGAGCGTTGAGGGTAGTTTCAATTCATTTGATGCGGCAGATGAACTGATCGAAGAATTAAA

 Met Ile Gln Leu Gln Leu Asn Phe Ile Lys Ser Val Glu Gly Ser Phe Asn Ser Phe Asp Ala Ala Asp Glu Leu Ile Glu Glu Leu Lys
                                                                                                 1080
AAAGATACCTCCTGATGACTTCTATGAATTGTTTCTCAGAATATCAGAAATATCCTTTGAGGAATTTGATAGTGAGGGAAACACAGTGGA

 Lys Ile Pro Pro Asp Asp Phe Tyr Glu Leu Phe Leu Arg Ile Ser Glu Ile Ser Phe Glu Glu Phe Asp Ser Glu Gly Asn Thr Val Glu
                                                                                                 1170
GAACGTAGAAGGTAATGTATATAAAATACTGTCATACTTGGAACAGTATCGAAGGGGTGACTTTGATCTAAGCTTAAAGGGGTTCTAGGC
                                                                          Hind III ↑   -------
 Asn Val Glu Gly Asn Val Tyr Lys Ile Leu Ser Tyr Leu Glu Gln Tyr Arg Arg Gly Asp Phe Asp Leu Ser Leu Lys Gly Phe END
                                                                                                 1260
TCCGTTAAAGGATGAAGCATATGCCGAGAAAGATGTATAGTTGTGACTTTGAGACAACTACTAAAGTGGAAGACTGTAGGGTATGGGCGT
      ----- Met Lys His Met Pro Arg Lys Met Tyr Ser Cys Asp Phe Glu Thr Thr Thr Lys Val Glu Asp Cys Arg Val Trp Ala

                                                                                                 1350
ATGGTTATATGAATATAGAAGATCACAGTGAGTACAAAATAGGTAATAGCCTGGATGAGTTTATGGCGTGGGTGTTGAAGGTACAAGCTG
Tyr Gly Tyr Met Asn Ile Glu Asp His Ser Glu Tyr Lys Ile Gly Asn Ser Leu Asp Glu Phe Met Ala Trp Val Leu Lys Val Gln Ala


ATCTATATTTCCATAACCTCAAATTTGACGGAGCTTTTATCATTAACTGGTTGGAACGTAATGGTTTTAAGTGGTCGGCTGACGGA
Asp Leu Tyr Phe His Asn Leu Lys Phe Asp Gly Ala Phe Ile Ile Asn Trp Leu Glu Arg Asn Gly Phe Lys Trp Ser Ala Asp Gl..
```

Fig. 3. Nucleotide sequence of genes 3 and 4. The sequence has been taken from the work of Escarmís and Salas (1982). The deduced amino acid sequence for proteins p4, p3, and part of p2 is given and the probable Shine-Dalgarno sequences for proteins p4 and p3 is underlined with *solid lines*. The two more likely Shine-Dalgarno sequences for protein p2 are underlined with *discontinuous lines*. The mutation in mutants *sus*4(56) and *sus*3(91) is shown

acids to protein p4, involved in the control of Φ29 DNA late transcription (Sogo et al. 1979), and that for the 266 amino acids protein to p3, the 5′-linked protein (Salas et al. 1978). The nucleotide sequence determined here is the same as that of a region at the left end of Φ29 DNA sequenced by Yoshikawa and Ito (1982).

As can be also seen in Fig. 3, at about ten nucleotides before the initiating triplets for proteins p3 and p4, a Shine-Dalgarno sequence is found which shows a strong complementarity with the sequence of the 3′ end of *B. subtilis* 16S rRNA (Murray and Rabinowitz 1982), this being a common feature for the efficient translation of *B. subtilis* mRNAs (McLaughlin et al. 1981).

The sequence of the adenovirus 5′-terminal protein has recently been reported (Smart and Stillman 1982; Aleström et al. 1982). It has no similarity to that of protein p3.

3 Mechanism of Φ29 DNA Replication

3.1 Structure of Protein-free and Protein-containing Replicative Intermediates of Φ29 DNA

To determine the origin as well as the mechanism of Φ29 DNA replication, we isolated Φ29 DNA replicative intermediates and analyzed them by electron microscopy.

Protein-free replicative intermediates were isolated by sucrose-gradient centrifugation followed by treatment with pronase and CsCl equilibrium density-gradient centrifugation. The material with a density higher than that of mature Φ29 DNA consisted of replicative intermediates, as analyzed by electron microscopy. Two major types of molecules were found (see Fig. 4) (Inciarte et al. 1980). One consisted of unit-length duplex Φ29 DNA with one single-stranded branch at a random position (type I). The length of the single-stranded branches was similar to that of one of the double-stranded regions. The other type of molecules consisted of unit-length DNA with one double-stranded and one single-stranded region extending a variable distance from one end (type II). Partial denaturation of type II molecules showed that replication was initiated with a similar frequency from either DNA end. These findings indicated that Φ29 DNA replication occurs by a mechanism of strand displacement and that replication starts, nonsimultaneously, from either DNA end, as in the case of adenovirus (Lechner and Kelly 1977). A similar conclusion for Φ29 DNA replication was reached by Harding and Ito (1980).

Since Φ29 DNA replication does not seem to occur through covalent circular structures or concatemeres, and the sequence at the ends is not compatible with the formation of a hairpin that would prime initiation

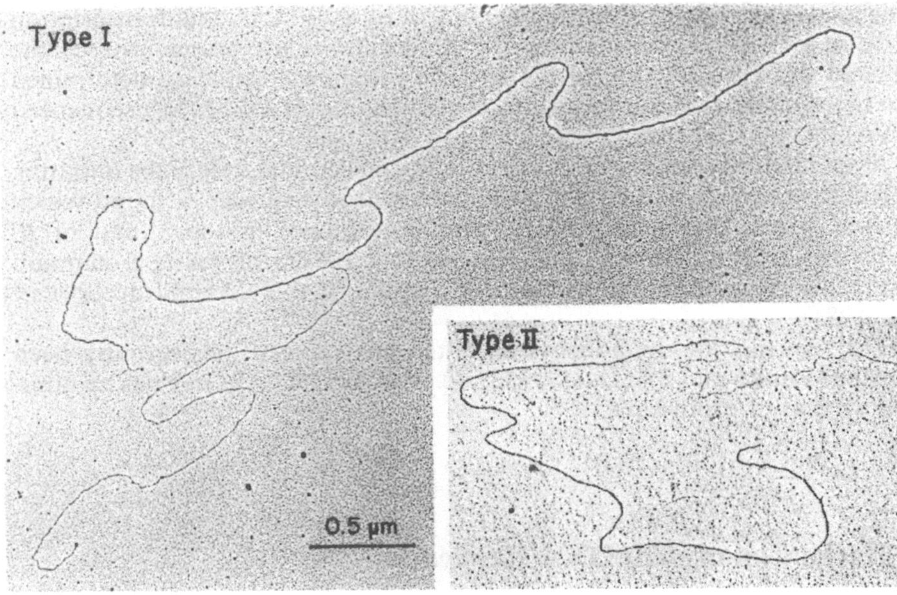

Fig. 4. Electron micrographs of type I and type II replicative Φ29 DNA molecules. The BAC protein-free spreading technique was used as described (SOGO et al. 1982). The electron micrographs were taken by J.M. SOGO

of replication (CAVALIER-SMITH 1974), the question arises of how the 5′ ends are primed, i.e., how the free 3′OH group needed by the DNA polymerase is provided. A model, first proposed by REKOSH et al. (1977) to solve this problem in the initiation of adenovirus DNA replication, assumes that a newly synthesized molecule of the terminal protein interacts with the parental protein linked to the DNA, with the inverted terminal repetition or with both and primes replication by reaction with the 5′-terminal dNTP giving rise to the formation of a protein-dNMP covalent complex, that provides the free 3′OH group needed for elongation by the DNA polymerase (INCIARTE et al. 1980; HARDING and ITO 1980).

In agreement with this model, we have found that Φ29 DNA replicative intermediates isolated without treatment with proteolytic enzymes have protein p3 associated at the ends of the parental and daughter DNA strands of the replicative type I and type II molecules (SOGO et al. 1982).

3.2 In Vivo Role of Proteins p2 and p3 in Initiation of Φ29 DNA Replication and of Proteins p5 and p6 in Elongation Steps

The Φ29 genes 2, 3, 5, 6, and 17 are involved in the synthesis of the viral DNA (CARRASCOSA et al. 1976; HAGEN et al. 1976). Since *ts* mutants in genes 2, 3, 5, and 6 are available (TALAVERA et al. 1971; MORENO et al.

1974), *B. subtilis* was infected at 30 °C with these mutants and after a short pulse the infected bacteria were shifted up to 42 °C. At different times after the shift-up the labeled DNA was analyzed by alkaline sucrose-gradient centrifugation (MELLADO et al. 1980). Unit-length DNA was formed shortly after infection with mutant *ts*3(132), suggesting a role of protein p3 in the initiation of replication consistent with the priming model proposed for the initiation of Φ29 DNA replication. A similar result was obtained after infection with mutant *ts*2(98). In contrast, mutant *ts*5(17) did not produce unit-length viral DNA even at very late times after the shift-up, suggesting that this mutation affects an elongation (or maturation) step. Mutant *ts*6(1360) gave rise to unit-length phage DNA at very late times after the shift-up, suggesting that the mutation might affect an elongation or maturation process but that either it is leaky or its function may be partially replaced by a bacterial function.

4 Initiation of Φ29 DNA Replication In Vitro

4.1 Formation of a Covalent Complex Between the Terminal Protein p3 and 5′dAMP

Although the in vivo evidence was consistent with the priming model for the initiation of Φ29 DNA replication, it was important to obtain in vitro support for the model, specifically to show the formation of a protein p3-dAMP covalent complex in vitro.

When extracts from Φ29-infected *B. subtilis* were incubated with [α-^{32}P]dATP in the presence of ATP, a labeled protein was found with the electrophoretic mobility of p3 (PEÑALVA and SALAS 1982). The reaction product was resistant to treatment with micrococcal nuclease, phosphatase, and RNases A and T1 and sensitive to proteinase K. Incubation of the ^{32}P-labeled protein with piperidine under conditions in which the Φ29 DNA-protein p3 linkage is hydrolyzed released 5′-dAMP. The reaction with [α-^{32}P]dATP was strongly inhibited by anti-p3 serum. Acid hydrolysis of the ^{32}P-labeled protein yielded phosphoserine under conditions in which this product was formed from the Φ29 DNA-protein p3 complex. Partial hydrolysis with V8 protease of the ^{32}P-labeled protein and ^{35}S-labeled protein p3 produced a common peptide having the same electrophoretic mobility in SDS-polyacrylamide gels. All these results indicated that a covalent complex between protein p3 and 5′-dAMP is formed in vitro.

The initiation complex (protein p3-dAMP) formed in the presence of 0.5 μ*M* [α-^{32}P]dATP could be elongated by addition of 40 μ*M* dNTPs (PEÑALVA and SALAS 1982). After treatment with piperidine, the product elongated in the presence of 2′,3′-dideoxycytidine 5′-triphosphate released the expected oligonucleotides, 9 and 12 bases long, taking into account the sequence at the left and right Φ29 DNA ends, respectively.

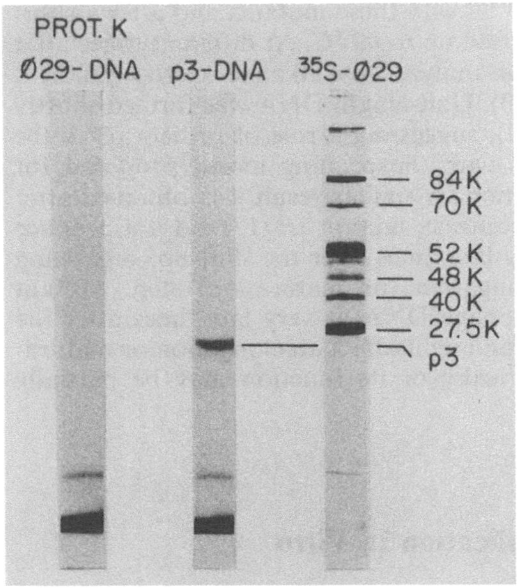

Fig. 5. Requirement of Φ29 DNA-protein p3 as template for the in vitro formation of protein p3-dAMP initiation complex. The DNA-free extracts prepared from Φ29-infected *B. subtilis* were incubated with [α-^{32}P]dATP as described (PEÑALVA and SALAS 1982), using proteinase-K-treated Φ29 DNA or Φ29 DNA–protein p3 complex as template. The samples were processed and subjected to SDS-polyacrylamide gel electrophoresis as described (PEÑALVA and SALAS 1982). ^{35}S-Φ29 proteins were used as a molecular weight marker. The experiment was carried out by M.A. PEÑALVA

4.2 Template Requirements for the Formation of the Initiation Complex

When endogenous Φ29 DNA-protein p3 was eliminated from the extracts by DEAE-cellulose chromatography, no formation of protein p3-dAMP initiation complex was obtained unless supplemented with exogenous Φ29 DNA-protein p3 (Fig. 5). Proteinase-K-treated Φ29 DNA was inactive as template (Fig. 5). These results are in agreement with the transfection of competent *B. subtilis* by Φ29 DNA-protein p3 but not by proteinase-K-treated Φ29 DNA (HIROKAWA 1972; SALAS et al. 1983). To study whether or not an intact Φ29 DNA molecule was needed, the Φ29 DNA–protein p3 complex was digested with various restriction endonucleases, such as *Bst*EII or *Cla*I (which produce one cut in Φ29 DNA), *Bcl*I (two cuts), *Eco*RI (four cuts) or *Hind*III (13 cuts) (Fig. 2). The mixture of the fragments was used as a template for the in vitro formation of the initiation complex. None of these treatments essentially affected the formation of the protein p3-dAMP complex (GARCÍA et al., in press).

To test whether a single terminal fragment of Φ29 DNA could act as template for the formation of the initiation complex, the isolated left and right fragments *Eco*RI A (9000 bp) and C (1800 bp), both containing protein p3 at the end, as well as an internal fragment, *Eco*RI B (5490 bp), were used as template for the formation of the initiation complex. Fragments *Eco*RI A and C gave rise to the formation of 61% and 25% respectively, of the p3-dAMP complex produced by intact *Φ*29 DNA-protein p3. As expected, the internal fragment *Eco*RI B was completely inactive. In a similar experiment using the terminal left and right fragments *Bcl*I C (73 bp) and A (12000 bp) respectively, the amount of p3-dAMP produced was 25% and 48% of that synthesized by intact *Φ*29 DNA–protein p3 complex (García et al., in press). These results indicate (a) that the isolated left or right protein p3-containing fragments can serve as template in the initiation reaction and (b) that there seems to be a size effect, i.e., longer fragments produce a higher amount of formation of p3-dAMP complex than shorter fragments, although the effect is not very drastic. To study the size effect further, the left fragment *Eco*RI A was cut with the restriction nucleases *Hha*I, *Bcl*I, *Taq*I, or *Mnl*I, which produce terminal fragments of 175, 73, 42, and 26 bp respectively. The formation of the p3-dAMP complex was 74%, 51%, 32%, and 27% respectively of that obtained with fragment *Eco*RI A as template. The right fragment *Eco*RI C was cut with *Hind*III, *Acc*I, *Hinf*I, *Taq*I, or *Rsa*I (terminal fragments of 269, 125, 59, 52, and 10 bp respectively), and the values obtained for the formation of the p3-dAMP complex were 181%, 157%, 36%, 29%, and 14% of that obtained with fragment *Eco*RI C, the last value being only slightly higher than the background obtained in the absence of DNA (García et al., in press). The above results indicate that a 26-bp fragment from the left end or a 52-bp fragment from the right end are still active in the formation of the initiation complex, but the initiation reaction decreased drastically when a 10-bp fragment from the right end was used. However, when an excess of lambda DNA digested with *Rsa*I was added to the mixture of fragments produced by treatment of the *Φ*29 DNA fragment *Eco*RI C with *Rsa*I, a marked increase in the formation of the initiation complex was obtained after ligation, suggesting that the size of the fragment, and not a specific DNA sequence after the 10 terminal base pairs at the right end, is important for the initiation reaction (García et al., in press).

4.3 Role of Proteins p3 and p2 in Formation of the Protein p3-dAMP Covalent Complex

To study the requirements for the in vitro formation of the protein p3-dAMP complex, extracts from *B. subtilis* infected with *Φ*29 mutants in genes 2, 3, 5, 6, and 17, involved in DNA synthesis, have been used. As expected, extracts from *sus*3-infected cells were completely inactive in the formation of the initiation complex. Extracts from *B. subtilis* infected with two different *sus* mutants in gene 2 were also inactive in the formation of the p3-dAMP complex, but could be complemented by addition of the *sus*3 extracts (Blanco et al. 1983). These results indicated that protein p2, as well as p3, is required in the formation of the initiation complex. Similar experiments using mutants in genes 5, 6, and 17 suggested that the products of these genes are not required in the initiation reaction (Blanco et al. 1983).

5 Overproduction and Purification of the Initiation Proteins p3 and p2

5.1 Cloning and Expression in *Escherichia coli* of Genes 3 and 2

To study the function of proteins p3 and p2 in the initiation of *Φ*29 DNA replication we undertook the cloning of these genes with the aim of overproducing the proteins for their purification.

A *Φ*29 DNA fragment containing gene 3 and the early genes 4, 5, and most of gene 6 was cloned (García et al. 1983a) in the pBR322 derivative plasmid pKC30 (Shimatake and Rosenberg 1981) under the control of the P_L promoter of bacteriophage λ. Four polypeptides of M_r 27000, 18500, 17500, and 12500 were labeled with [^{35}S]methionine after heat induction of the *E. coli* transformed with the recombinant plasmid pKC30Al, accounting for about 15% of the de novo synthesized protein. The 27000 protein, which represents about 3% of the de novo protein synthesis, has been characterized as p3 by radioimmunoassay. The protein p3 synthesized in *E. coli* was active in the in vitro formation of the initiation complex p3-dAMP when supplemented with extracts from *B. subtilis* infected with a *sus*3 mutant (García et al. 1983a).

A *Hin*dIII-*Bcl*I fragment of *Φ*29 DNA containing gene 2 has been cloned in the *Hin*dIII-*Bam*HI sites of plasmid pBR322. To overproduce protein p2, the *Eco*RI-*Pst*I fragment of the recombinant plasmid containing the *Φ*29 DNA insert was further cloned into the *Eco*RI-*Pst*I sites of plasmid pPLc28 (Remaut et al. 1981), under the control of the P_L promoter of phage λ. A protein of M_r 68000, the size expected for protein p2, which was synthesized after heat induction of *E. coli* transformed with the recombinant plasmid pLBw2 accounted for about 2% of the de novo protein synthesis and was active in the in vitro formation of the initiation complex when complemented with extracts from *B. subtilis* infected with a *sus*2 mutant (L. Blanco, J.A. García, and M. Salas, unpublished results).

5.2 Purification of Proteins p3 and p2

Protein p3 has been purified from *E. coli* harboring the gene 3-containing recombinant plasmid pKC30A1 by a procedure that includes DEAE-cellulose chromatography, precipitation with polyethyleneimine, and phosphocellulose chromatography. The protein p3 obtained was essentially homogeneous and was active in the formation of the initiation complex when complemented with extracts from *sus*3-infected *B. subtilis*. No DNA polymerase or ATPase activities were present in the purified protein (Prieto et al., submitted).

The purification of protein p2 from *E. coli* harboring the gene 2-containing recombinant plasmid pLBw2 is under way. A DNA polymerase activity copurifies with protein p2 (L. BLANCO, J.A. GARCIA, J.M. LAZARO and M. SALAS, unpublished results).

5.3 In Vitro Mutagenesis of Gene 3: Effect of Specific Mutations on In Vitro Formation of the Initiation Complex

The Φ29 DNA *Hin*dIII G fragment contains gene 4 and most of gene 3, except the last 15 nucleotides coding for the five C-terminal amino acids. This fragment was cloned in the *Hin*dIII site of pBR322, giving rise to a recombinant that produced a mutated protein p3, p3′, in which the five C-terminal amino acids of protein p3 (*ser-leu-lys-gly-phe*) would be replaced by *ser-phe-asp-ala-val-val-tyr-his-ser* (MELLADO and SALAS 1982). To obtain a high level of expression of the mutated protein p3′, the *Eco*RI-*Bam*Hl fragment from the above recombinant was cloned in the *Eco*RI-*Bam*Hl sites of plasmid pPLc28 (REMAUT et al. 1981) under the control of the P_L promoter of phage λ. After heat induction of *E. coli* transformed with the recombinant plasmid pRMw51, two proteins with the electrophoretic mobility expected for p3′ and p4 were labeled with [^{35}S]methionine, and accounted for 6% and 30% respectively of the total de novo protein synthesis. By performance of a similar cloning using the *Hin*dIII G fragment from a *sus*3 and a *sus*4 mutant, the two proteins were characterized as p3′ and p4 (MELLADO and SALAS 1982).

Protein p3′ was active in the in vitro formation of the initiation complex (p3′-dAMP), although the specific activity was about 15% of that obtained with intact protein p3 (MELLADO and SALAS, in press).

The Φ29 DNA *Hin*dIII G fragment was also cloned in the pBR322 derivative plasmid pKTH601, which has the *Hin*dIII-*Bam*H1 fragment of the former replaced by a synthetic stop oligonucleotide with *Hin*dIII-*Bam*H1 linkers (kindly donated by Dr. R.F. Pettersson). This cloning gave rise to a recombinant in which the last three C-terminal amino acids of protein p3, *lys-gly-phe*, would be replaced by *leu-ile-asp*. As before, to obtain a high level of expression of the mutated protein p3″ the *Eco*RI-*Bam*H1 fragment from the above recombinant was cloned in the *Eco*RI-*Bam*H1 sites of plasmid pPLc28, under the control of the P_L promoter of phage λ, resulting in the recombinant plasmid pRMt121 (MELLADO and SALAS, in press). After heat induction, the mutated protein p3″ was labelled with [^{35}S]methionine and found to account for 8% of the total de novo protein synthesis. This protein was also active in the in vitro formation of the initiation complex, although the specific activity was about 12% of that obtained with intact protein p3 (MELLADO and SALAS, in press).

We intend to produce deletions at the C-terminal of protein p3 to determine how many amino acids can be eliminated without total loss of the protein activity in the initiation of replication. The effect of mutations at the N-terminal of p3 as well as at internal positions will be also studied.

The serine residue in protein p3 through which the linkage to 5′-dAMP is carried out is presently being studied with a view to introducing mutations in the neighboring region by oligonucleotide-directed mutagenesis and studying their effect on the function of protein p3 in the formation of the initiation complex.

6 A New Mechanism for Initiation of *Φ*29 and Adenovirus DNA Replication: A Comparison

As already pointed out, adenovirus and *Φ*29 DNAs are similar in that both have a specific viral protein covalently linked at the two 5′ ends through a phosphoester bond between serine and dCMP (for adenovirus) or dAMP (for *Φ*29). Both DNAs have inverted terminal repetitions, six nucleotides long in *Φ*29 and about 100 nucleotides long in adenovirus, and both start replication at the ends and replicate by a similar mechanism of strand displacement, giving rise to the same types of replicating molecules. Also, the two proteins at the ends of the DNA interact with each other, producing circular DNA structures and aggregates.

In agreement with the model proposed for the initiation of adenovirus and *Φ*29 DNA replication in which the terminal protein acts as a primer by reaction with the 5′-terminal dNTP, a covalent complex between protein p3 and dAMP (Peñalva and Salas 1982; Shih et al. 1982; Watabe et al. 1982) and between a M_r 80000 precursor of the adenovirus terminal protein and dCMP (Lichy et al. 1981; Pincus et al. 1981; Challberg et al. 1982; Tamanoi and Stillman 1982; de Jong et al. 1982) has been found. Unlike the adenovirus system, no precursor for the terminal protein exists in *Φ*29 (Escarmís and Salas 1982).

The adenovirus and *Φ*29 in vitro systems for the initiation of replication require the DNA-protein complex as template and are inactive with protease-treated DNA. However, while the adenovirus protein-dCMP complex is formed with piperidine-treated DNA which produces a DNA without the residual peptide that remains after the protease treatment (Tamanoi and Stillman 1982), essentially no initiation complex is formed with piperidine-treated *Φ*29 DNA (García et al., in press). Denatured pronase-treated adenovirus DNA is active as a template in the formation of the initiation complex (Ikeda et al. 1982; Tamanoi and Stillman 1982), but denatured proteinase-K-treated *Φ*29 DNA is not. The adenoviral protein-primed initiation of DNA chains in vitro is also active with single-stranded *Φ*X174 DNA (Ikeda et al. 1982) but no initiation reaction occurs with this template in the *Φ*29 system (García et al., in press).

Treatment of adenovirus DNA-protein complex with several restriction nucleases still supports the initiation reaction (Horwitz and Ariga 1981; Tamanoi and Stillman 1982). Similarly, treatment of the *Φ*29 DNA–protein p3 complex with several restriction endonucleases produces terminal

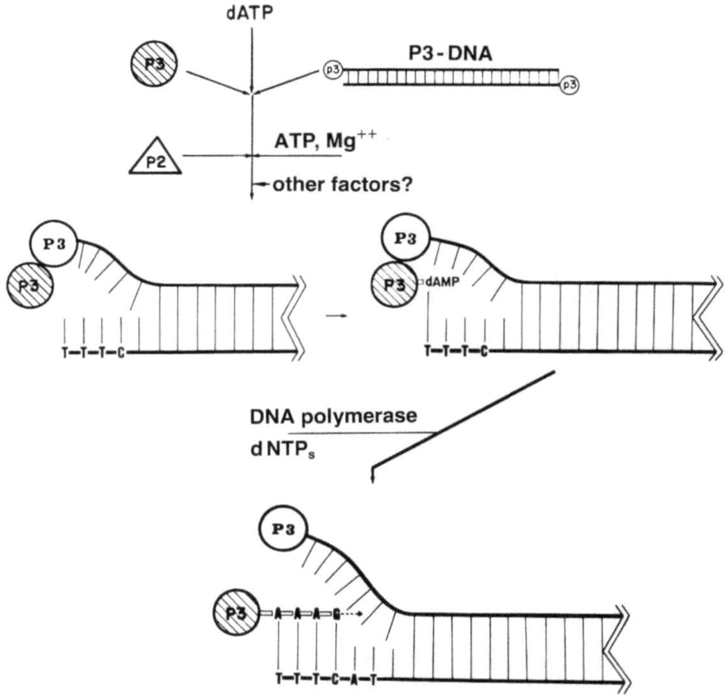

Fig. 6. Model for the initiation of phage Φ29 DNA replication

fragments still active for the formation of the initiation complex. Moreover, isolated left or right protein-containing terminal Φ29 DNA fragments, as small as 26 and 52 bp long respectively, are active in the formation of the p3-dAMP complex (García et al., in press). Similarly, a fragment from the left end of adenovirus DNA cloned in pBR322 was active in the formation of the initiation complex when the recombinant plasmid DNA was linearized at a position close to the adenovirus origin of replication (Tamanoi and Stillman 1982; van Bergen et al. 1983).

The adenovirus terminal protein precursor has been purified in a functional form and has been shown to copurify with a DNA polymerase activity (Enomoto et al. 1981). More recently, the two proteins have been separated and it has been shown that they are both involved in the initiation of adenovirus DNA replication (Lichy et al. 1982; Stillman et al. 1982). In addition, a host factor has been purified and shown to stimulate the formation of the adenovirus protein-dCMP complex in vitro, being an absolute requirement for the reaction when the adenovirus DNA binding protein is present (Nagata et al. 1982).

The terminal protein p3 of Φ29 has been purified to homogeneity in a functional form and no DNA polymerase activity is present in the p3 preparation. The product of the Φ29 gene 2, also required for the formation of the protein p3-dAMP complex, is being purified and a DNA polymerase activity copurifies with it. Fig. 6 shows a summary of the model proposed for the initiation of Φ29 DNA replication. A newly synthesized molecule of protein p3 interacts with the parental protein p3, and possibly also with the inverted terminal repetition, and is positioned at the end(s) of the DNA to start replication by reaction with dATP in the presence of protein p2 and ATP. A covalent protein p3-dAMP complex is formed and provides the free 3′OH group needed for elongation. Whether another viral and/or host factor besides protein p2 is also involved in the initiation of Φ29 DNA replication remains to be elucidated.

Acknowledgments. Different aspects of the Φ29 work presented here were carried out by L. Blanco, C. Escarmís, J.A. García, J.M. Hermoso, M.R. Inciarte, J.M. Lázaro, R.P. Mellado, J. Ortín, R. Pastrana, M.A. Peñalva, I. Prieto, J.M. Sogo, C. Vásquez, and E. Viñuela. This research has been supported by grant 1 R01 GM27242 from the National Institutes of Health and by grants from the Comisión Asesora para el Desarrollo de la Investigación Científica y Técnica and the Fondo de Investigaciones Sanitarias.

References

Aleström P, Akusjärvi G, Pettersson M, Pettersson U (1982) DNA sequence analysis of the region encoding the terminal protein and the hypothetical N-gene product of adenovirus type 2. J Biol Chem 257:13492–13498

Blanco L, García JA, Peñalva MA, Salas M (1983) Factors involved in the initiation of phage Φ29 DNA replication in vitro: requirement of the gene 2 product for the formation of the protein p3-dAMP complex. Nucleic Acids Res 11:1309–1323

Brown DT, Westpahl M, Burlingham BT, Winterhoff U, Doerfler W (1975) Structure and composition of the adenovirus type 2 core. J Virol 16:366–387

Carrascosa JL, Camacho A, Moreno F, Jiménez F, Mellado RP, Viñuela E, Salas M (1976) *Bacillus subtilis* phage Φ29: characterization of gene products and functions. Eur J Biochem 66:229–241

Carusi EA (1977) Evidence for blocked 5′ termini in human adenovirus DNA. Virology 76:390–394

Cavalier-Smith T (1974) Palindromic base sequences and replication of eukaryote chromosome ends. Nature 250:467–470

Challberg MD, Ostrove JM, Kelly TJ Jr (1982) Initiation of adenovirus DNA replication: detection of covalent complexes between nucleotide and the 80 kilodalton terminal protein. J Virol 41:265–270

De Jong PJ, Kwant MM, van Driel W, Jansz HS, van der Vliet PC (1982) The ATP requirements of adenovirus type 5 DNA replication and cellular DNA replication. Virology 124:45–58

Desiderio SV, Kelly TJ Jr (1981) Structure of the linkage between adenovirus DNA and the 55000 molecular weight terminal protein. J Mol Biol 145:319–337

Enomoto T, Lichy JH, Ikeda JE, Hurwitz J (1981) Adenovirus DNA replication in vitro: purification of the terminal protein in a functional form. Proc Natl Acad Sci USA 78:6779–6783

Escarmís C, Salas M (1981) Nucleotide sequence at the termini of the DNA of *Bacillus subtilis* phage Φ29. Proc Natl Acad Sci USA 78:1446–1450

Escarmís C, Salas M (1982) Nucleotide sequence of the early genes 3 and 4 of bacteriophage Φ29. Nucleic Acids Res 10:5785–5798

García JA, Pastrana R, Prieto I, Salas M (1983a) Cloning and expression in *Escherichia coli* of the gene coding for the protein linked to the ends of *Bacillus subtilis* phage Φ29 DNA. Gene 21:65–76

García E, Gómez A, Ronda C, Escarmís C, López R (1983b) Pneumococcal bacteriophage Cp-1 contains a protein tightly bound to the 5′ termini of its DNA. Virology 128:92–104

Garcia JA, Peñalva MA, Blanco L, Salas M, Template requirements for the initiation of phage Φ29 DNA replication in vitro. Proc Natl Acad Sci USA, in press

Hagen EW, Reilly BE, Tosi ME, Anderson DL (1976) Analysis of gene function of bacteriophage Φ29 of *Bacillus subtilis*: identification of cistrons essential for viral assembly. J Virol 19:501–517

Harding NE, Ito J (1980) DNA replication of bacteriophage Φ29: characterization of the intermediates and location of the termini of replication. Virology 104:323–338

Harding NE, Ito J, David GS (1978) Identification of the protein firmly bound to the ends of bacteriophage Φ29 DNA. Virology 84:279–292

Hermoso JM, Salas M (1980) Protein p3 is linked to the DNA of phage Φ29 through a phosphoester bond between serine and 5′-dAMP. Proc Natl Acad Sci USA 77:6425–6428

Hirokawa H (1972) Transfecting deoxyribonucleic acid of *Bacillus* bacteriophage Φ29 that is protease sensitive. Proc Natl Acad Sci USA 69:1555–1559

Horwitz MS, Ariga H (1981) Multiple rounds of adenovirus DNA synthesis in vitro. Proc Natl Acad Sci USA 78:1476–1480

Ikeda JE, Enomoto T, Hurwitz J (1982) Adenoviral protein-primed initiation of DNA chains in vitro. Proc Natl Acad Sci USA 79:2442–2446

Inciarte MR, Salas M, Sogo JM (1980) Structure of replicating DNA molecules of *Bacillus subtilis* bacteriophage Φ29. J Virol 34:187–199

Ito J (1978) Bacteriophage Φ29 terminal protein: its association with the 5′ termini of the Φ29 genome. J Virol 28:895–904

Lechner RL, Kelly TJ Jr (1977) The structure of replicating adenovirus 2 DNA molecules. Cell 12:1007–1020

Lichy JH, Horwitz MS, Hurwitz J (1981) Formation of a covalent complex between the 80000-dalton adenovirus terminal protein and 5′-dCMP in vitro. Proc Natl Acad Sci USA 78:2678–2682

Lichy JH, Field J, Horwitz MS, Hurwitz J (1982) Separation of the adenovirus terminal protein precursor from its associated DNA polymerase: role of both proteins in the initiation of adenovirus DNA replication. Proc Natl Acad Sci USA 79:5225–5229

McLaughlin JR, Murray CL, Rabinowitz JC (1981) Initiation factor-independent translation of mRNAs from Gram-positive bacteria. J Biol Chem 256:11283–11291

Mellado RP, Salas M (1982) High-level synthesis in *Escherichia coli* of the *Bacillus subtilis* phage Φ29 proteins p3 and p4 under the control of phage lambda P_L promoter. Nucleic Acids Res 10:5773–5784

Mellado RP, Salas M. Initiation of phage Φ29 DNA replication by the terminal protein modified at the carboxyl end. Nucleic Acids Res, in press

Mellado RP, Peñalva MA, Inciarte MR, Salas M (1980) The protein covalently linked to the 5′ termini of the DNA of *Bacillus subtilis* phage Φ29 is involved in the initiation of DNA replication. Virology 104:84–96

Moreno F, Camacho A, Viñuela E, Salas M (1974) Suppressor-sensitive mutants and genetic map of *Bacillus subtilis* bacteriophage Φ29. Virology 62:1–16

Murray CL, Rabinowitz JC (1982) Nucleotide sequences of transcription and translation initiation regions in *Bacillus* phage Φ29 early genes. J Biol Chem 257:1053–1062

Nagata K, Guggenheimer RA, Enomoto T, Lichy JH, Hurwitz J (1982) Adenovirus DNA replication in vitro: identification of a host factor that stimulates synthesis of the preterminal protein-dCMP complex. Proc Natl Acad Sci USA 79:6438–6442

Ortín J, Viñuela E, Salas M, Vásquez C (1971) DNA-protein complex in circular DNA from phage Φ29. Nature 234:275–277

Peñalva MA, Salas M (1982) Initiation of phage Φ29 DNA replication in vitro: formation of a covalent complex between the terminal protein, p3, and 5′-dAMP. Proc Natl Acad Sci USA 79:5522–5526

Pincus S, Robertson W, Rekosh D (1981) Characterization of the effect of aphidicolin on

adenovirus DNA replication: evidence in support of a protein primer model of initiation. Nucleic Acids Res 9:4919–4938
Rekosh DMK, Russell WC, Bellett AJD (1977) Identification of a protein linked to the ends of adenovirus DNA. Cell 11:283–295
Remaut E, Stanssens P, Fiers W (1981) Plasmid vectors for high-efficiency expression controlled by the P_L promoter of coliphage lambda. Gene 15:81–93
Robinson AJ, Younghusband HB, Bellett AJD (1973) A circular DNA-protein complex from adenoviruses. Virology 56:54–69
Ruben M, Bacchetti S, Graham F (1983) Covalently closed circles of adenovirus 5 DNA. Nature 301:172–174
Salas M, Mellado RP, Viñuela E, Sogo JM (1978) Characterization of a protein covalently linked to the 5′ termini of the DNA of *Bacillus subtilis* phage *Φ*29. J Mol Biol 119:269–291
Salas M, García JA, Peñalva MA, Blanco L, Prieto I, Mellado RP, Lázaro JM, Pastrana R, Escarmís C, Hermoso JM (1983) Requirements for the initiation of phage *Φ*29 DNA replication in vitro primed by the terminal protein. In: Cozzarelli NR (ed) Mechanisms of DNA replication and recombination. Liss, New York (UCLA symposia on molecular and cellular biology, new series, vol 10.)
Shih M, Watabe K, Ito J (1982) In vitro complex formation between bacteriophage *Φ*29 terminal protein and deoxynucleotide. Biochem Biophys Res Commun 105:1031–1036
Shimatake H, Rosenberg M (1981) Purified λ regulatory protein cII positively activates promoters for lysogenic development. Nature 292:128–132
Smart JE, Stillman BW (1982) Adenovirus terminal protein precursor. Partial amino acid sequence and the site of covalent linkage to virus DNA. J Biol Chem 257:13499–13506
Sogo JM, Inciarte MR, Corral J, Viñuela E, Salas M (1979) RNA polymerase binding sites and transcription map of the DNA of *Bacillus subtilis* phage *Φ*29. J Mol Biol 127:411–436
Sogo JM, García JA, Peñalva MA, Salas M (1982) Structure of protein-containing replicative intermediates of *Bacillus subtilis* phage *Φ*29 DNA. Virology 116:1–18
Stillman BW, Tamanoi F, Mathews MB (1982) Purification of an adenovirus-coded DNA polymerase that is required for initiation of DNA replication. Cell 31:613–623
Talavera A, Jiménez F, Salas M, Viñuela E (1971) Temperature-sensitive mutants of bacteriophage *Φ*29. Virology 46:586–595
Tamanoi F, Stillman BW (1982) Function of adenovirus terminal protein in the initiation of DNA replication. Proc Natl Acad Sci USA 79:2221–2225
Tolun A, Aleström P, Pettersson U (1979) Sequence of inverted terminal repetitions from different adenoviruses: demonstration of conserved sequences and homology between SA7 termini and SV40 DNA. Cell 17:705–713
Van Bergen BMG, van der Ley PA, van Driel W, van Mansfeld ADM, van der Vliet PC (1983) Replication of origin containing adenovirus DNA fragments that do not carry the terminal protein. Nucleic Acids Res 11:1975–1989
Watabe K, Shih MF, Sugino A, Ito J (1982) In vitro replication of bacteriophage *Φ*29 DNA. Proc Natl Acad Sci USA 79:5245–5248
Yanofsky S, Kawamura F, Ito J (1976) Thermolabile transfecting DNA from temperature-sensitive mutant of phage *Φ*29. Nature 259:60–63
Yehle CO (1978) Genome-linked protein associated with the 5′ termini of bacteriophage *Φ*29 DNA. J Virol 27:776–783
Yoshikawa H, Ito J (1981) Terminal proteins and short inverted terminal repeats of the small *Bacillus* bacteriophage genomes. Proc Natl Acad Sci USA 78:2596–2600
Yoshikawa H, Ito J (1982) Nucleotide sequence of the major early region of bacteriophage *Φ*29. Gene 17:323–335
Yoshikawa H, Friedmann T, Ito J (1981) Nucleotide sequences at the termini of *Φ*29 DNA. Proc Natl Acad Sci USA 78:1336–1340

The Messenger RNAs from the Transforming Region of Human Adenoviruses

ULF PETTERSSON[1], ANDERS VIRTANEN[1], MICHEL PERRICAUDET[2] and GÖRAN AKUSJÄRVI[1]

1 Introduction

The adenoviruses are attractive to molecular biologists not because of their clinical significance, but rather as a useful tool in fundamental biological research. Both the papovaviruses and the adenoviruses have been extensively used as model systems to study the organization and expression of eukaryotic genes. Many properties unique to the eukaryotic organisms were in fact first observed in studies of animal DNA viruses and later shown to be shared by the more complex eukaryotes.

The small DNA viruses have in addition played an important role in cancer research in that they provide one of the simplest model systems, allowing the identification of genes and gene products involved in transfor-

1 Departments of Medical Genetics and Microbiology, The Biomedical Center, Box 589, S-751 23 Uppsala, Sweden
2 Institut de Recherches, Scientifiques sur le Cancer, 7, Rue Guy-Mocquet, F-94800 Villejuif

Current Topics in Microbiology and Immunology, Vol. 109

mation. It was discovered as long as two decades ago (HUEBNER et al. 1962; TRENTIN et al. 1962) that adenoviruses cause tumors when injected into newborn hamsters and rats. Since then a vast amount of knowledge has been accumulated concerning the structure and expression of adenovirus transforming genes. Adenovirus research, like many other kinds of biological research, has made a dramatic leap forward during the past years because of the rapid progress in DNA technology. cDNA copies of many spliced adenovirus mRNAs have been cloned and their exact structures have been determined at the molecular level. In this review we will summarize our current knowledge of the mRNAs which are transcribed from the transforming region of human adenoviruses.

2 The Transforming Region

By analyzing cell lines transformed by human Ad2 it was shown that all cell lines possessed the left-hand end of the viral DNA, in some cases as little as 14%. This suggests that the genes required for maintenance of transformation reside in the left part of the genome.

The finding of GRAHAM and VAN DER EB (1973) that cells can be transformed by naked DNA using the now well-known calcium phosphate coprecipitation technique represented a breakthrough, since it made it possible to introduce specific restriction enzyme fragments into cells and thus to assay their transforming capacity. Using this experimental approach it was shown that transformation can indeed be achieved by fragments containing only the left-hand end of Ad5 DNA representing as little as 8% of the viral genome (GRAHAM et al. 1974a, b; VAN DER EB et al. 1977). Partial or incomplete transformation can also be achieved with the 4.5% *Hpa*I E fragment of Ad5 DNA (HOUWELING et al. 1980), although the transformation efficiency is lower than with larger fragments. The *Hpa*I E-transformed cells also show a difference in morphology compared to cells transformed by larger fragments (see VAN DER EB and BERNARDS, in this volume). As will be discussed below, the leftmost 11.2% contains all the genetic information required for transformation.

3 The mRNAs from the Transforming Region of Human Ad2

3.1 The Transcription Units

The transforming region of adenovirus includes three transcription units – E1A, E1B, and the transcription unit for polypeptide IX (WILSON et al. 1979a). The E1A region is located between map coordinates 1.4 and 4.5

Table 1. Major landmarks in Ad2, E1A, E1B and polypeptide X

Landmark	Position[a]
TATA box for E1A mRNAs	467
Cap site for E1A mRNAs	498
Initiator AUG for E1A polypeptides	559
Donor splice site for E1A 9S mRNA	636
Donor splice site for E1A 12S mRNA	973
Donor splice site for E1A 13S mRNA	1111
Acceptor splice site for E1A mRNAs	1226
Terminator UGA for 9S mRNA polypeptide	1313
Terminator UAA for 12S and 13S mRNAs polypeptides	1540
AAUAAA for E1A mRNAs	1608
Poly(A) addition site for E1A mRNAs	1630
TATA box for E1B mRNAs	1669
Cap site for E1B mRNAs	1699
Initiator AUG for E1B 21K polypeptide	1711
Initiator AUG for E1B 55K polypeptide	2016
Terminator UGA for E1B 21K polypeptide	2236
Donor splice site for E1B 13S mRNA	2249
Terminator UGA for E1B 55K polypeptide	3501
Donor splice site for E1B 22S mRNA	3504
TATA box for polypeptide IX mRNA	3545
Cap site for polypeptide IX mRNA	3576
Acceptor splice site for E1B mRNAs	3589
Initiator AUG for polypeptide IX	3600
Terminator UAA for polypeptide IX	4020
AAUAAA for E1B and polypeptide IX mRNAs	4029
Poly(A) addition site for E1B and polypeptide IX mRNAs	4061

[a] Positions refer to the first nucleotide in each structure: for splice donor and acceptor sites, the last or the first nucleotide of each exon. Numbering is according to the sequence of GINGERAS et al. (1982)

and region E1B between coordinates 4.8 and 11.2. The transcription unit for polypeptide IX overlaps completely with region E1B and is located between coordinates 9.9 and 11.2 (Fig. 1).

3.2 The mRNAs from Region E1A

The mRNAs transcribed from the transforming region of Ad2 and Ad5 have been studied by a variety of methods such as S1 endonuclease mapping (BERK and SHARP 1978), electron microscopy (CHOW et al. 1979; KITCHINGMAN and WESTPHAL 1980), DNA sequencing (GINGERAS et al. 1982; VAN ORMONDT et al. 1980b), molecular cloning (PERRICAUDET et al. 1979; PERICAUDET et al. 1980a), and primer extension (VIRTANEN and PETTERSSON 1983). The conclusion drawn from these studies is that three mRNAs, a 9S, a 12S, and a 13S species, are generated from region E1A (Fig. 1). All three species appear to have common 5′ as well as 3′ ends and they differ

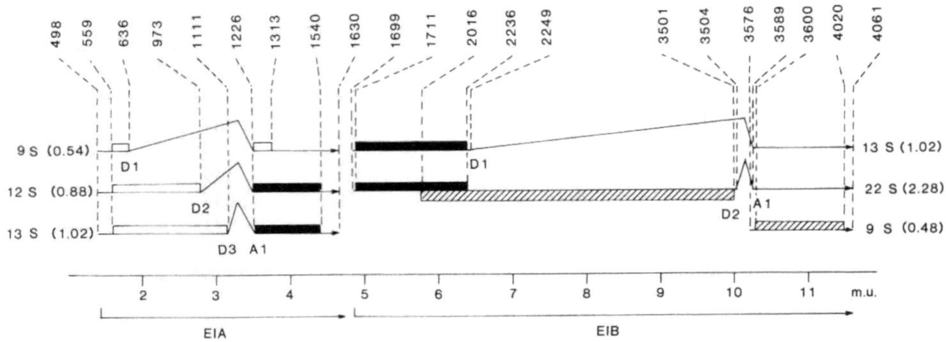

Fig. 1. A schematic drawing which illustrates the mRNAs which are transcribed from the r-strand of region E1. The three different reading frames are indicated with *open, filled*, and *hatched boxes*. Donor (*D*) and acceptor (*A*) sites for splicing are also shown. The numbers on top of the figure refer to positions in the sequence of GINGERAS et al. (1982). The numbers in parentheses indicate the length in kilobases of each individual mRNA

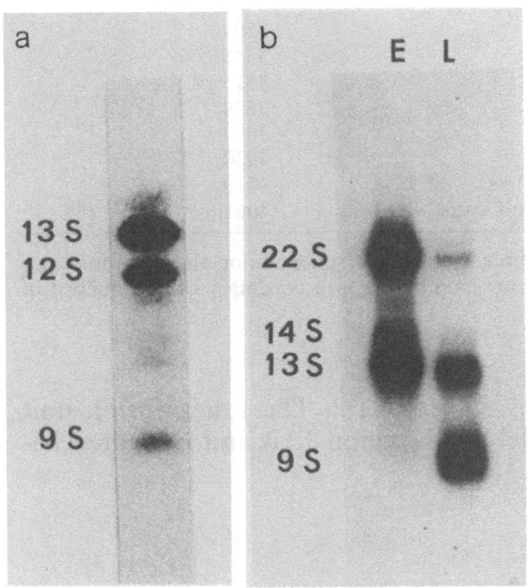

Fig. 2a, b. Major mRNA species from regions E1A and E1B. **a** Analysis of early mRNAs transcribed from region E1A by neutral S1 analysis. (Data from SVENSSON et al. 1983) **b** Analysis of mRNAs transcribed from region E1B by electrophoresis in agarose gels. The RNAs were transferred to nitrocellulose and hybridized with nick-translated *Sac*I (5.0–10.3) probe. Lane *E*, early RNA; lane *L*, late RNA

by the size of the intervening sequences, which are excised during mRNA maturation (Figs. 1, 2a). They are in all likelihood processed from a common precursor which thus can be spliced in three alternative fashions using a common acceptor site. The common cap site for the E1A mRNAs in Ad2 is located 498 nucleotides from the left-hand end (BAKER and ZIFF 1980) and is, as expected, preceded by a TATA motif (VAN ORMONDT et al. 1980b). The common poly(A) addition site for E1A mRNAs is located at nucleotide 1630 on the Ad2 genome (PERRICAUDET et al. 1979) and is preceded by the hexanucleotide AATAAA, which precedes the poly(A) addition site in most eukaryotic mRNAs (PROUDFOOT and BROWNLEE 1976). The splicing events which generate the 12S and 13S mRNAs occur within the coding region of these mRNAs and result in the deletion of a markedly A·T-rich region which contains many stop codons. In this way the coding capacity of region E1A is used in the most efficient way. The polypeptides which are specified by the 12S and 13S mRNAs in Ad2 will be 242 and 288 amino acids long with molecular weights of 26K and 32K respectively, assuming that the AUG closest to the capped 5′ end is used for initiation of translation. The two polypeptides are predicted to have identical N- and C-terminal ends, the only difference between them being a deletion of 46 internal amino acids from the polypeptide specified by the shorter mRNA (Fig. 1). The polypeptide specified by the 9S mRNA starts at the same AUG triplet used for initiation in the 12S and 13S mRNAs but terminates in a different reading frame (Fig. 1), giving the protein product a predicted molecular weight of 6.1K (VIRTANEN and PETTERSSON 1983).

3.3 The mRNAs from Region E1B

Early after infection two major mRNAs, a 13S and a 22S species, are transcribed from a common promoter in region E1B (BERK and SHARP 1978; CHOW et al. 1979; PERRICAUDET et al. 1980a; KITCHINGMAN and WESTPHAL 1980). As is the case for the E1A mRNAs, the E1B mRNAs are completely overlapping with identical 5′ and 3′ ends (Fig. 1). The 13S and the 22S mRNAs are thus likely to be formed by differential splicing of a common RNA precursor. A TATA motif is, as expected, located a short distance upstream from the cap site (BAKER and ZIFF 1980; VAN ORMONDT et al. 1980b). The DNA sequence which covers the 13S and 22S mRNAs contains two long, open translational reading frames (BOS et al. 1981) with the possibility of encoding a 55K and a 21K polypeptide. Both reading frames can be used in their entirety in the 22S mRNA, since both terminate before the splice present in this mRNA (Fig. 1). Experimental evidence exists that both a small and a large polypeptide can be translated from the 22S mRNA (BOS et al. 1981; LUPKER et al. 1981; VAN DER EB, personal communication) thus making the 22S mRNA polycistronic. The open translational reading frame which specifies the 21K polypeptide can also be utilized in its entirety in the 13S mRNA (Fig. 1).

The E1B 13S and 22S mRNAs are unique among eukaryotic mRNAs in that two different AUGs can be selected for translation. In the 22S mRNA both AUGs appear to be used, whereas in the 13S mRNA the first AUG seems to be chosen. It is not yet known whether translation also begins at the second AUG in the 13S mRNA; if this were so, it would encode a 10K polypeptide by using sequences on both sides of the splice.

In addition to the 22S and the 13S mRNAs ESCHE et al. (1980) have reported the existence of an additional 14S mRNA in region E1B by in vitro translation of size-fractionated mRNA. A corresponding mRNA has also been identified by analysis of the E1B mRNAs by a RNA blotting procedure (Fig. 2B) (VIRTANEN and PETTERSSON, in preparation). The precise structure of this mRNA is not yet known, although it seems to encode a polypeptide slightly larger than that encoded by the 13S mRNA (ESCHE et al. 1980).

3.4 The mRNA for Polypeptide IX

The polypeptide IX mRNA differs from all the other adenovirus mRNAs in that it is known to mature without splicing (ALESTRÖM et al. 1980). Structural studies have shown that the cap site for the polypeptide IX mRNA is located in the intervening sequence common to the 22S and the 13S mRNAs of region E1B (Fig. 1) (ALESTRÖM et al. 1980; GINGERAS et al. 1982; VAN ORMONDT et al. 1980b). The nucleotide sequence reveals the presence of a TATA motif approximately 30 nucleotides upstream from the cap site (ALESTRÖM et al. 1980; VAN ORMONDT et al. 1980b), suggesting that the synthesis of this mRNA is regulated from a separate promoter. The polypeptide IX mRNA overlaps with the 13S and the 22S mRNAs and shares its poly(A) addition site with the other E1B mRNAs (Fig. 1) (ALESTRÖM et al. 1980).

The fact that polypeptide IX, which is a structural polypeptide of the virion, is controlled by a promoter located within the transforming region of adenovirus may suggest that it has additional functions besides being a structural polypeptide. Against this notion speaks the observation that the deletion mutant d1313, which lacks the gene for polypeptide IX, is still viable, although the virions are less heat-stable than wild-type virions (COLBY and SHENK 1981). However, the mutant can only be propagated in cells which already contain integrated viral sequences. Therefore the integrated gene for polypeptide IX may produce sufficient amounts of the protein to carry out a hypothetical catalytic function, although the protein is not produced in quantities large enough for assembly into virions.

Since the gene for polypeptide IX is located outside the minimum region required for transformation there is no reason to suspect that it plays an important role in adenovirus transformation, and experimental evidence suggests that it is not even expressed in transformed cell lines (PERSSON et al. 1978; LEWIS and MATHEWS 1981).

3.5 Transcripts from the l-Strand of the Transforming Region

The mRNA species which have been discussed above are all transcribed from promoters located on the r-strand. KATZE et al. (1982) have recently reported that an adenovirus-specific 11K polypeptide is translated from a mRNA originating from the l-strand of region E1. In vitro translation studies of size-fractionated mRNA suggest that the l-strand-specific mRNA sediments around 22S, and more detailed mapping studies indicate that it spans the border between regions E1A and E1B. Neither the promoter for this mRNA nor its precise molecular structure has been established so far. The function of the 11K polypeptide is, moreover, unknown, and genetic studies do not suggest a role for this mRNA in transformation.

4 Kinetics of Region E1 Expression

During the infectious cycle the five early transcription units are expressed with different kinetics (NEVINS et al. 1979). Region E1A, which enhances the rate of initiation of transcription from the other early transcription units, is the first to be expressed, and transcripts from this region can be detected as early as 45 min after infection (p.i.). The rate of transcription from the E1A promoter reaches a maximum by 3–4 h p.i. and is then followed by a slight decline. Transcription from region E3 and E4 begins around 1.5 h p.i. and reaches a maximal rate around 3 h p.i., while region E2 and E1B are activated around 3 h p.i. and reach their maximum rates of transcription 6–7 h p.i. With the onset of DNA replication, transcription from all early promoter sites increases three- to tenfold, probably due to an increase in number of DNA templates (SHAW and ZIFF 1980). The early viral transcription units, which are stimulated approximately 50-fold by the E1A region (NEVINS 1981), reach a steady-state level of approximately 500–1000 copies per cell early after infection (FLINT and SHARP 1976).

Early after infection the E1A 9S mRNA represents only about 5% of the level of the E1A 12S and 13S mRNAs. Late after infection, in contrast, the 9S mRNA becomes the most abundant E1A mRNA (SPECTOR et al. 1978; CHOW et al. 1979; WILSON et al. 1979b; SVENSSON et al. 1983). The shift in the steady-state levels of the E1A mRNAs could be explained by an increase in the cytoplasmic half-life of the E1A 9S mRNA relative to the 12S and 13S mRNAs. Alternatively, the shift may be due to a change in a sequential splicing reaction late during the infectious cycle, as will be discussed below.

A similar preferential accumulation of the shorter spliced mRNA (13S) from region E1B has been demonstrated late after infection (SPECTOR et al. 1978; CHOW et al. 1979; WILSON and DARNELL 1983). By 20 h p.i. approximately 20 times as much 13S mRNA as 22S mRNA has accumulated in the cytoplasm, whereas almost equal amounts are found early after infection. The change in abundancy of cytoplasmic E1B mRNA can in part be explained by an increase in the cytoplasmic stability of the 13S mRNA compared with the 22S mRNA. Early after infection both mRNAs have a half-life of 15–20 min but the half-life of the 13S mRNA is increased five- to tenfold, late after infection (WILSON and DARNELL 1981). The change in the abundancy of the E1B mRNAs is dependent on the synthesis of late adenovirus proteins (BABICH and NEVINS 1981).

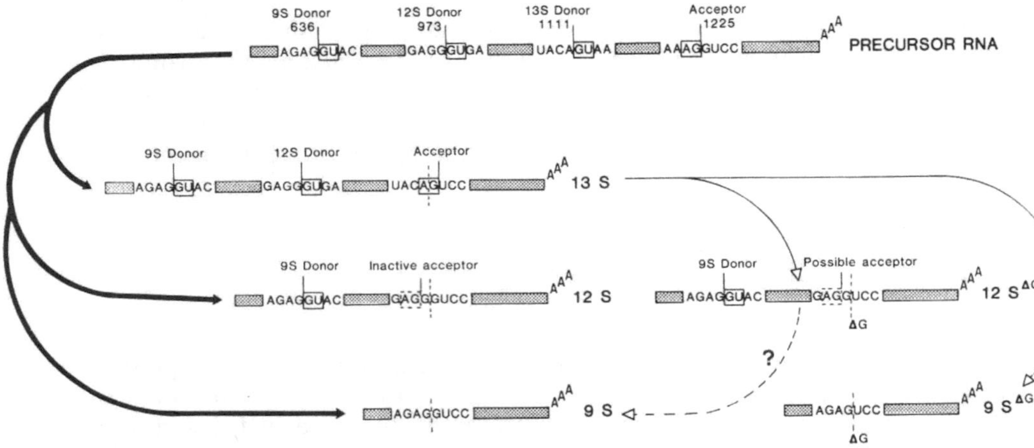

Fig. 3. RNA splicing pathways for region E1A mRNAs. *Broad arrows* show the experimentally found pathways in generation of the E1A 13S, 12S, and 9S mRNAs. *Thin arrows* indicate hypothetical ways to further splice the E1A 13S mRNA to defective 12S and 9S RNAs lacking one G-residue across the splice junction (*G*). The *dashed arrow* shows how a correct 9S RNA could in theory be generated by further splicing of the defective 12S RNA. The GU-AG dinucleotides defining the splice sites of the E1A RNAs are in *boxes*. The *dashed lines* denote the splice junctions in the resulting mRNA structures

The promoter for the polypeptide IX gene is not active in transformed cells, and expression appears to require the synthesis of viral proteins (PERSSON et al. 1978; LEWIS and MATHEWS 1981). The polypeptide IX mRNA is, however, expressed before and in the absence of DNA replication.

5 Nonsequential Splicing of E1A mRNAs

Transcription both early and late after adenovirus infection produces a complicated set of RNA structures. Each transcription unit generates multiple RNAs differing from each other only in the size and number of introns removed by splicing. The overlapping nature of adenovirus mRNAs has raised the question of whether RNA splicing should be envisaged as a multistep reaction whereby an intron is removed in two or more successive stages, thus generating functional splicing intermediates.

The processing pathways for the E1A mRNAs have been studied by using a transient expression assay of recombinant plasmids containing the E1A region (SVENSSON et al. 1983). The most interesting conclusion that can be drawn from this study is that in the absence of adenovirus-specific gene products, splicing of the E1A mRNAs is nonsequential, i.e., the correct 13S, 12S, and 9S RNAs are generated by separate splicing events using the colinear nuclear transcript as the only precursor RNA (Fig. 3). The

13S mRNA can act as a precursor for further splicing, generating 12S and 9S RNAs; however, these RNAs have an aberrant structure, both lacking one G-residue at the splice junction (Fig. 3). From these results it has been proposed (SVENSSON et al. 1983) that splicing and transport of the E1A mRNAs are two tightly coupled events which prevent the further processing of the 13S mRNA during the normal infection with wild-type virus. If the splicing of the precursor takes place at the nuclear membrane and the products are immediately transported to the cytoplasm, no spliced RNA would be present in the nucleus, thereby preventing a second round of processing from occurring.

An interesting feature of the E1A RNAs is their kinetics of appearance, as described in Sect. 4. The 9S mRNA, which early after infection is barely detectable, becomes the most prominent E1A mRNA late in the infection cycle. Since the relative frequency of the E1A mRNA species present after transfection resembles that in the early phase, it might be speculated that the change in abundancy occurs as the result of virus-induced modifications of the splicing enzymes. For example, if the mechanism of splicing changes to a sequential splicing reaction late in the infectious cycle, the increased level of 9S RNA could be explained by a processing of the 13S mRNA through an intermediate stage consisting of a defective 12S RNA to a correct 9S mRNA (Fig. 3). Virus-induced changes of the splicing machinery have been demonstrated for the major late adenovirus transcription unit (AKUSJÄRVI and PERSSON 1981; NEVINS and WILSON 1981).

6 Relationship Between mRNAs and Adenovirus T Antigens

6.1 T Antigens Encoded by Region E1A

The 12S and 13S mRNAs from region E1A appear to give rise to several related polypeptides ranging in size between 35K and 53K (ESCHE et al. 1980; HALBERT et al. 1979; HARTER and LEWIS 1978; JOCHEMSEN 1981; JOCHEMSEN et al. 1981; LEWIS et al. 1976, 1980; LUPKER et al. 1981). By in vitro translation of size-fractionated RNA, it has been shown that the 13S mRNA encodes polypeptides of molecular weights of 41K and 53K, whereas the 12S mRNA yields polypeptides of 35K and 47K (ESCHE et al. 1980; HALBERT et al. 1979; RICCIARDI et al. 1981). By two-dimensional gel electrophoresis these components can be resolved into additional species (HARTER and LEWIS 1978). The results are interpreted to mean that both the 12S and 13S mRNAs from region E1A give rise to two or more polypeptides, presumably by modification in the C-terminal part (RICCARDI et al. 1981). Tryptic peptide analysis and amino acid radiosequence analysis have shown that all polypeptides encoded by the 12S and 13S mRNAs are related, as was expected from the structure of the mRNAs (GREEN et al. 1979;

Green et al. 1980; Halbert and Raskas 1982; Harter and Lewis 1978; Levinson et al. 1976; Smart et al. 1981). The molecular weights of the E1A polypeptides as predicted from the DNA sequence are 26K and 33K (Perricaudet et al. 1979; van Ormondt et al. 1980b). There is thus a considerable discrepancy between the predicted and the estimated molecular weights of the polypeptides. The difference can probably be attributed in part to the comparatively high proline content in both polypeptides.

The protein product translated from the 9S mRNA of region E1A has been estimated to be a 28K polypeptide (Esche et al. 1980; Spector et al. 1980), exceeding the predicted molecular weight of 6.1K by far (Virtanen and Pettersson 1983), and further work is necessary to explain this discrepancy.

6.2 T Antigens Encoded by Region E1B

Polypeptides with molecular weights of 21K and 55K are translated from different but overlapping reading frames in region E1B (Bos et al. 1981) and are therefore totally unrelated, as has been verified by tryptic peptide analysis (Green et al. 1980; Halbert and Raskas 1982). As mentioned above, the 22S mRNA is polycistronic and has been shown to yield both the 21K and the 55K polypeptide by in vitro translation (Lupker et al. 1981; van der Eb, personal communication). Another prominent polypeptide with an estimated molecular weight of 18K has been assigned to region E1B (Green et al. 1979; Green et al. 1980; van der Eb et al. 1979). In contrast to the 21K polypeptide, this polypeptide shares many of its tryptic peptides with the 55K polypeptide. The exact structure of its mRNA has not yet been deduced and it cannot be excluded that it represents a degradation product of the 55K polypeptide.

7 Postulated Role of the Different mRNAs in Adenovirus Transformation

The phenotypic properties connected with transformation appear to be associated with the E1B region, with either the product of the 13S mRNA, the 21K polypeptide, or the N-terminal end of the 55K polypeptide. This conclusion rests on the observation that cells transformed by the *Hin*dIII G fragment (0–8.0 map units) of Ad5 express the full transformed phenotype. Fragment *Hin*dIII G includes the entire coding region for the 21K protein but only the N-terminal end of the 55K polypeptide (Fig. 1). Analysis of T antigens in *Hin*dIII G-transformed cells shows, as expected, that the 21K protein is present but not the 55K protein (van der Eb et al.

1979; JOCHEMSEN et al. 1981). A certain correlation has, however, been obtained between the presence of the 55K sequences and tumorigenicity in the case of Ad12 (VAN DEN ELSEN et al. 1982; M. PERRICAUDET, unpublished information). These observations suggest that the 21K polypeptide, or alternatively the N-terminal end of the 55K polypeptide, is responsible for the major transforming properties of adenoviruses. Region E1A is sufficient for immortalization of cells, and the products of the 12S and 13S mRNAs are thus likely to be responsible for this property. Genetic experiments imply that the product of the 13S mRNA is responsible for both the regulation of early gene expression and for the first step in adenovirus transformation (BERK et al. 1979; JONES and SHENK 1979). Since a product from region E1A is necessary for the activation of the other early transcription units, including the E1B region, the main role of region E1A in transformation could be to enhance the expression of region E1B. In support of this notion VAN DEN ELSEN et al. (1982) have reported that fragments which contain the E1B region but lack parts of the E1A region are unable to transform cells. This model is, however, likely to be an oversimplification.

8 Adenovirus Mutants Defective in Region E1 Expression

Studies using adenovirus mutants with lesions in region E1 have been very informative and made it possible to assign at least one function to the gene products encoded by region E1. One of the most significant results which stem from studies of region E1 mutants is that one or more products from this region control the expression of other early adenovirus genes (BERK et al. 1979; JONES and SHENK 1979). More detailed studies indicate that the viral product inactivates a cellular protein that either degrades viral mRNA (KATZE et al. 1981) or prevents viral transcription from the other early regions (NEVINS 1981). This function has been ascribed to the product of the 13S mRNA since the mutant, designated Ad5 hr1 (HARRISON et al. 1977; BERK et al. 1979; RICCIARDI et al. 1981), which expresses the product of the 12S mRNA but not the 13S mRNA product is unable to stimulate the expression of the other early regions during infection of HeLa cells. The conclusion is, moreover, corroborated by studies using the mutant Ad2 pm975 (MONTELL et al. 1982). This mutant is unable to make the E1A 12S mRNA but synthesizes the 13S mRNA and replicates normally in HeLa cells. This result demonstrates that the 12S mRNA is dispensible for replication of adenovirus in tissue culture cells, which is not an unexpected finding since all genetic information contained in the 12S mRNA is also present in the 13S mRNA.

Studies on the E1 mutants have also given interesting information as to the maturation pathways of region E1 mRNAs. The mutant pm975 has a single-point mutation located precisely at the donor site which is required for splicing of the 12S mRNA. The mutation alters the GT dinucleotide located on the 5′ side of the intron that is removed from the 12S mRNA to a GG dinucleotide. Since the mutant was constructed in such a way that the protein product of the 13S mRNA is unaltered, the results of MONTELL et al. (1982) demonstrate that a single base change is sufficient to render the splice site inactive.

The mutant hr440 constructed by SCOLNICK (1981) contains several single-point mutations two of which are located near the 5′ splice junction of the 12S mRNA. At this position the mutant sequence reads GG GTGAATAG instead of the wild-type sequence GG GTGAGGAG. This mutant does not produce the E1A 12S mRNA, suggesting that the mutation alters a part of the intron which is required for splicing.

The d1502 mutant constructed by CARLOCK and JONES (1981) is also interesting with regard to splicing. A 340-bp-long segment has been deleted which include the common acceptor site for splicing of the E1A mRNAs. This mutant is of course defective for replication in HeLa cells, since it does not produce functional E1A products. It does, however, produce significant amounts of unspliced E1A mRNA, which demonstrates that splicing is not a prerequisite for transport of mRNA to the cytoplasm.

9 A Comparison of mRNAs Transcribed from the Transforming Regions of Human Adenoviruses from Different Oncogenic Subgroups

Since the human adenoviruses are known to differ with regard to their oncogenic potential it is of particular interest to compare the structures of the mRNAs which are transcribed from the transforming region of highly oncogenic (Ad12), weakly oncogenic (Ad7), and nononcogenic adenoviruses (Ad2, Ad5).

The overall organization of mRNAs encoded by the region E1A of Ad2, Ad7 and Ad12 is very similar (SUGISAKI et al. 1980; VAN ORMONDT et al. 1980a). In members of all three oncogenic subgroups, this region is transcribed into three overlapping mRNAs with identical 5′ and 3′ ends which differ in the size of their intervening sequences. For Ad2, Ad5, Ad7, and Ad12 it has been shown that the splicing event deletes from the primary transcript a markedly A·T-rich region containing many termination codons. For all serotypes analyzed so far, it has been shown that the splicing events in region E1A utilizes different donor sites (D1, D2, and D3), but uses one single acceptor site (A1; Fig. 1). A sequence of nine nucleotides around the the acceptor site is identical in Ad2, Ad7, and Ad12, and has thus been extremely well conserved in adenovirus evolution (PERRICAUDET et al. 1980b). This is also true for the donor site D3, where eleven nucleotides are identical in Ad5, Ad7, and Ad12. In contrast, donor site D2, which is used to mature the 12S mRNA, varies with regard to the nucleotide

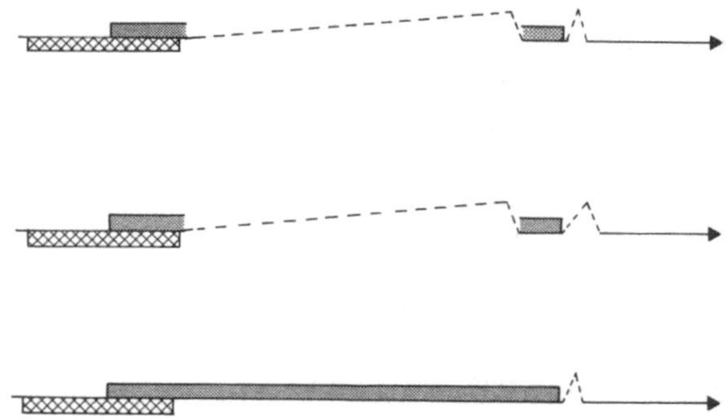

Fig. 4. Structure of the spliced mRNAs from region E1B of ad12. The two open translational reading frames are indicated (▩, ▤). [Data from VIRTANEN et al. (1981)]

sequence, as well as to the exact location in different serotypes. In Ad2, donor site D2 has been displaced 45 nucleotides downsteam relative to its position in Ad7 and Ad12.

The main consequence of the altered splice signals in region E1A is a change in the size of the intervening sequences removed from the 12S mRNA: 255, 187, and 168 nucleotides long in Ad2, Ad7, and Ad12 respectively.

A third E1A mRNA, the 9S mRNA, has been detected in large quantities late after lytic infection with Ad2 and Ad7. This mRNA is spliced between the acceptor site A1 and donor site D1 (Fig. 1). In both Ad2 and Ad7 it is known that the splice in the 9S mRNA causes a different kind of frame shift than the splice in the 12S and 13S mRNAs, giving the protein product of the 9S mRNA a unique C-terminal end (DIJKEMA et al. 1980; VIRTANEN and PETTERSSON 1983).

The 13S mRNA intron in region E1A is very A·T rich in Ad2, Ad7, and Ad12, but its sequence has apparently diverged considerably during adenovirus evolution. In addition, the regions which flank the poly(A) addition site in region E1A have diverged considerably and the hexanucleotide AAUAAA is the only conserved feature. In contrast, the nucleotide sequences which flank the cap site of the E1A mRNAs are highly conserved; a stretch of 40 nucleotides corresponding to the 5′ nontranslated region and the promoter of the E1A mRNAs is identical in Ad2, Ad7, and Ad12.

For region E1A the overall DNA sequence homology between any pair of serotypes, representing subgroups A, B, or C, is approximately 50% (MACKEY et al. 1979), and from a direct sequence comparison it is not possible to establish a closer relationship between any two subgroups. Analysis of the structures at donor and acceptor sites for splicing suggest, however, that subgroups A and B are more closely related to each other than to the nononcogenic serotypes (PERRICAUDET et al. 1980b; VIRTANEN et al. 1982).

The major open translational reading frames identified in Ad2 and Ad5 are also present in the weakly oncogenic Ad7 and the highly oncogenic Ad12 (Fig. 4). The proteins encoded by region E1A of different subgroup members have very similar amino acid compositions, suggesting that they all serve the same functions (VAN ORMONDT et al. 1980a). In accordance with this it has been demonstrated that Ad12 can complement mutants of Ad5 with defects in the transforming region (ROWE and GRAHAM 1981), and it has also been shown that the E1B region of Ad12 can be activated by the E1A region of Ad5 in transfection experiments (HOUWELING et al. 1980; VAN DEN ELSEN et al. 1982).

The E1B regions in Ad5, Ad7, and Ad12 also have many features in common (VAN ORMONDT et al. 1980b; PERRICAUDET et al. 1980a; BOS et al. 1981; KIMURA et al. 1981; DIJKEMA et al. 1982; GINGERAS et al. 1982; VIRTANEN et al. 1982; VAN ORMONDT and HESPER 1983). In all cases there exist a large (22S) and a small (13S) mRNA which in two separate but overlapping reading frames can give rise to a large (54–55K) and a small (19–21K) polypeptide. In addition, the gene for polypeptide IX can be identified in human adenoviruses representing all oncogenic subgroups. An interesting difference is an apparent lack in Ad7 of a TATA motif adjacent to the cap site of the polypeptide IX mRNA (DIJKEMA et al. 1982). The structure of the 22S mRNA appears to be the same for the nononcogenic Ad5 and the highly oncogenic Ad12 (VIRTANEN et al. 1982). The intron of the 22S mRNA in different subgroup members is better conserved than the E1A introns, probably because it contains the promoter and cap site for the polypeptide IX mRNA.

The structure of the 13S mRNA from region E1B, in contrast, differs significantly between oncogenic and nononcogenic serotypes (Figs. 1 and 4). In cells infected with Ad12, the 13S mRNA with a single large splice appears to be absent (VIRTANEN et al. 1982). Instead, two additional mRNAs have been found, both of which contain two intervening sequences. Both intervening sequences are located after the termination codon which ends the reading frame encoding the 19K polypeptide, and will hence cause no obvious change in the coding properties of the 13S mRNA. The difference between the two types of 13S mRNA present in Ad12-infected cells, is that two slightly different acceptor sites are used to splice out one of the introns, thus giving rise to two mRNAs which differ in their 3′ noncoding regions (VIRTANEN et al. 1982). The two alternative splices in the 13S mRNAs of Ad12 are both located in the 3′ noncoding part of the mRNA, which is unusual. The biological function of the 3′ noncoding region of eukaryotic mRNAs is poorly understood. It is conceivable that sequences present in this part of mRNAs influence the transport or the stability of the mRNAs.

The part of the E1 region which shows the greatest variation between members of the different oncogenic subgroups is located in the C-terminal of the E1B 21K polypeptide and the N-terminal of the E1B 55K polypeptide. It cannot yet be decided whether the difference in oncogenicity between different adenovirus serotypes is related to a critical difference in protein structure or to the slight difference in mRNA structure.

Acknowledgments. The authors are indebted to Marianne Gustafson and Jeanette Backman for patient and skilful secretarial assistance.

References

Akusjärvi G, Persson H (1981) Controls of RNA splicing and termination in the major late adenovirus transcription unit. Nature 292:420–426

Aleström P, Akusjärvi G, Perricaudet M, Mathews MB, Klessig DF, Petterson U (1980) The gene for polypeptide IX of adenovirus type 2 and its unspliced messenger RNA. Cell 19:671–681

Babich A, Nevins JR (1981) The stability of early adenovirus mRNA is controlled by the viral 72K DNA binding protein. Cell 26:371–379

Baker CC, Ziff EB (1980) Biogenesis, structures and sites encoding the 5′-termini of adenovirus-2 mRNAs. Cold Spring Harbor Symp Quant Biol 44:415–428

Berk AJ, Sharp PA (1978) Structure of adenovirus 2 early mRNAs. Cell 14:695–711

Berk AJ, Lee F, Harrison T, Williams J, Sharp PA (1979) Pre-early adenovirus 5 gene product regulates synthesis of early viral messenger RNAs. Cell 17:1935–1944

Bos JL, Polder LJ, Bernards R, Schrier PI, van den Elsen PJ, van der Eb AJ, van Ormondt H (1981) The 2.2kb Elb mRNA of human ad12 and ad5 codes for two tumor antigens starting at different AUG triplets. Cell 27:121–131

Carlock LR, Jones N (1981) Synthesis of an unspliced cytoplasmic message by an adenovirus 5 deletion mutant. Nature 294:572–574

Chow LT, Broker TR, Lewis JB (1979) Complex splicing patterns of RNAs from the early regions of adenovirus-2. J Mol Biol 134:265–303

Colby WW, Shenk T (1981) Adenovirus type 5 virions can be assembled in vivo in the absence of detectable polypeptide IX. J Virol 39:977–980

Dijkema R, Dekker BMM, van Ormondt H, de Waard A, Maat J, Boyer HW (1980) Gene organization of the transforming region of weakly oncogenic adenovirus type 7: the E1A region. Gene 12:287–299

Dijkema R, Dekker BMM, van Ormondt H (1982) Gene organization of the transforming region of adenovirus type 7 DNA. Gene 18:143–156

Esche H, Mathews MB, Lewis JB (1980) Proteins and messenger RNAs of the transforming region of wild-type and mutant adenoviruses. J Mol Biol 142:399–417

Flint SJ, Sharp PA (1976) Adenovirus transcription. V. Quantitation of viral RNA sequences in adenovirus 2 infected and transformed cells. J Mol Biol 106:749–771

Gingeras TR, Sciaky D, Gelinas RE, Bing-Dang J, Yen C, Kelly M, Bullock P, Parsons B, O'Neill K, Roberts RJ (1982) Nucleotide sequences from the adenovirus-2 genome. J Biol Chem 257:13475–13491

Graham FL, van der Eb AJ (1973) A new technique for assay infectivity of human adenovirus DNA. Virology 52:456–467

Graham FL, Abrahams PJ, Mulder C, Heinjeker HL, Warnaar SO, de Vries FAJ, Fiers W, van der Eb AJ (1974a) Studies on in vitro transformation by DNA and DNA fragments of human adenoviruses and simian virus 40. Cold Spring Harbor Symp Quant Biol 39:637–750

Graham FL, van der Eb AJ, Heijneker HL (1974b) Size and location of the transforming region in human adenovirus type 5 DNA. Nature 251:687–691

Green M, Wold WSM, Brackman KH, Cartas MA (1979) Identification of families of overlapping polypeptides coded by early region 1 of human adenovirus type 2. Virology 97:275–286

Green M, Wold WSM, Brackman KH, Cartas MA (1980) Studies of early proteins and transformation proteins of human adenoviruses. Cold Spring Harbor Symp Quant Biol 44:457–470

Halbert DN, Raskas HJ (1982) Tryptic and chymotryptic methionine peptide analysis of the in vitro translation products specified by the transforming region of adenovirus type 2. Virology 116:406–418

Halbert DN, Spector DJ, Raskas HJ (1979) In vitro translation products specified by the transforming region of adenovirus type 2. J Virol 31:621–629

Harrison T, Graham F, Williams J (1977) Host range mutants of adenovirus type 5 defective for growth in HeLa cells. Virology 77:319–329

Harter ML, Lewis JB (1978) Adenovirus type 2 early proteins synthesized in vitro and in

vivo: identification in infected cells of the 38000- to 50000-molecular-weight protein encoded by the left end of the adenovirus type 2 genome. J Virol 26:736–749
Houweling A, van den Elsen PJ, van der Eb AJ (1980) Partial transformation of primary rat cells by the leftmost 4.5% fragment of adenovirus 5 DNA. Virology 105:537–550
Huebner RJ, Rowe WP, Lane WT (1962) Oncogenic effects in hamsters of human adenovirus type 12 and 18. Proc Natl Acad Sci USA 48:2051–2058
Jochemsen H (1981) Studies on the transforming genes and their products of human adenovirus types 12 and 5. Thesis, University of Leiden
Jochemsen H, Hertoghs JJL, Lupker JH, Davis A, van der Eb AJ (1981) In vitro synthesis of adenovirus type 5 T antigens. II. Translation of virus-specific RNA from cells transformed by fragments of adenovirus type 5 DNA. J Virol 37:530–534
Jones N, Shenk T (1979) An adenovirus type 5 early gene functions regulates expression of other early viral genes. Proc Natl Acad Sci USA 76:3665–3669
Katze MB, Persson H, Philipson L (1981) Control of adenovirus early gene expression – posttranscriptional control mediated by both viral and cellular gene-products. Mol Cell Biol 9:807–813
Katze MG, Persson H, Philipson L (1982) A leftward reading transcript from the transforming region of adenovirus DNA encodes a low molecular weight polypeptide. EMBO Journal 1:783–790
Kimura T, Sawada Y, Shinagawa M, Shimizu Y, Shiroki K, Shimojo H, Sugisaki H, Takanami M, Vemizu Y, Fujinaga K (1981) Nucleotide sequence of the transforming region of E1b of adenovirus type 12 DNA: structure and gene organization and comparison with those of adenovirus type 5 DNA. Nucleic Acids Res 9:6571–6589
Kitchingman GR, Westphal H (1980) The structure of adenovirus 2 early nuclear and cytoplasmic RNAs. J Mol Biol 137:23–48
Levinson A, Levine AJ (1977) The isolation and identification of the adenovirus group C tumor antigens. Virology 76:1–11
Levinson A, Levine AJ, Anderson S, Osborn M, Rosenwirth B, Weber K (1976) The relationship between group C adenovirus tumor antigen and the adenovirus single-strand DNA binding protein. Cell 7:575–584
Lewis JB, Mathews MB (1981) Viral messenger RNAs in six lines of adenovirus-transformed cells. Virology 115:345–360
Lewis JB, Atkins JF, Baum PR, Solem R, Gesteland RB, Anderson CW (1976) Location and identification of the genes for adenovirus type 2 early polypeptides. Cell 7:141–151
Lewis JB, Esche H, Smart JE, Stilman BW, Harter ML, Mathews MB (1980) Organization and expression of the left third of the genome of adenovirus. Cold Spring Harbor Symp Quant Biol 44:493–508
Lupker JH, Davis A, Jochemsen H, van der Eb AJ (1981) In vitro synthesis of adenovirus type 5 T antigens. I. Translation of early region 1-specific RNA from lytically infected cells. J Virol 37:524–529
Mackey J, Wold W, Rigden P, Green M (1979) Transforming Region of Group A, B, and C Adenoviruses: DNA homology studies with twenty-nine human adenovirus serotypes. J Virol 29:1056–1064
Montell C, Fisher EF, Caruthers MH, Berk AJ (1982) Resolving the functions of overlapping viral genes by site-specific mutagenesis at a mRNA splice site. Nature 295:380–384
Nevins JR (1981) Mechanism of activation of early viral transcription by the adenovirus E1A gene products. Cell 26:213–220
Nevins JR, Wilson MC (1981) Regulation of adenovirus-2 gene expression at the level of transcriptional termination and RNA processing. Nature 290:113–118
Nevins JR, Ginsberg HS, Blanchard JM, Wilson MC, Darnell JE (1979) Regulation of the primary expression of early adenovirus transcription units. J Virol 32:727–733
Perricaudet M, Akusjärvi G, Virtanen A, Pettersson U (1979) Structure of two spliced mRNAs from the transforming region of human subgroup C adenoviruses. Nature 281:694–696
Perricaudet M, Le Moullec JP, Pettersson U (1980a) The predicted structure of two adenovirus T-antigens. Proc Natl Acad Sci USA 77:3778–3782
Perricaudet M, Le Moullec J-M, Tiollais P, Pettersson U (1980b) Structure of two adenovirus type 12 transforming polypeptides and their evolutionary implications. Nature 288:174–176

Persson H, Pettersson U, Mathews MB (1978) Synthesis of a structural adenovirus polypeptide in the absence of viral DNA replication. Virology 90:67–79
Proudfoot NJ, Brownlee GG (1976) 3′ Non-coding region sequences in eucaryotic messenger RNA. Nature 263:211–214
Ricciardi RP, Jones RL, Cepko CL, Sharp PA, Roberts BE (1981) Expression of early adenovirus genes requires a viral encoded acidic polypeptide. Proc Natl Acad Sci USA 78:6121–6125
Rowe DT, Graham FL (1981) Complementation of adenovirus type 5 host range mutants by adenovirus type 12 in coinfected HeLa and BHK-21 cells. J Virol 38:191–197
Scolnik D (1981) An adenovirus mutant defective in splicing RNA from early region 1A. Nature 291:508–510
Shaw AR, Ziff EB (1980) Transcripts from the adenovirus-2 major late promoter yield a singe family of a coterminal mRNAs during early infection and five families at late times. Cell 22:905–916
Smart JE, Lewis JB, Mathews MB, Harter ML, Andersson CW (1981) Adenovirus type 2 early proteins. Assignment of the early region 1A proteins synthesized in vivo and in vitro to specific mRNAs. Virology 112:703–713
Spector DJ, McGrogan M, Raskas HJ (1978) Regulation of the appearance of cytoplasmic RNAs from region 1 of the adenovirus genome. J Mol Biol 126:395–414
Spector DJ, Crossland LD, Halbert DN, Raskas HJ (1980) A 28K polypeptide is the translation product of 9S RNA encoded by region 1A of adenovirus 2. Virology 102:218–221
Sugisaki H, Sugimoto K, Takanami M, Shiroki K, Saito I, Shimiojio H, Sawada Y, Uemizu Y, Uesugi S-I, Fujinaga K (1980) Structure and gene organization in the transforming *Hin*dIII-G fragment of ad12. Cell 20:777–786
Svensson C, Pettersson U, Akusjärvi G (1983) Splicing of adenovirus 2 early region 1A mRNAs is non-sequential. J Mol Biol 165:475–479
Trentin JJ, Yabe Y, Taylor G (1962) The quest for human cancer viruses. Science 137:835–841
Van den Elsen P, de Pater S, Houweling A, van der Veer I, van der Eb A (1982) The relationship between region E1a and E1b of human adenoviruses in cell transformation. Gene 18:175–185
Van der Eb A, Bernards R (to be published) Transformation and oncogenecity by adenoviruses. In: Doerfler W (ed) The molecular biology of adenoviruses 2. Curr Top Microbiol Immunol 110
Van der Eb AJ, Mulder C, Graham FL, Houweling A (1977) Transformation with specific fragments of adenovirus DNAs. I. Isolation of specific fragment with transforming activity of adenovirus 2 and 5 DNA. Gene 1:115–132
Van der Eb AJ, van Ormondt H, Schrier PI, Lupker JH, Jochemsen H, van den Elsen PJ, de Leys RJ, Maat J, van Beveren CP, Dijkema R, de Waard A (1979) Structure and function of the transforming genes of human adenoviruses and SV40. Cold Spring Harbor Symp Quant Biol 44:383–399
Van Ormondt H, Hesper B (1983) Comparison of the nucleotide sequences of early region E1b DNA of human adenovirus types 12, 7 and 5 (subgroups A, B and C) Gene 21:217–226
Van Ormondt H, Maat J, Dijkema R (1980a) Comparison of nucleotide sequences of the early E1a regions for subgroups A, B and C of human adenviruses. Gene 12:63–76
Van Ormondt H, Maat J, Van Beveren CP (1980b) The nucleotide sequence of the transforming region E1 of adenovirus type 5 DNA. Gene 11:299–399
Virtanen A, Pettersson U (1983) The molecular structure of the 9S mRNA from early region 1A of adenovirus 2. J Mol Biol 165:496–499
Virtanen A, Pettersson U, LeMoullec JM, Tiollais P, Perricaudet M (1982) Different mRNA structures are transcribed from the transforming region of highly and non-oncogenic human adenoviruses. Nature 295:705–707
Wilson MC, Darnell JE Jr (1981) Control of mRNA concentration by differential cytoplasmic half-life: adenovirus mRNAs from transcription units 1A and 1B. J Mol Biol 148:231–251
Wilson MC, Fraser N, Darnell J (1979a) Mapping of RNA initiation sites by high doses of UV irradiation. Evidence for three independent promoters within the left 11% of the Ad-2 genome. Virology 94:175–184
Wilson MC, Nevins JR, Blanchard JM, Ginsberg HS, Darnell JE Jr (1979b) The metabolism of mRNA from the transforming region of adenovirus type 2. Cold Spring Harbor Symp Quant Biol 44:447–455

In Vitro Transcription of Adenovirus Genes

Roberto Weinmann[1], Steven Ackerman[1], David Bunick[2],
Michael Concino[1], and Ruben Zandomeni[1]

1 Introduction

Before the development of recombinant DNA technology, DNA tumor viruses provided a convenient source of DNA for analysis of a limited number of genes in eukaryotic cells. The virus can be purified in large amounts and the DNA extracted from the virions is free of cellular DNA sequences. The lytic viral cycle affects cellular metabolism in a dramatic way, but does not alter the levels of cellular RNA polymerases I, II, and III or induce a virus-coded one (Weinmann et al. 1976). The genetics of adenovirus was well developed, with deletion and temperature-sensitive mutants (For review see Shenk and Williams, Vol. 111, (in press). The advent of restriction enzymes allowed the direct correlation of mutations with gene products and specific DNA regions. The transcribed regions of the genome were first identified by hybridization kinetic analysis of RNA products. Combined with size analysis and hybridization to separated DNA strands, the polarities of the 3′ and 5′ ends of mRNAs were determined. At least half

1 The Wistar Institute of Anatomy and Biology, Philadelphia, PA 19104, USA
2 Biology Department, University of North Carolina, Chapel Hill, NC 27514, USA

Abbreviations. DRB, 5,6-dichloro-1β-D-ribofuranosylbenzimidazole, DRB^{R}, DRB-resistant, MNNG, *N*-methyl-*N*-nitro-*N*-nitrosoguanidine, PIX, protein IX.
Positive numbers (+1, etc.) refer to the position of transcription initiation on the corresponding template, negative numbers refer to upstream sequences

Current Topics in Microbiology and Immunology, Vol. 109

of the late-infected cell viral transcripts are derived from the major late promoter. A common undecanucleotide present at the 5′ end of all adenovirus structural protein mRNAs was the first suggestion for a common transcriptional start site (GELINAS and ROBERTS 1977). Studies using UV-induced inactivation of the transcriptional unit further suggested a common origin for all these transcripts (GOLDBERG et al. 1977). Comparison of nuclear with cytoplasmic RNA sequences by R-loop hybridization and electron microscopy, S1 nuclease (BERK and SHARP 1978) and sequence analysis revealed the amazing phenomenon of splicing. The major late promoter of the virus is where most late transcripts are initiated (BERGET et al. 1977). Studies using adenovirus-infected HeLa cell nuclei indicated that this sequence of DNA was coding for the 5′ end of the mRNA, since the sequence derived from the RNA fingerprint could be accurately aligned with the DNA sequence (EVANS et al. 1977; ZIFF and EVANS 1978). An A·T-rich sequence (Goldberg-Hogness box) was identified 25–30 nucleotides upstream from the 5′ end of the mRNA, in analogy to other eukaryotic genes (ZIFF and EVANS 1978). No transcripts were detected upstream of the cap, suggesting that the cap was probably the site of initiation of RNA synthesis. Rapid capping of the RNA in vivo did not allow the isolation of the diagnostic 5′ polyphosphate ends (SALDITT-GEORGIEFF et al. 1980). The knowledge of the in vivo transcription start site is essential to develop in vitro systems that faithfully reproduce the in vivo situation. Initiation of transcription was first shown to occur faithfully in isolated nuclei using the adenovirus major late promoter (ZIFF and EVANS 1978; MANLEY et al. 1979). Cell-free in vitro transcription systems have been developed for the RNA polymerase III-transcribed VA RNA genes of adenovirus (WU 1978). These small virus-coded RNAs are essential for late viral protein synthesis (THIMMAPPAYA et al. 1983) and can assemble both in vivo and in vitro into ribonucleoprotein particles reacting with lupus antisera (LERNER and STEITZ 1981). An RNA polymerase III transcription system and deletion mutagenesis were used to demonstrate that an internal control region is required for faithful transcription of the VA RNA gene (FOWLKES and SHENK 1980; GUILFOYLE and WEINMANN 1981). Based on the soluble RNA polymerase III transcription systems, a HeLa cell extract containing the high-speed supernatant proteins (S-100) and exogenous calf thymus RNA polymerase II was developed by WEIL et al. (1979). A cruder whole cell extract which did not require the addition of exogenous RNA polymerase II was developed by MANLEY et al. (1980). The assay utilizes the DNA template truncated at a specific site by a restriction enzyme. The RNA is of discrete size only if initiated at a specific point on the template and obligatorily terminates at the site of restriction enzyme template truncation. Size analysis of the RNA products on polyacrylamide gels and extrapolation from the restriction enzyme site indicate that the 5′ end corresponds to the in vivo 5′ end. Some characteristics common to these systems are:

1. The very high protein concentration of the active extracts (5–25 mg/ml). The purified RNA polymerase II is necessary but not sufficient for faithful transcription initiation. Other factors contained in the extract are required.

2. The narrow range of extract and template concentrations under which activity can be detected, suggesting the presence of transcription inhibitory and stimulatory factors. Some of these factors bind to the DNA in an unspecific manner, as demonstrated by the fact that 50–75% of the DNA template in the reaction can be substituted by the homopolymer dI-dC (HONDA et al. 1980).

3. The low efficiency of the system. At most, 1–2.5% of the templates give RNA runoffs. This is much lower than other in vitro transcription systems, such as those developed for RNA polymerase III and I (WU 1978; GRUMMT 1981).

The multiplicity of factors has been partially sorted out by direct chromatographic fractionation of the extracts (MATSUI et al. 1981; SAMUELS et al. 1982; DYNAN and TJIAN 1983). Purification eliminates many of the proteins and enzymes which are probably not involved directly in transcription. However, purification to homogeneity of many of these proteins has not yet been achieved, due to problems of the functional assay, abundance, yield, stability, and recovery.

These crude transcriptional systems generate runoff RNAs that are fully capped and methylated. They are able to transcribe not only the adenovirus major late promoter gene, but also other adenovirus early as well as cellular genes. Runoff transcripts from E1A, E1B, PIX, E3 and E4 were obtained (MANLEY et al. 1980; LEE and ROEDER 1981; FIRE et al. 1981). In the case of region 4, heterogeneity of the 5′ end caps was found in vivo. A major U start is separated by five Us (which can also serve as weak start sites) from the other major A start (BAKER and ZIFF 1981). The TATAA sequence for this gene is long. This start-site heterogeneity is also reflected in the in vitro transcription system (LEE and ROEDER 1981; FIRE et al. 1981), probably because alternate positions in the TATA box allow multiple start sites, maintaining a constant 25-nucleotide (n) alignment between them (BAKER and ZIFF 1981). Transcription from E2 and IVA2 genes was very low or nonexistent. Careful examination of the DNA sequence indicated that these sequences upstream of the cap site lacked the typical TATAA or Goldberg-Hogness consensus sequences (BAKER and ZIFF 1981). The genes for adenovirus regions E2 and IVA2 were later transcribed in vitro, but at very low efficiencies (MATHIS et al. 1981; MATSUI 1982). The finding that uninfected cell extracts were able to transcribe from the major late promoter prompted a reexamination of the sequence of events following viral infection. Careful analysis indicated that indeed the late promoter is also active at early times after infection, giving RNAs with the same 5′ ends as at late times (CHOW et al. 1979; LEWIS and MATHEWS 1980; SHAW and ZIFF 1980). An examination of several adenovirus early, intermediate, and late promoters with extracts prepared from uninfected and infected cells revealed no gross regulatory feature retained by the extracts (FIRE et al. 1981). Only for the PIX promoter, a gene activated at intermediate to late times after infection, and for the major late promoter, some preferential transcription with infected vs uninfected cell extracts was detected (FIRE et al. 1981).

We will describe here our experiments designed to establish the coincidence of the cap site with the site of initiation of transcription by direct isolation of the polyphosphate end. In the course of this work we have discovered a requirement for hydrolysis of the β-γ bond of ATP for specific initiation of RNA polymerase II-mediated transcription (Bunick et al. 1982). We have used the in vitro transcription system to establish that DRB inhibits specific RNA polymerase II transcripts at the same concentrations as in vivo (Zandomeni et al. 1982). Furthermore, and in contrast to results obtained by others in vivo, we have established that DRB does affect RNA initiation but not elongation (Zandomeni et al. 1983). DRB^R cell mutants and biochemical fractionation are being used to elucidate the mechanism of action of this compound (Mittleman et al. 1983).

2 Results

2.1 The Cap Site Is the Site of Transcription Initiation

We have analyzed the E4 promoter of adenovirus, which is transcribed from the right end of the virus in a leftward direction, in an in vitro transcriptional system. This promoter was used to demonstrate the coincidence of the transcription start site with the actual site where the cap of the mRNA is located by isolating the respective 5′ end triphosphate. The assay for in vitro transcription contains the adenovirus *Eco*RI C fragment inserted into pBR322 (generously donated by Dr. J. Nevins, Rockefeller University). It has been shown that both in vivo and in vitro, at least six different initiation sites are utilized. These sites, shown schematically in Fig. 1 a, are the five different U-residues and the A-residue. Most of the cap sites detected both in vivo and in vitro (80–90%) are the first U-residue and the A-residue (Baker and Ziff 1981; Fire et al. 1981). If the DNA template is truncated at the *Sma*I site, major runoff transcripts of 245 and 240 n in length are obtained as shown in Fig. 1 b, lane 1, corresponding to the major U and A cap sites respectively. It is not easy to determine that transcription does indeed start at the cap site, since the transcriptional extracts contain phosphohydrolase, guanyltransferase, and all the methylases and *S*-adenosylmethionine required for efficient RNA capping and methylation (Shatkin 1976; Hagenbüchler and Schibler 1981). Nucleotide analogues containing β-γ imido bonds are resistant to the action of the phosphohydrolase (Yount et al. 1971), but have normal α-β bonds which allow ready incorporation into RNA chains and can thus be used in our in vitro assay (Howard and de Crombrugghe 1976; Bunick et al. 1982). To avoid capping of the U-initiated RNA, the β-γ imido analogue of UTP, UMP-P*N*P, was used for in vitro transcription. When the reactions are performed in the presence of UMP-P*N*P, transcripts initiated with a U-residue cannot be capped due to blocking of the phosphohydrolase activity by the imido analogue. As shown in Fig. 1 b, lane 2, no quantitative difference between the fully capped 240-n A-initiated transcripts and the UMP-P*N*P 245-n U-initiated transcripts can be detected. A slight slowdown of the rate of elongation has

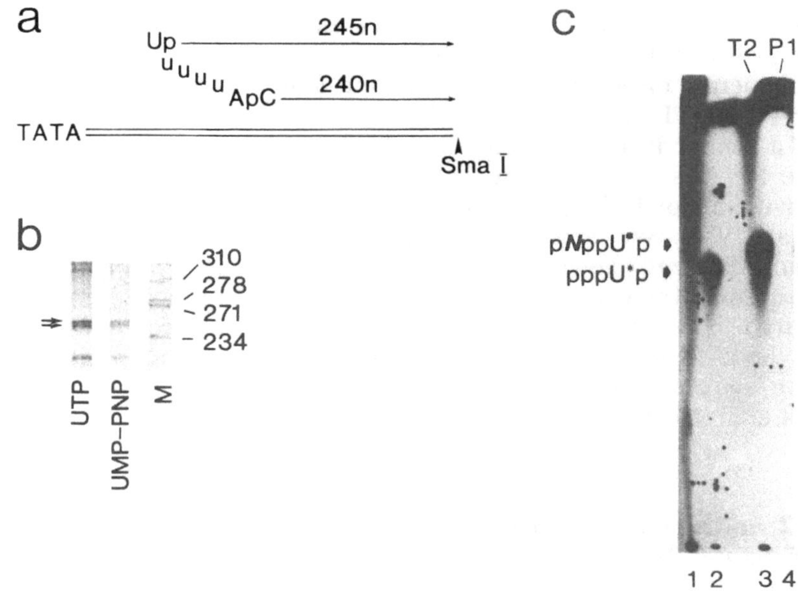

Fig. 1a–c. 5′-End analysis of adenovirus E4 in vitro RNA transcripts. **a** Schematic of the E4 template used for in vitro transcription, truncated with the restriction enzyme *Sma*I. The major U and A initiations give runoffs of 245 and 240 n respectively, while the minor U initiations give RNAs of intermediate sizes. **b** A portion of an autoradiogram from a 10% polyacrylamide-urea gel with the pair of 245 and 240 n runoff RNAs indicated by the *double arrows*. Reactions contained 4 m*M* creatine phosphate and 150 μ*M* of all ribonucleotide triphosphates, except for cold UTP and UMP-P*N*P, which were kept at 50 μ*M* each. The label (α[^{32}P]UTP) was 500 μCi (50 μ*M*) in each 250-μl reaction. Each reaction contained 50% of a whole cell extract prepared as described by Manley et al. (1980) and 25 μg/ml of the truncated DNA. After 1 h incubation at 30 °C, the RNAs were phenol-extracted and ethanol-precipitated. The first lane shows the RNAs synthesized in the presence of UTP, the second in the presence of UMP-P*N*P. The sizes of the *Φ*X-174 *Hae*III DNA markers displayed in lane M are indicated on the *right*. **c** An autoradiogram of EIV RNA made in the presence of UMP-P*N*P, eluted from a gel as in Fig. 2, lane 2, and digested with T2 (lane 3) and P1 (lane 4) nucleases for 5′-end analysis. The digests were run on PEI plates with 0.6 *M* phosphate buffer. Lanes 1 and 2 contain the tetraphosphates indicated by the *arrows*, made with *E. coli* RNA polymerase on a poly dAT template using α[^{32}P]ATP as the radioactive precursor and digested with ribonuclease T2

occurred for both U- and A-initiated transcripts with the imido analogue, but there is no preferential inhibition of the synthesis of the presumably uncapped U-initiated RNAs, in comparison to the A-initiated and capped transcripts. To further establish that these RNAs are indeed uncapped, we proceeded to isolate the 5′-end tetraphosphate. Since labeled UMP-P*N*P was not available, we proceeded to label the tetraphosphate via its nearest neighbor, α[^{32}P]UTP. After T2 digestion (Fig. 1c, lane 3) a tetraphosphate, p*N*ppUp^{32}, could be isolated from the 245-n transcript eluted from a polyacrylamide gel like the one in Fig. 1b. Capped T2 oligonucleotides resulting from A initiations remain unlabeled, while those that result from incorporation of α[^{32}P]UTP at the 5′ end migrate with the front of the solvent. The structure of the nucleotide tetraphosphate was confirmed by removal of the 3′-labeled phosphate with P1 nuclease (Fig. 1c, lane 4) and by comigration with $pppUp^{32}$ and $pNppUp^{32}$ markers made with *Escherichia coli* RNA polymerase on a poly dAT alternating template in the presence of UMP-P*N*P or UTP and $\alpha[^{32}P]$ATP (Fig. 1c, lanes 1 and 2). Fingerprint analysis of a similar RNA indicated that all oligonucleotides are identical whether the RNA is synthesized with UTP or with UMP-P*N*P precursors, except for the major U-capped oligonucleotide, which differs in mobility (Bunick et al. 1982). We conclude that this 5′-oligonucleotide contains the polyphosphate end shown in Fig.1c, and thus capping is not required for transcription. Moreover, capping occurs on the 5′-nucleotide where transcription starts, thus indicating the equivalence of start and cap sites. Methylation at the 2′-O ribose position of the starting nucleotide must require guanylation, since the tetraphosphate could be released with T2 ribonuclease (Bunick et al. 1982, 1983).

2.2 Initiation of Transcription Requires Hydrolysis of the β-γ Bond of ATP

Initial experiments on the use of the β-γ imido triphosphates were performed using the adenovirus major late promoter. These experiments failed because when the initiating nucleotide was A, substituting the ATP for AMP-P*N*P resulted in failure of transcription (Bunick et al. 1982). This was not a failure due to capping, since in the case of adenovirus early region 4 (Fig. 1) and murine leukemia virus LTR (Bunick et al. 1983), the respective U- or G-initiated uncapped transcripts could be isolated. Moreover, as shown in Fig. 2, the U-initiated transcripts of E4 or PIX also require the hydrolysis of the β-γ bond of ATP, since they cannot be synthesized when ATP is substituted with the imido analogue (lanes 3 and 8). AMP-P*N*P can be used by either *E. coli* RNA polymerase or purified HeLa cell RNA polymerase II in a system that lacks transcription fidelity (Howard and de Crombrugghe 1976; Bunick et al. 1982). We have shown that elongation of preinitiated transcripts can occur in the presence of AMP-P*N*P, but transcription initiation (as determined by the formation of a ternary complex) is blocked by the imido-ATP analogue (Bunick et al. 1982). We conclude that hydrolysis of the β-γ bond of ATP is required for faithful transcription initiation, for a function distinct from polymerization of nucleotides into RNA.

2.3 Effect of DRB on In Vitro Transcription

DRB is an adenosine analogue which has been widely used to analyze cellular and adenovirus-specific transcripts (see review by Tamm et al. 1983). It enters rapidly into cells and dramatically inhibits mRNA synthesis. Short

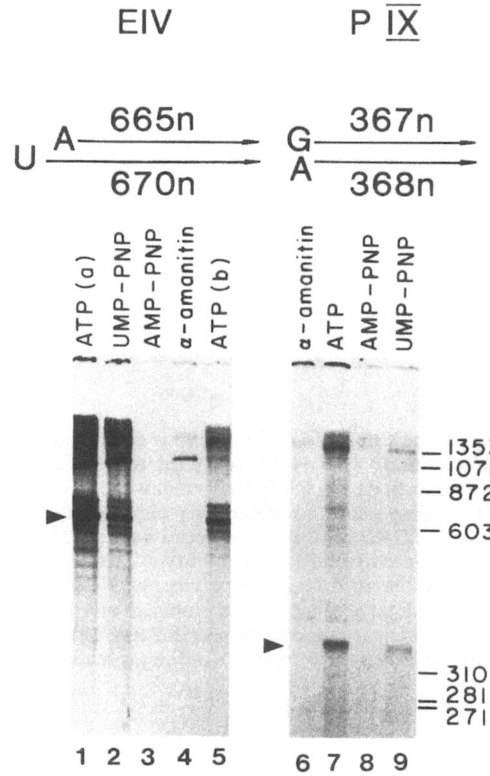

Fig. 2. Effect of imidonucleotide diphosphate analogues on specific RNA polymerase II transcription. The *upper part* of the figure shows a diagram of the two adenovirus templates corresponding to E4 and PIX. The E4 template, corresponding to the *Eco*RI C fragment (89.7–100 map units) cloned in pBR322 and the PIX templates corresponding to the *Hin*dIII C fragment (8–17 map units) were cut with restriction enzymes *Hin*dIII and *Sma* respectively. The sizes of the expected runoff RNAs are shown in the *upper panels* and indicated by the *arrows* in the autoradiogram of a 5% gel shown in the *lower panels*. Incubations were as indicated in Fig. 1, but each reaction was in a volume of 25 μl with 3 μCi/reaction (5 μ*M*) of α[^{32}P]GTP (410 Ci/m*M*). Control reactions with ATP at 100 μ*M* (a; lanes 1 and 7) or at 50 μ*M* (b; lane 5). In lanes 2 and 9, UMP-P*N*P was used instead of UTP; in lanes 3 and 8, AMP-P*N*P was used instead of ATP. Lanes 4 and 6 show reactions in the presence of 1 μg/ml α-amanitin, enough to completely inhibit transcription by RNA polymerase II

RNAs, synthesized in the presence of DRB, are correctly initiated and capped (Fraser et al. 1979).

The mechanism of action of DRB remained controversial because in vivo data suggested that it either acted on mRNA transcription initiation (Egyhazi 1974) or induced or enhanced premature termination (Fraser et al. 1979; see review by Tamm et al. 1983). When DRB is tested in a transcription assay with the adenovirus major late promoter we have found that the concentrations required for inhibition of specific RNA polymerase II transcripts are similar to those required to inhibit mRNA synthesis in vivo (Fig. 3) (Zandomeni et al. 1982). Thus, the faithful transcription systems provide the first in vitro model where the action of DRB can be studied at the molecular level, since no effect on the purified RNA polymerase II (under nonspecific transcription conditions) can be detected. Moreover, and as summarized in Fig. 3, the nucleoside is more active than the DRB nucleotide monophosphate and the triphosphate. No effect on RNA polymerase III can be detected at the concentrations which affect RNA polymerase II-specific transcription (Fig. 3, ■–■), as in vivo. It was suggested by various investigators that DRB either induced or enhanced premature termination (Fraser et al. 1978, 1979; reviewed by Tamm et al.

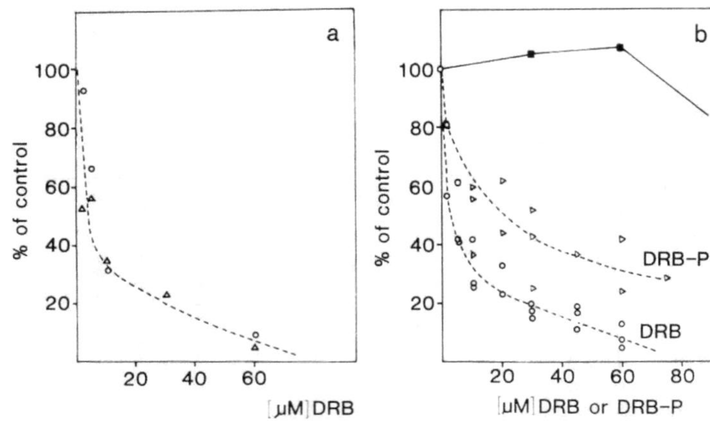

Fig. 3a, b. Effect of DRB on in vivo and in vitro transcription. **a** Incorporation of [5-^{3}H]uridine (13 Ci/mmol) into HeLa cell total RNA. Cells preincubated for 1 h in the presence of the concentrations of DRB indicated were labeled for 15 (o-----o) or 30 (Δ-----Δ) min with [^{3}H]uridine and the radioactivity incorporated into RNA determined. No corrections for the effect of DRB on the uridine pool were made because these are only significant at the higher DRB concentrations. **b** Summary of results obtained in in vitro incubations either for VA RNA transcribed by RNA polymerase III (*filled squares*) or for RNA polymerase II transcribing the adenovirus major late promoter or the human ε globin gene (*open symbols*). The *triangles* represent measurements of specific RNA runoffs in the presence of 5′-DRB monophosphate and the *circles* measurements in the presence of DRB (see ZANDOMENI et al. 1982)

1983), or, in contrast, that short RNAs made in the absence of DRB continued to be synthesized in its presence (SEHGAL et al. 1979; TAMM et al. 1980; MONTANDON and ACHESON 1982). Since these conclusions derived from in vivo labeling experiments were open to several interpretations, we decided to analyze these phenomena in the in vitro system. In the experiments summarized in Fig. 3b and in ZANDOMENI et al. (1982), we could not detect any short discrete transcripts that accounted for all the RNA polymerase II-specific transcription being shifted by increasing DRB concentrations to a different molecular weight category. The absence of short transcripts could be explained by a failure of detection, since a molecule 50 n long would contain 10 times less radioactive nucleotides. To increase the sensitivity of the assay, reduce the background, and be able to detect smaller molecules, we performed the experiment shown in Fig. 4. The RNA transcribed in vitro from a pBR322 vector containing the adenovirus major late promoter was selected by hybridization to single-stranded M-13 phage DNA containing the coding strand of the same gene from coordinates −50 to +500 (in relation to the cap site). The in vitro reactions were controls (Fig. 4, lane a), DRB at 10 µ*M* (b) or 60 µ*M* (c), or in the presence of enough α-amanitin (1 µg/ml) to completely inhibit RNA polymerase II activity (d). In the first panel, aliquots of the total reaction are shown. Similar aliquots were selected by hybridization with the single-stranded M-13 recombinant DNA probe. Panels A and B display autoradiograms at two different expo-

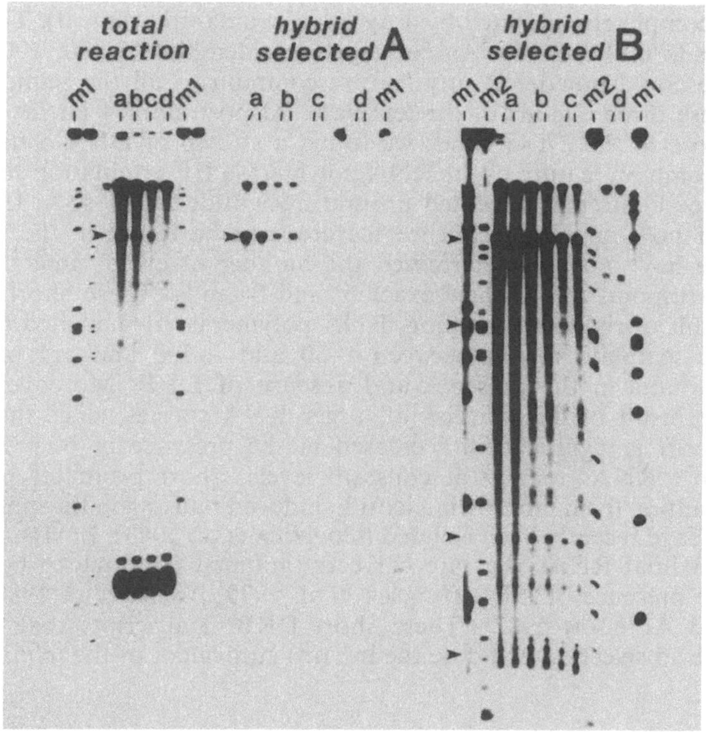

Fig. 4. No effect of DRB on premature termination. Transcription reactions similar to those in Fig. 2 contained the adenovirus major late promoter template *Sma*F (11–18 map units) inserted into the plasmid pBR322 and cut with the restriction enzyme *Sma*I. The 563-n runoff is indicated by the *arrow* in the first panel. In all three panels, lanes labeled *a* were controls with ethanol, lanes labeled *b* contained 10 μM DRB, lanes labeled *c* contained 60 μM DRB, and lanes labeled *d* contained 1 µg/ml α-amanitin. The in vitro transcription reactions were either analyzed directly (*first panel*) or after hybrid selection (*second and third panels*) in 5% acrylamide gels. We used a single-stranded M-13 phage recombinant containing DNA complementary to the rightward reading strand (generously donated by R. Roberts) to select the RNA from nucleotide −50 to +500 [in relation to the transcription start site; Gingeras et al. (1982)]. Duplicate aliquots of the hybrid-selected RNA are displayed in the second and third panels. The autoradiographic exposure of the A panel is the same as in the first panel (overnight) but the B panel is 20 times longer. The arrow points indicate either the position of the runoff RNAs (536 n) or discrete DRB-resistant RNA species 270, 100 and 50–60 n in length. The m1 markers are the *Hae*III-cut *Φ*X-174 end-labeled DNAs as in Fig. 5. The m2 markers are *Hae*III-*Hha*I-cut *Φ*X-174 DNA fragments with sizes of 614, 490, 445, 366, 305, 300, 277, 271, 241, 201, 150, 143, 123, 101, 88, 72, 57, 54, 50, 43, 35, 28 and 21 n

sures of the same gel of hybrid-selected RNA. The background is almost eliminated by hybridization selection, and the inhibition of specific runoff transcription induced by DRB is clearly evident (compare lanes b and c to a). A prolonged exposure (20 times longer, panel B) shows that shorter DRB^R RNAs are synthesized both in the presence and absence of DRB. These short RNAs are made by RNA polymerase II since their synthesis is completely inhibited by 1 μg/ml of α-amanitin (lane d). The more intense bands of DRB^R RNAs correspond to lengths of 270, 100, and 50–60 n. We could not detect any feature common to all the sequences located at these three regions of the template. At position 264 on the DNA template (close to the 270-n band) we found a stretch of DNA coding for five Us, a common feature of the RNA polymerase III termination signals. A similar three-U stretch is located around nucleotide 96 (close to the 100-n band), but no equivalent sequence features can be found in the 50–60-n region. We have not yet determined the number of events that these transcripts correspond to, or their exact 5′ and 3′ ends. These short RNAs are the result of rightward reading, RNA polymerase II-mediated transcription on the template region between −50 and +500 and are present in equal amounts in the presence and absence of DRB. Moreover, they are not the result of degradation of larger RNA species, since the specific RNA runoff is significantly decreased in the presence of 60 μ*M* DRB but the short RNAs remain at constant levels. Short promoter proximal RNAs resulting from limiting nucleotide-induced pausing in the major late promoter have recently been isolated (COPPOLA et al. 1983). In vivo, short promoter proximal RNAs that are correctly initiated continue to be synthesized in the presence of DRB (FRASER et al. 1979; TAMM et al. 1980; MONTANDON and ACHESON 1982). These short DRB^R transcripts that we detect after hybrid selection could be the in vitro equivalent of the in vivo species.

2.4 Transcription of Adenovirus Protein IX mRNA

From inside the coding region of adenovirus E1B, a short unspliced mRNA coding for the viral structural PIX is initiated (ALESTROM et al. 1980). The polyadenylation site is shared with the E1B mRNA polyadenylation site. This mRNA is transcribed at intermediate to late times after viral infection and seems to be one that responds differentially to transcription by infected cell extracts in vitro. It had been reported that in vivo, transcription of this mRNA and the synthesis of the corresponding PIX protein was resistant to DRB (VENNSTRÖM et al. 1979). To test if transcription of this gene was differentially resistant to DRB we performed the experiments shown in Fig. 5. DNA templates included in these reactions are the *Hin*dIII C fragment of Ad2, containing the PIX promoter, and the human ε globin *Bam*HI fragment, containing its own promoter. Increasing DRB concentrations to 60 μ*M* results in parallel inhibition of the ε globin and PIX transcripts. Thus, the DRB sensitivity of the PIX runoff transcript is similar to other promoters tested in vitro, in contrast to what occurs in vivo (ALESTROM et al. 1980). The remaining PIX mRNA detected in vivo in the presence of DRB (ALESTROM et al. 1980), could be due to the presence of heterogeneously sized short RNAs which are DRB resistant, some of them as long as the original mRNA. The short length of the transcript could explain the residual DRB resistant translation detected.

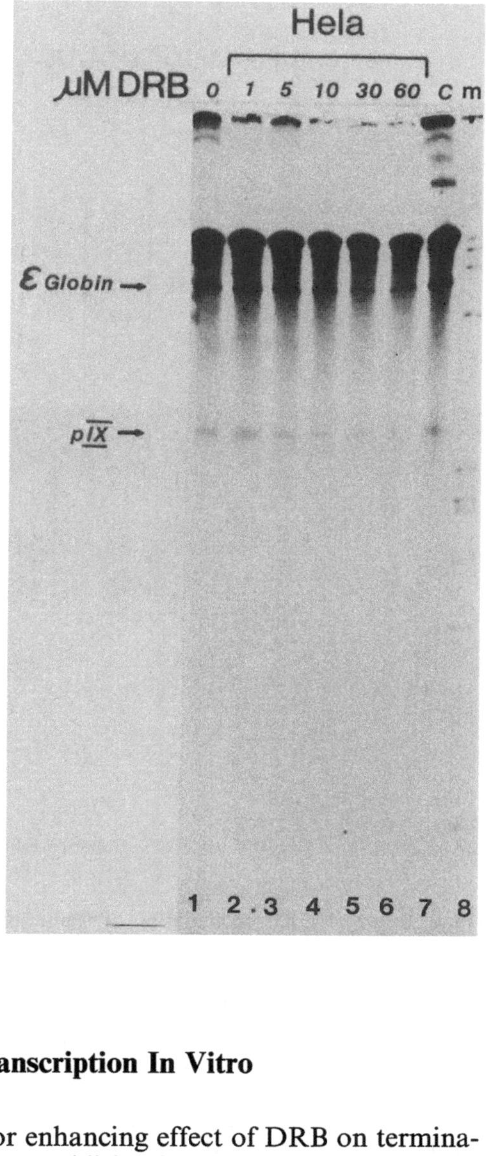

Fig. 5. Effect of DRB on the transcription of the protein IX and the globin promoters. In this in vitro reaction two templates were mixed. One, containing the 367–368 n RNA runoff of the PIX gene, is the same used in Fig. 2, but at 12.5 µg/ml. The other, also at 12.5 µg/ml, was the human ε globin *Bam*HI fragment, which gives a runoff 760 n in length when the vector is cut with *Sal*I. This is a chimeric runoff, with the first 460 n of the human insert and the last 300 n of pBR322. All reactions contained 5 µ*M* α[^{32}P]CTP (410 Ci/mM) as the radioactive precursor. Lane 7 is a control. Lanes 1–6 contain the concentrations of DRB indicated in the figure, all with the same amount of ethanol (0.5%) which is used as a solvent for the DRB. Lane 8 contains the *Hae*III-cut ΦX-174 size markers 1353, 1078, 872, 603, 310, 281, 271, 234, 194, and 118 n in length. An autoradiogram of the 5% polyacrylamide urea gel is illustrated

2.5 DRB Affects Initiation of Transcription In Vitro

Since we did not find any inducing or enhancing effect of DRB on termination, we performed an experiment to establish whether DRB was affecting transcription initiation (Zandomeni et al. 1983). The adenovirus major late promoter template used is shown at the top of Fig. 6. Transcription was initiated with an A followed by a C. Reactions were performed as indicated in the table and the autoradiograms of the resulting gels are shown in the lower part of the figure. Regular 30-min transcription reactions with or without 0.5% ethanol (the solvent used for the DRB) are shown in lanes 1 and 2 respectively. Lane 3 contains a reaction with 60 µ*M* DRB (0.5% ethanol) which shows the almost complete inhibition of synthesis

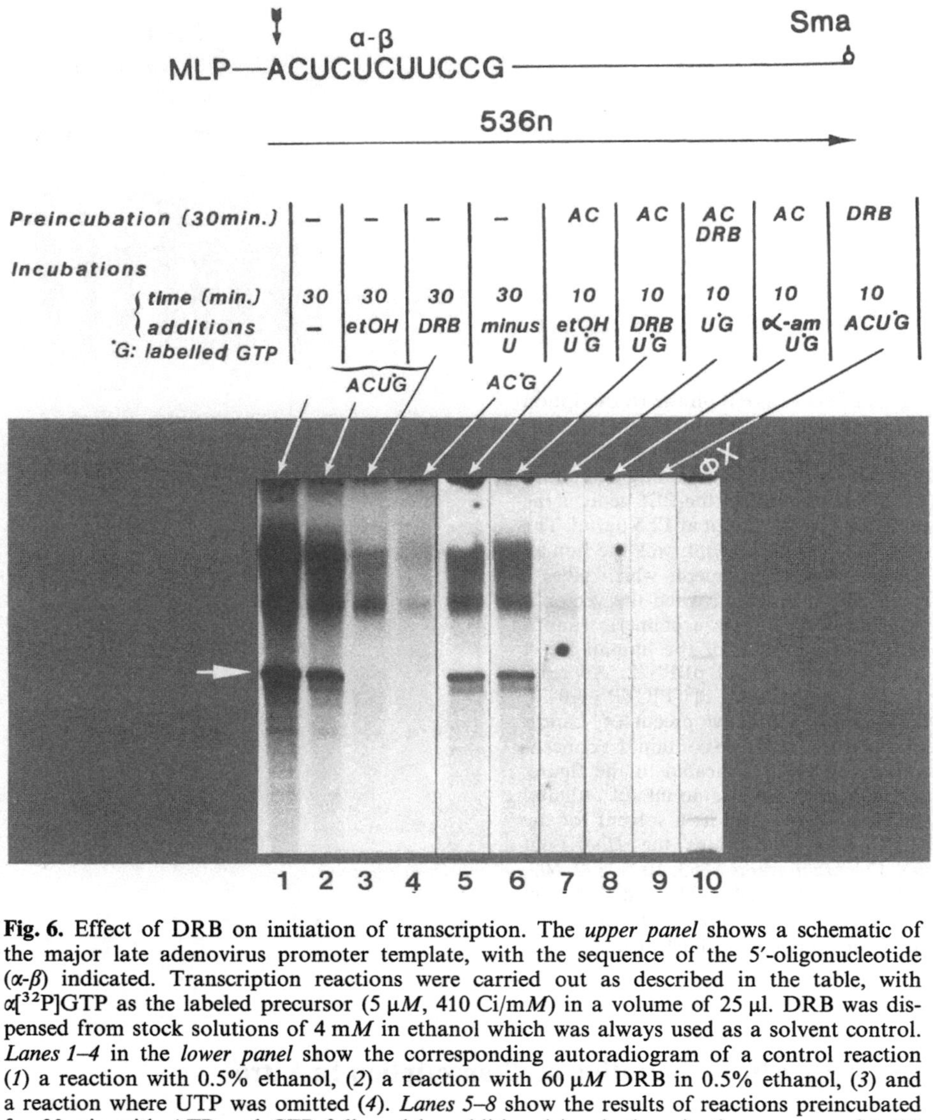

Fig. 6. Effect of DRB on initiation of transcription. The *upper panel* shows a schematic of the major late adenovirus promoter template, with the sequence of the 5′-oligonucleotide (α-β) indicated. Transcription reactions were carried out as described in the table, with α[^{32}P]GTP as the labeled precursor (5 μ*M*, 410 Ci/m*M*) in a volume of 25 μl. DRB was dispensed from stock solutions of 4 m*M* in ethanol which was always used as a solvent control. *Lanes 1–4* in the *lower panel* show the corresponding autoradiogram of a control reaction (*1*) a reaction with 0.5% ethanol, (*2*) a reaction with 60 μ*M* DRB in 0.5% ethanol, (*3*) and a reaction where UTP was omitted (*4*). *Lanes 5–8* show the results of reactions preincubated for 30 min with ATP and CTP followed by additional incubations in the presence of 0.5% ethanol, UTP, and α[^{32}P]GTP (*5*), DRB in 0.5% ethanol, UTP, and α[^{32}P]GTP (*6*) and α-amanitin, UTP, and α[^{32}P]GTP (*8*). In *lane 7* the DRB was added during the preincubation with ATP and CTP followed by addition of UTP and α[^{32}P]GTP for another 10 min. *Lane 9* displays the products of a reaction preincubated only with DRB followed by a 10-min incubation with all four nucleoside triphosphates. *Lane 10* shows the ΦX-174 *Hae*III-cut size markers described in Fig. 5

of the 536-n runoff RNA indicated by the arrow. Lane 4 shows that the extract used in this experiment is dependent on exogenous triphosphates, since omission of UTP does not result in runoff RNA synthesis. In lane 5 the reaction was preincubated for 30 min in the presence of A and C triphosphates to form a ternary initiation complex. Incubation was continued for 10 min after addition of U and labeled G triphosphate, giving a good level of RNA runoff. Lane 6 shows the results of a similar preincubation experiment, except that in this case DRB was added together with the radioactive label. A similar level of 536-n RNA can be detected, indicating that elongation can proceed efficiently in the presence of DRB. However, if DRB is present during the preincubation period (lane 7), no RNA runoff can be detected afterwards. This indicates that DRB is affecting the formation of the ternary complex. That the runoff synthesized after preincubation is indeed mediated by RNA polymerase II is shown in lane 8 by its α-amanitin sensitivity (LINDELL et al. 1970). Preincubating the extract with the DNA template in the presence of DRB and then adding all four ribonucleotide triphosphates does not result in the synthesis of runoff either, indicating that the presence of A and C triphosphates is necessary to form the ternary complex. Fingerprint analysis of the RNA synthesized in the presence of DRB after preincubation indicates that the labeled portion extends all the way to the 5′-oligonucleotide (ZANDOMENI et al. 1983). This supports the notion that the RNA polymerase II moves, at the most, just a few nucleotides from the transcription start site during the preincubation with A and C. The experiments described above provide an unequivocal demonstration that DRB acts on transcription initiation without affecting RNA elongation, probably before the formation of the first dinucleotide bond.

2.6 Generation of DRB-Resistant Cell Mutants

The first DRB^R cell mutants were isolated by GUPTA and SIMINOVITCH (1980). These cells were able to grow in lethal concentrations of DRB but the molecular basis for the resistance was not analyzed. FUNANAGE (1982) described several Chinese hamster ovary DRB^R cell lines which showed some DRB resistance in a crude in vitro transcriptional system. Since we were able to show in the in vitro transcriptional system that DRB acted at the same concentrations as in vivo, we proceeded to isolate a DRB^R cell mutant (MITTLEMAN et al. 1983). HeLa cells were mutagenized using MNNG, and one of the cell colonies able to grow in 60 μ*M* DRB, concentration that completely kills HeLa cells, was isolated. The mutation is stable since the DRB resistance can transfer from one cell to another by DNA-mediated transfection. Growth in the absence of DRB for up to 2 months did not result in a detectable rate of reversion.

The cells look normal and grow about 25% slower in the presence of DRB. They can support adenovirus infection and multiplication in the presence of 60 μ*M* DRB. The rate of uptake of [^{3}H]uridine is affected in the mutant to the same extent as the parental cells, but RNA synthesis, as measured by pulse labeling of total RNA, shows an inhibition midpoint at 30 μ*M* DRB, compared with 2.5 μ*M* DRB for HeLa cells. More importantly, transcriptionally active cell extracts were prepared from the DRB^R-1 mutant cells. The transcriptional efficiency of these extracts is similar to those extracts prepared from HeLa cells. In the presence of increasing concentrations of DRB, extracts from the mutant show a biphasic inhibition profile. Half of the activity has the same DRB sensitivity as the parental cells and half is partially resistant to higher DRB concentrations. Thus DRB resistance is a property not of the specific templates, but of the transcriptional extract, and the DRB^R-1 mutation is in either the RNA polymerase II or one of the factors affecting specific transcription. Since the partially purified RNA polymerase II is completely insensitive to DRB, we believe that the mutation affects either one of the factors required or their site of interaction for transcriptional specificity in vitro. Other mutants in the RNA polymerase II itself, such as human α-amanitin-resistant cells, have been isolated and characterized (Shander et al. 1982).

2.7 Formation of Ternary Complexes

Another approach to the analysis of the factor(s) involved in specific gene transcription is the purification of the components of the gene transcriptional machinery. Conventional biochemical fractionation has been used to partially separate some of the components involved in giving specificity to the RNA polymerase II in these in vitro assays (Matsui et al. 1981; Samuels et al. 1982; Dynan and Tjian 1983). The problems for this approach are the low abundance and instability of these factors. Moreover, the runoff assay cannot distinguish between factors required for initiation and those required for elongation. We have developed a simplified assay which is able to detect factors required for initiation, without a requirement for RNA elongation (Ackerman et al. 1983). We monitor the formation of ternary complexes between a specific template, RNA polymerase II, transcription factors contained in the whole cell extract, and radioactive nucleotide precursors. The ternary complexes are separated from the other components of the in vitro reaction by electrophoresis on agarose gels. The RNA polymerase II–template–nascent RNA complex remains intact in the sarkosyl-EDTA buffer used to stop the reaction and separate the complexes. An example of an autoradiogram of a dried agarose gel analysis of the adenovirus major late promoter is shown in Fig. 7. The adenovirus DNA restriction enzyme fragment *Sma*F 2400 bp, is indicated by the arrow and is clearly separated from the 9500-bp pBR313 vector. The 5′-end sequence of the major late promoter-specific RNA is shown in the upper part of Fig. 6. Only the nucleotide combinations compatible with the DNA sequence result in efficient labeling of the 2400-bp adenovirus DNA ternary complex. The fungal toxin α-amanitin, which inhibits transcription after

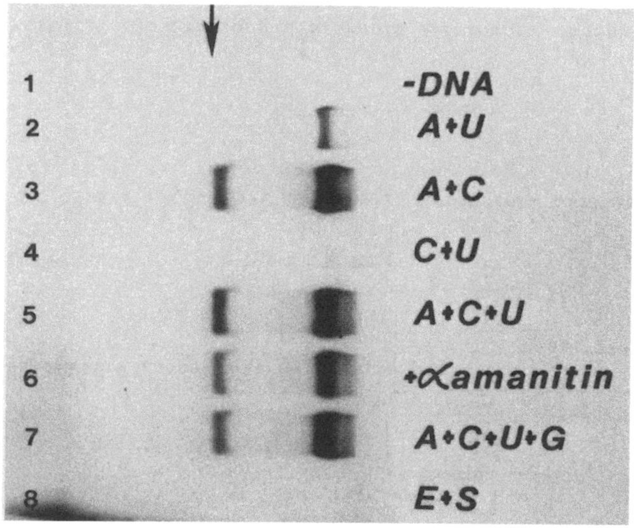

Fig. 7. Formation of ternary transcription complexes. The template DNA used was the *Sma*F (11–18 map units) of Ad2 inserted into the plasmid pBR313 (WEIL et al. 1979) digested with restriction enzyme *Sma*I. Transcription reactions were set up as described for Fig. 2 and stopped after 20 min by adding EDTA to 10 m*M* and sarkosyl to 0.25%. The pBR313 moiety (9500 bp in length) was separated from the adenovirus insert (2400 bp in length) and from other reaction components by electrophoresis in 2% agarose–0.25% sarkosyl gels containing 40 m*M* TRIS acetate, pH 8.3, 1 m*M* EDTA. An autoradiogram of the dried gel is shown, the *arrow* indicating the position of the viral DNA insert. *Lane 1* is without DNA; *lane 2* contains ATP and labeled UTP; *lane 3*, ATP and labeled CTP; *lane 4*, UTP and labeled CTP; *lane 5*, ATP, UTP, and labeled CTP; *lane 6*, as lane 5 but with 1 μg/ml α-amanitin; *lane 7*, as lane 5 with GTP. *Lane 8* is similar to lane 5, but EDTA (10 m*M*) and sarkosyl (0.25%) were added before the whole cell extract

the first dinucleotide, slightly reduced complex formation (lane 6), as expected from its site of action (VAISIUS and WIELAND 1982). Omission of the DNA (lane 1) or preincubation with sarkosyl (lane 8), which inhibits initiation, result in no complexes being formed. Thus the template, the transcriptional machinery, and the correct sequence alignment on the DNA are essential for ternary complex formation.

2.8 Transcriptional Signals

Analysis of the signals for transcription was facilitated by developments in the manipulation of recombinant DNA and by the development of in vitro systems that allow facile assay of the effects of these template alterations. Deletion analysis of the major late promoter (CORDEN et al. 1980; HU and MANLEY 1981) and of the E3 promoter (LEE et al. 1982) indicated that indeed the TATA box itself was required for both in vitro and in

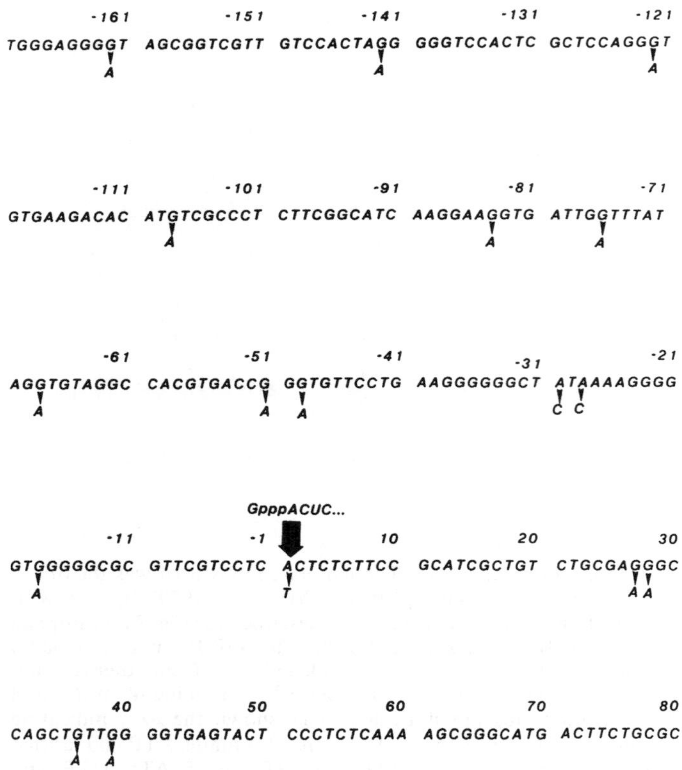

Fig. 8. Point mutants in the adenovirus major late promoter. The changes are indicated below the normal sequence, with position +1 symbolizing the transcription start site. Mutants at positions −30, −28 (CONCINO et al. 1983), and +1 were generated using synthetic oligonucleotides, while all the others were induced with hydroxylamine on the single-stranded DNA. Mutants GA −108 and GA −68 have lower DNA optima, GA −49 has a promoter with increased affinity, AC −30 and AT +1 reduce the level of transcription to 50% in comparison with wild type, and AC −28 to 20%

vivo expression of these sequences. A series of point mutations in this region of the genome have also suggested that they play an important role in determining the level of transcription from the major late adenovirus promoter (CONCINO et al. 1983). We have used synthetic oligonucleotides and region-specific point mutagenesis in an M-13 recombinant to obtain the major late promoter mutants indicated in Fig. 8. The single mutation which affects transcription in a most dramatic way is the AC-28 change in the TATA box, which reduces transcription to 20% of the wild type (CONCINO et al. 1983). Changes between −108 and +1 affect the level of transcription (slightly), or the optimal DNA concentration required for maximal expression, or the avidity of the template for some of the factors required for transcription (M. Concino and R. Weinmann, in preparation).

2.9 Mechanism of Transcription Initiation

Although the use of in vitro systems and cloned adenovirus genes has not allowed us to unveil all features of regulation of viral gene expression, important insights into the biochemical reactions involved in the mechanism of mammalian cell transcription by RNA polymerase II for both viral and cellular genes have been gained. First, it has been shown that hydrolysis of the β-γ of ATP is required for formation of the ternary transcriptional complex. Second, capping reactions are not required for in vitro transcription, although the cap is added at the 5′-polyphosphate containing nucleotide where transcription is initiated. Third, DRB inhibits initiation of transcription of full-size runoff RNAs, as defined by the formation of the ternary complex. However, some initiation events that might be faithful remain, producing short RNAs even in the presence of DRB. These results are in conflict with the reports on DRB-induced enhancement of premature transcription termination. However, DRB seems to inhibit transcription initiation of the *Chironomus* Balbiani ring 75S RNA (EGYHAZY et al. 1982) and no short DRB resistant RNAs were found in this system. Moreover, short RNAs that continue to be made in the presence of DRB are also made in its absence in mammalian cells (TAMM et al. 1980; FRASER et al. 1979; MONTANDON and ACHESON 1982). A speculative model summarizing our current understanding of the mechanism of transcription initiation is presented below (Fig. 9).

We assume that some of the transcription factors and the RNA polymerase II isolated by fractionation of whole cell or S-100 extracts interact with some DNA sequences, at −100, −70 (CAAT box), and/or the TATA box (see review by BREATNACH and CHAMBON 1981). It was suggested by DAVISON et al. (1983) that some of the factors (contained in the phosphocellulose C and in the DEAE 0.35 *M* fractions) interact first. Additional interaction with the RNA polymerase II results in commitment to transcription of a specific gene (DAVISON et al. 1983). This step would not require hydrolysis of the β-γ bond of ATP, as suggested by W. Dynan and R. Tjian (personal communication). Some of the transcriptional factors seem to confer some selectivity to SV40 promoters over adenovirus promoters (DYNAN and TJIAN 1983). In the next step, hydrolysis of the β-γ bond of ATP is required for unwinding of the DNA double helix, conformational transitions of the RNA polymerase, or phosphorylation or stabilization into what now seems to be an "open" transcriptional complex. In bacterial systems, the "open" complex has been defined by kinetic parameters like half-life, temperature dependence, rifampicin sensitivity, etc. (CHAMBERLIN 1976). The first and second nucleotide triphosphates of the specific transcript are now entering the complex. Once the nucleotides line up with the complementary strand on the template and the first dinucleotide bond is made, we define it as a ternary complex. DRB acts at some step prior to ternary complex formation (ZANDOMENI et al. 1983). To fully explain the presence of DRB resistant transcripts, we assume a second type of initiation event. These initiation events lack the factor which confers DRB sensitivity and can occur in the

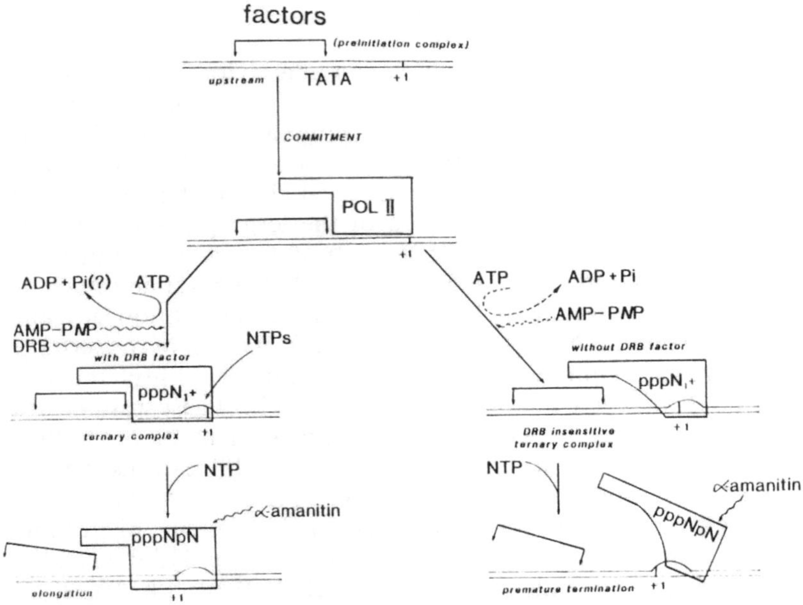

Fig. 9. Schematic of the events surrounding specific RNA polymerase II-specific transcription. Factors (probably phosphocellulose fraction C and DEAE fraction; Davison et al. 1983) interact with the DNA to form a preinitiation complex. The polymerase II is required to interact with the factors and DNA promoter sequences. Addition of nucleotides allows full engagement and the generation of an "open" complex. Two kinds of open complex are postulated. In one kind (*left*) the addition of the second nucleotide is sensitive to DRB and full-length RNA runoffs are obtained. We have demonstrated that hydrolysis of the β-γ bond of ATP is required for the formation of this kind of complex. The second kind (*right*) does not involve the factor(s) which confer(s) DRB sensitivity. The defective open complexes result in abnormal and/or early termination of transcription of these complexes. These are DRB insensitive, and an ATP hydrolysis requirement has not yet been demonstrated. The short RNAs described in Fig. 5 are derived from this kind of initiation event. The inhibitor of elongation α-amanitin (Lindell et al. 1970) affects RNA synthesis starting from the second dinucleotide bond and thereafter (Vaisius and Wieland 1982)

presence or absence of DRB. These initiation events would be partially defective, since the polymerase II seems to have a tendency to fall off or pause, accounting for the short RNAs observed. The assumption that β-γ ATP hydrolysis is required for this branch of the pathway is under investigation at this point. Ternary complexes are fairly stable and can be isolated on gels (Ackerman et al. 1983). The fungal toxin α-amanitin inhibits RNA elongation after the formation of the first dinucleotide bond in the ternary complex. Inhibition of elongation of RNA beyond the second nucleotide was shown using purified RNA polymerase II in vitro (Vaisius and Wieland 1982). Dissection of a series of steps involved in transcription initiation will now be possible using specific inhibitors such as DRB, AMP-PNP, and α-amanitin in combination with specific fractions required for transcription, and by the use of mutants in the DNA templates. Identification of the proteins responsible for specific transcription activity has to await the purification of the transcriptional factors. Using pure proteins will allow footprinting and crosslinking experiments to elucidate the nature of the specific interactions required for each one of these steps. Isolation of ternary complexes and analysis with template point mutants will help disclose the exact sites where these interactions occur. Finally, cellular mutants in the transcriptional apparatus will allow the isolation of the genes responsible for the α-amanitin- and DRB-resistant elements. Detailed information on proteins. DNA, and nucleotide interactions will allow us to understand more clearly the molecular details that lead to regulation of gene expression at the transcriptional level.

Acknowledgments. This work was supported by grants CA-21124 and AI-13231 to RW and by institutional grant CA-10815. SA and MC were supported by training grant 2T32 CA-09171 and DB by training grant 5T32 07922, all from the National Institutes of Health.

References

Ackerman S, Bunick D, Zandomeni R, Weinmann R (1983) RNA polymerase II ternary transcription complexes generated in vitro. Nucleic Acids Res 11:6041–6069

Alestrom P, Akusjärvi G, Perricaudet M, Matthews MB, Klessig DF, Pettersson U (1980) The gene for polypeptide IX of adenovirus type 2 and its unspliced mRNA. Cell 19:671–681

Baker CC, Ziff EB (1981) Promoters and heterogenous 5′ termini of mRNAs of adenovirus 2. J Mol Biol 149:189–221

Berget SM, Moore C, Sharp PA (1977) Spliced segments at the 5′ terminus of adenovirus late mRNA. Proc Natl Acad Sci USA 74:3171–3175

Berk AJ, Sharp PA (1978) Structure of the adenovirus 2 early mRNAs. Cell 14:695–711

Breatnach R, Chambon P (1981) Organization and expression of eukaryotic split genes coding for proteins. Annu Rev Biochem 50:349–383

Bunick D, Zandomeni R, Ackerman S, Weinmann R (1982) Mechanism of RNA polymerase II-specific initiation of transcription in vitro: ATP requirement and uncapped runoff transcripts. Cell 29:877–886

Bunick D, Zandomeni R, Ackerman S, Weinmann R (1983) Mechanism of initiation of RNA polymerase II. XIII Int Cancer Congress. Part B, Biology of Cancer (1) 9–18, Alan R Liss, Inc

Chamberlin MJ (1976) RNA polymerase: an overview. In: Losick R, Chamberlin M (eds) RNA polymerase. Cold Spring Harbor Laboratories, Cold Spring Harbor, pp 17–67

Chow L, Broker TR, Lewis JB (1979) Complex splicing patterns of RNAs from early regions of adenovirus 2. J Mol Biol 134:265–303

Concino M, Goldman RA, Carruthers M, Weinmann R (1983) Point mutations of the adenovirus major late promoter with different transcriptional efficiencies in vitro. J Biol Chem 258:8493–8496

Coppola JA, Field AS, Luse DS (1983) Promoter proximal pausing by RNA polymerase II in vitro: transcripts shorter than 20 nucleotides are not capped. Proc Natl Acad Sci USA 80:1251–1255

Corden J, Wasylyk B, Buchwalder A, Sassone-Corsi P, Kedinger C, Chambon P (1980) Promoter sequences of eukaryotic protein coding genes. Science 209:1406–1414

Davison BL, Egly J-M, Mulvihill ER, Chambon P (1983) Formation of stable preinitiation complexes between eukaryotic class B transcription factors and promoter sequences. Nature 301:680–686

Dynan WS, Tjian R (1983) Isolation of transcription factors that discriminate between different promoters recognized by RNA polymerase II. Cell 32:669–680

Egyhazi E (1974) A tentative initiation inhibitor of chromosomal hn RNA synthesis. J Mol Biol 84:173–183
Egyhazi E, Pigon A, Rydlander L (1982) DRB inhibits the rate of transcription initiation in intact *Chironomus* cells. Eur J Biochem 122:445–451
Evans RM, Fraser N, Ziff E, Weber J, Wilson M, Darnell JE (1977) The initiation sites for RNA transcription in Ad2 DNA. Cell 12:733–739
Fire A, Baker CC, Manley JL, Ziff EB, Sharp PA (1981) In vitro transcription of adenovirus. J Virol 40:703–719
Fowlkes DM, Shenk T (1980) Transcriptional control regions of the adenovirus VA I RNA gene. Cell 22:405–413
Fraser N, Sehgal PB, Darnell JE (1978) DRB-induced premature termination of late adenovirus transcription. Nature 272:590–593
Fraser N, Sehgal PB, Darnell JE (1979) Multiple discrete sites for premature RNA chain termination late in Ad2 infection: enhancement by DRB. Proc Natl Acad Sci USA 76:2571–2575
Funanage VL (1982) Isolation and characterization of DRB^R mutants of the CHO cell line. Mol Cell Biol 2:467–477
Gelinas RE, Roberts RJ (1977) One predominant 5′-undecanucleotide in adenovirus 2 late messenger RNAs. Cell 11:533–544
Gingeras TR, Sciaky D, Gelinas RE, Bing-Dong J, Yen CE, Kelly MM, Bullock PA, Parsons BL, O'Neill KE, Roberts RJ (1982) Nucleotide sequences from the adenovirus-2 genome. J Biol Chem 257:13475–13491
Goldberg S, Weber J, Darnell JE Jr (1977) The definition of a large viral transcription unit late in Ad2 infection of HeLa cells: mapping by effects of ultraviolet irradiation. Cell 10:617–621
Grummt I (1981) Specific transcription of mouse rDNA in a cell-free systems that mimics control in vivo. Proc Natl Acad Sci USA 78:727–731
Guilfoyle R, Weinmann R (1981) The control region for transcription of the adenovirus VA RNA. Proc Natl Acad Sci USA 78:3378–3382
Gupta RS, Siminovitch L (1980) DRB resistance in Chinese hamster and human cells: genetic and biochemical characterization of the selection system. Somatic Cell Genet 6:151–169
Hagenbüchler O, Schibler U (1981) Mouse β-globin and adenovirus 2 major late transcripts are initiated at the cap sites in vitro. Proc Natl Acad Sci USA 78:2283–2286
Honda H, Kaufman RJ, Manley J, Gefter M, Sharp PA (1980) Transcription of SV40 in a HeLa whole cell extract. J Biol Chem 256:478–482
Howard BH, de Crombrugghe B (1976) ATPase activity required for termination of transcription by the *Escherichia coli* protein factor Rho. J Biol Chem 251:2520–2524
Hu S-L, Manley J (1981) Identification of the DNA sequence required for initiation of transcription in vitro from the major promoter of adenovirus-2. Proc Natl Acad Sci USA 78:879–883
Lee DC, Roeder RG (1981) Transcription of adenovirus type 2 genes in a cell-free system: apparent heterogeneity of initiation at some promoters. Mol Cell Biol 1:635–651
Lee DC, Roeder RG, Wold WSM (1982) DNA sequences affecting specific initiation of transcription in vitro from the EIII promoter of adenovirus 2. Proc Natl Acad Sci USA 79:41–45
Lerner MR, Steitz JA (1981) Snurps and scyrps. Cell 25:298–300
Lewis J, Mathews M (1980) Control of adenovirus early gene expression: a class of immediate early products. Cell 21:303–313
Lindell TJ, Weinberg F, Morris PW, Roeder RG, Rutter WJ (1970) Specific inhibition of nuclear RNA polymerase II by α-amanitin. Science 170:447–448
Manley JL, Sharp PA, Gefter ML (1979) RNA synthesis in isolated nuclei; in vitro initiation of the Ad2 major late mRNA precursor. Proc Natl Acad Sci USA 76:160–164
Manley JL, Fire A, Cano A, Sharp PA, Gefter ML (1980) DNA-dependent transcription of adenovirus genes in a soluble whole cell extract. Proc Natl Acad Sci USA 77:3855–3859
Mathis DJ, Elkaim R, Kedinger C, Sassone-Corsi P, Chambon P (1981) Specific in vitro initiation of transcription on the adenovirus type 2 early and late EII transcription units. Proc Natl Acad Sci USA 77:7102–7106

Matsui T (1982) In vitro accurate initiation of transcription on the adenovirus type 2 IVa2 gene which does not contain a TATA box. Nucleic Acids Res 10:789–7101

Matsui T, Segall J, Weil PA, Roeder RG (1981) Multiple factors required for accurate initiation of transcription by RNA polymerase II. J Biol Chem 255:11992–11996

Mittleman B, Zandomeni R, Weinmann R (1983) Mechanism of action of DRB II: A resistant human cell mutant with an altered transcriptional machinery. J Mol Biol 165:461–473

Montandon PE, Acheson NH (1982) Synthesis of prematurely terminated late transcripts of polyoma virus DNA is resistant to inhibition by DRB. J Gen Virol 59:367–376

Salditt-Georgieff M, Harpold M, Chen-Kiang S, Darnell JE Jr (1980) The addition of 5′ cap structures occurs early in hnRNA synthesis and prematurely terminated molecules are capped. Cell 19:69–78

Samuels M, Fire A, Sharp PA (1982) Separation and characterization of factors mediating accurate transcription by RNA polymerase II. J Biol Chem 257:14419–14427

Sehgal PB, Fraser NW, Darnell JE (1979) Early Ad2 transcription units: only promoter-proximal RNA continues to be made in the presence of DRB. Virology 94:185–191

Shander MTM, Croce CM, Weinmann R (1982) Human mutant cell lines with altered RNA polymerase II. J Cell Physiol 113:324–328

Shatkin AJ (1976) Capping of eukaryotic mRNAs. Cell 9:645–653

Shaw AR, Ziff EB (1980) Transcripts from the adenovirus-2 major late promoter yield a single family of 3′ coterminal mRNAs during early infection and five families at late times. Cell 22:905–916

Tamm I, Kikuchi T, Darnell JE, Salditt-Georgiev M (1980) Short capped hnRNA precursor chains in HeLa cells: continued synthesis in the presence of DRB. Biochem 19:2743–2748

Tamm I, Sehgal PB, Lamb RA, Goldberg AR (1983) Halogenated ribofuranosylbenzimidazoles. In: Becker Y (ed) Antiviral drugs and interferon: the molecular basis of their activity. Nijhoff, The Hague. In press

Thimmappaya B, Weinberger C, Schneider RJ, Shenk T (1983) Adenovirus VAI RNA is required for efficient translation of viral mRNAs at late times after infection. Cell 31:543–551

Vaisius AC, Wieland T (1982) Formation of a single phosphodiester bond by RNA-polymerase B from calf thymus is not inhibited by α-amanitin. Biochemistry 21:3097–3101

Vennström B, Persson H, Pettersson U, Philipson L (1979) A DRB-resistant adenovirus mRNA. Nucl Acids Res 7:1405–1418

Weil PA, Luse DS, Segall J, Roeder RG (1979) Selective and accurate initiation of transcription at the Ad2 major late promoter in a soluble system dependent on purified RNA polymerase II and DNA. Cell 18:469–484

Weinmann R, Jaehning JA, Raskas HJ, Roeder RG (1976) Viral RNA synthesis and levels of DNA-dependent RNA polymerases during replication of adenovirus 2. J Virol 17:114–126

Wu GJ (1978) Adenovirus DNA-directed transcription of 5.5S RNA in vitro. Proc Natl Acad Sci USA 75:2175–2179

Yount RG, Babcock D, Ballantyne W, Ojala D (1971) Adenylimidodiphosphate, an adenosine triphosphate analog containing a P-N-P linkage. Biochem 10:2484–2489

Zandomeni R, Mittleman B, Bunick D, Ackerman S and Weinmann R (1982) Mechanism of action of DRB: effect on in vitro transcription. Proc Natl Acad Sci USA 79:3167–3170

Zandomeni R, Bunick D, Ackerman S, Mittleman B (1983) Mechanism of action of DRB III. Effect on specific in vitro initiation of transcription. J Mol Biol 167:561–574

Ziff E, Evans RM (1978) Coincidence of the promoter and capped 5′ terminus of RNA from the adenovirus 2 major late transcription unit. Cell 15:1463–1475

Adenovirus Early Gene Regulation and the Adeno-associated Virus Helper Effect

WILLIAM D. RICHARDSON[1] and HEINER WESTPHAL[2]

1 Introduction

As one of the best studied eukaryotic viruses, adenovirus offers excellent opportunities for a thorough study of gene regulation at the molecular level. Evidence for an intricate system of viral regulatory functions mediating lytic growth or transformation is rapidly accumulating. Modern techniques of gene analysis have created experimental conditions for studying mechanisms of gene regulation in considerable detail. We will mention some of these new approaches before turning to the adeno-associated virus (AAV) helper effect, the topic of our own studies.

2 Adenovirus Gene Regulation: New Tools of Analysis

Progress in the understanding of adenovirus gene regulation has been closely coupled to the elucidation of the structure of viral DNA, RNA, and proteins. For instance, the phenomenon of RNA splicing was detected when

1 National Institute for Medical Research, The Ridgeway, Mill Hill, London NW7 1AA, United Kingdom
2 Laboratory of Molecular Genetics, National Institutes of Health, Bethesda, MD 20205, USA

Current Topics in Microbiology and Immunology, Vol. 109

the structure of adenovirus mRNAs were examined in the electron microscope (Berget et al. 1977; Westphal and Lai 1977; Chow et al. 1977). Similarly, the detection of terminal protein at the ends of virion DNA (Robinson et al. 1973) played an important part in establishing the first in vitro system of eukaryotic DNA replication (Challberg and Kelly 1979). The reader is referred to other chapters of this volume for complete, up-to-date information on the viral chromosome and pathways of gene expression and control during lytic growth or cell transformation. Here we wish to explore some new avenues of viral gene analysis.

Traditionally, viral gene functions are identified and examined with the help of mutants. Their central importance for the analysis of gene regulation is well documented throughout this book. We merely recall a few outstanding examples, such as H5*ts*125 (Ensinger and Ginsberg 1972), H5*ts*36 (Wilkie et al. 1973), H5*ts*149 (Ginsberg et al. 1974), H5*dl*312 (Jones and Shenk 1979), and H5*dl*330 (Thimmappaya et al. 1982), and the wealth of information these and other mutants have generated.

More recently, several laboratories have begun to dissect the viral genome and test the interactions of selected genes in their mammalian host cell environment. This is especially important in instances where two or more genes are involved in one regulatory event. The idea is, of course, not new. Even prior to the availability of restriction enzymes, cell lines were established that contained and expressed sequences from the transforming region of adenovirus (Graham and van der Eb 1973). Today virtually any viral sequence can be excised, propagated in the wild-type or mutated form in suitable vector systems, and reintroduced into the natural host cell by a variety of methods. For instance, the role of individual gene functions in cell transformation has been examined in transfection experiments using portions of the transforming region of the adenovirus genome cloned in prokaryotic vectors (Klessig et al. 1982; Bernards et al. 1982; van den Elsen et al. 1982). Regulation by the product of the adenovirus E1A gene is now also being studied this way (Bos and ten Wolde-Kraamwinkel 1983; Weeks and Jones 1983).

For studies of viral gene interactions, it is desirable to transfer into the host cell not only the genes of interest but also the corresponding gene products. A method uniquely suited for this purpose is the microinjection technique, which allows the transfer of macromolecules, such as DNA, RNA, or proteins, via glass capillaries into the cell compartment of choice, i.e., either the nucleus or cytoplasm. A widely used version of this method is demonstrated in Fig. 1. The tip of a fine capillary is filled with the sample to be injected. The capillary is connected to a syringe via air-filled tubing. Cells grown in a petri dish and covered with medium are placed in the light path of the microscope and the tip of the capillary is lowered into the medium. A micromanipulator guides the tip into the cell, and manual pressure on the syringe controls the amount of solution to be injected, usually about 10^{-11} ml. Change in appearance of the cell upon injection (see Fig. 1) serves as a visual control. Details of the method and its applica-

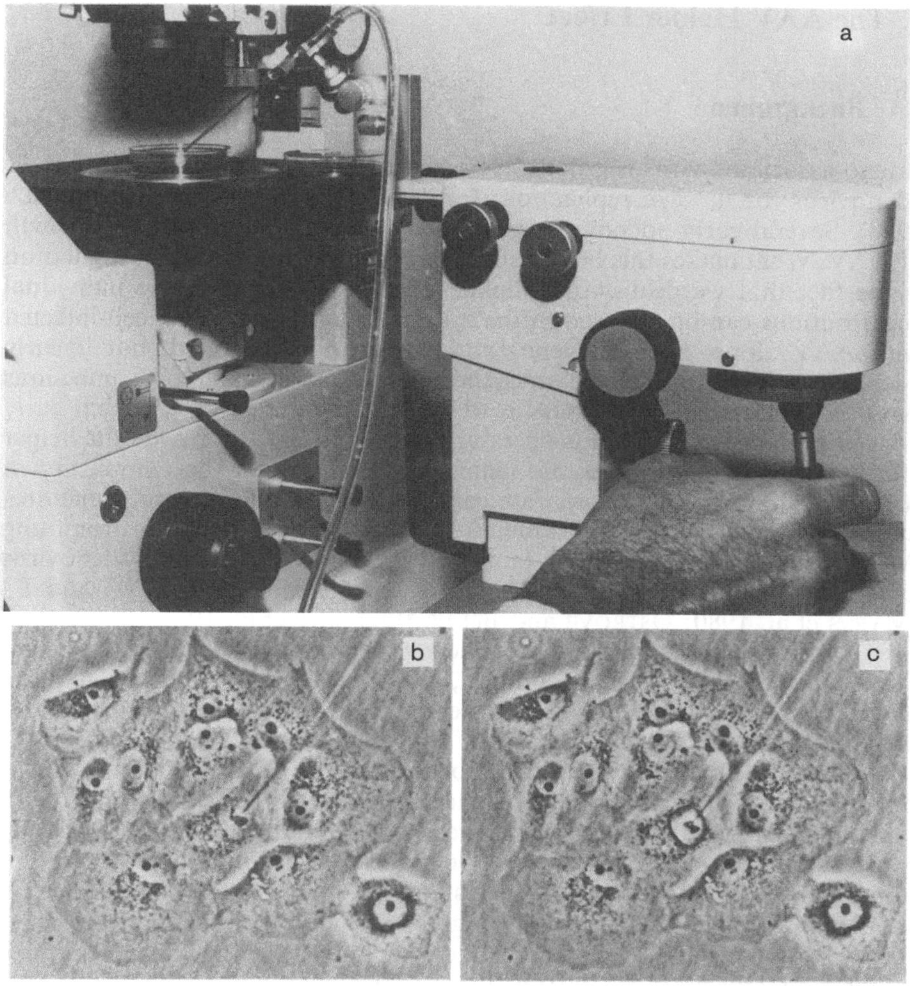

Fig. 1 a–c. The manual microinjection technique. **a** The equipment in use. On the specimen stage of an inverted phase contrast microscope is a petri dish containing the cells to be injected. The glass injection capillary is connected via a length of flexible plastic tube to a 50 ml syringe (not visible in the photograph) to which pressure is applied to expel liquid from the needle. The needle is guided to the cells by means of a micromanipulator. With practice, up to 300 cells can be injected in 15 min. **b** The view down the microscope as the injection needle (entering the field from the *top right*) approaches the nucleus of a cell. The needle has a tip diameter of approximately 0.5 μm. **c** The cell immediately after a successful nuclear injection. The refractive index of the nucleus, but not the cytoplasm, has changed due to dilution of its contents

tion may be found in GRAESSMANN et al. (1980). On the following pages we will demonstrate how complex pathways of adenovirus gene regulation can be analyzed with the help of this microinjection technique.

3 The AAV Helper Effect

3.1 Background

Adeno-associated virus is a defective parvovirus which depends on a helper adenovirus for its own replication (for review see WARD and TATTERSALL 1978). Several early adenovirus genes act together to promote the growth of AAV. What makes this helper effect so attractive for studies of regulation is the fact that we deal with unlinked adenovirus genes whose individual contributions can be assessed in the microinjection assay. In a cell infected with AAV alone, no viral gene expression can be detected, but if early adenovirus RNA is injected into the cytoplasm of such cells, infectious AAV is produced (RICHARDSON et al. 1980). While this tells us that early adenovirus gene expression is necessary and sufficient to provide the helper effect, a glance at the genomic map (Fig. 2) shows us the complexity of the problem. Each of the separate early regions encodes several functions, and all of these had to be examined for their potential role in promoting AAV growth. Studies with mutants of Ad2 and Ad5 or with adenovirus DNA fragments narrowed the choice to three regions, E1, E2A, and E4 (MYERS et al. 1980; OSTROVE and BERNS 1980; JAY et al. 1981; JANIK et al. 1981), with functions from each of these regions contributing to the helper effect (JANIK et al. 1981). In addition, JANIK et al. (1981) found a requirement for VA1, a small RNA polymerase III transcript made throughout the adenovirus life cycle.

What is the individual contribution of each of these functions? It was possible that each early region participated directly in the AAV growth cycle or alternatively, that one or more played an indirect role by regulating expression of the others. Answering these questions is difficult using conventional genetics, but poses a good test of the microinjection technique. It should be possible, by microinjecting gene products rather than the genes themselves, to bypass normal control events and so distinguish genes which interact directly with AAV from those which play a regulatory role. We purified mRNAs from E1, E2A, E3, and E4 by hybridizing cytoplasmic RNA from Ad2-infected cells to plasmid DNAs representing each region, and injected them in various combinations into AAV-infected cells (RICHARDSON and WESTPHAL 1981). Our experimental procedure, called the AAV assay, is briefly described in the legend to Fig. 3. We showed that injection of E4-specific mRNA was both necessary and sufficient for AAV antigen production and AAV DNA replication. In contrast, when we injected gene-sized DNA fragments of Ad2, we found that in addition to E4 it was necessary also to inject fragments encompassing E1 and E2A. These results infer that the requirement for E1 and E2A may be to activate expression of a direct-acting helper function encoded in E4. If so, what is the order of events in the regulatory scheme? We approached this question by injecting combinations of mRNA and DNA fragments into the same AAV-infected cells. Cells which received only E2A mRNA and the E4 DNA

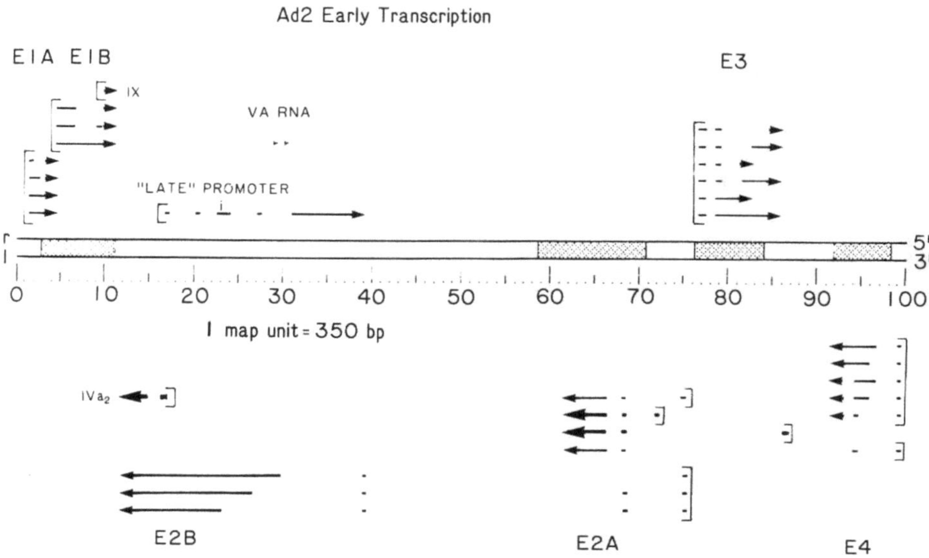

Fig. 2. Map of the Ad2 genome and the early transcripts, reprinted with permission from RICHARDSON and WESTPHAL (1981). This and subsequent figures combine data from several laboratories, reviewed by ZIFF (1980). The major early regions are numbered E1A to E4 and follow the nomenclature of KITCHINGMAN et al. (1977) and STILLMAN et al. (1981). *Brackets* denote promoter locations and *interrupted lines* represent spliced RNA transcripts. *Arrows* show the direction of transcription. RNAs mapping outside the main early regions include rare transcripts starting at the late promoter and including the "i" leader, and RNAs that begin to accumulate early, but specify late products such as proteins IVA2 and IX. The VA RNAs are small RNA polymerase III transcripts made throughout the virus life cycle

fragment were able to support AAV antigen production, as were cells which received E1 mRNA and a mixture of E2A and E4 DNAs. Taken together, the experiments described suggested an ordered sequence of gene activation: E1 → E2A → E4 → AAV. Our subsequent experiments, described below, have been aimed at obtaining further details of this scheme and assigning specific gene products to each step of the cascade. These later experiments, while supporting the regulatory sequence outlined above, provide evidence for alternative routes of gene activation, and illustrate the complexity of adenovirus early gene control.

3.2 Independent Roles for E1A and E1B

The experiments described above did not discriminate between E1A and E1B, since the DNA fragment we used for injection (0–29 on the Ad2 map) encompassed both regions. Since a functional E1A gene is required for activation of transcription of other early genes (JONES and SHENK 1979; BERK et al. 1979; NEVINS 1981), we speculated that this explained the re-

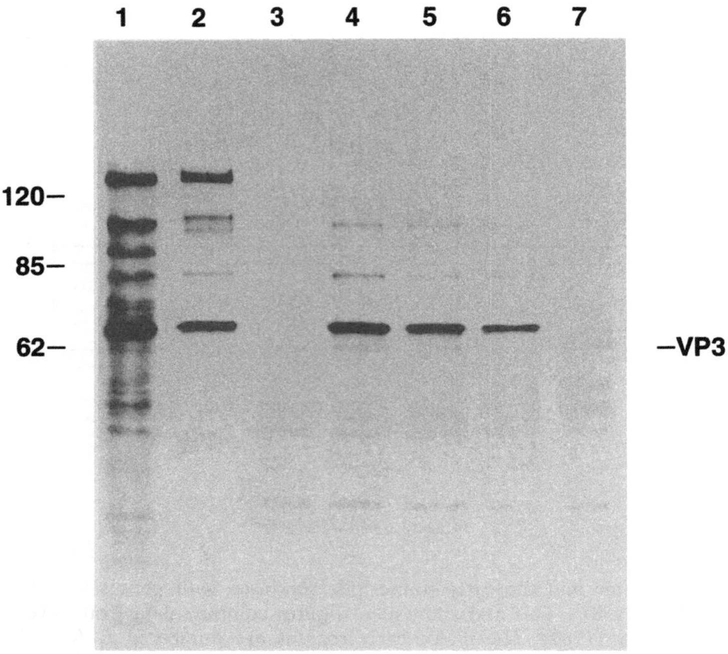

Fig. 3. AAV antigen synthesis in cells infected with AAV and injected with DNA fragments or coinfected with adenovirus. Experimental details are as follows. Permissive monolayer cells (Vero) were grown in petri dishes in isolated islands of around 200 cells. These were systematically injected in the nucleus with DNA solution. After a short recovery period in the incubator, AAV virus was added to the dish for 1 h. Next day the cells were labeled with [^{35}S]methionine, and immunoprecipitated AAV proteins were displayed on an SDS-polyacrylamide gel. Further details are given in RICHARDSON and WESTPHAL (1981). *Lane 1,* Ad2 late protein markers; *lane 2,* Ad2 coinfection; *lane 3,* AAV alone. *Lanes 4–7* are all from cells which received DNA fragments containing E2A (*Sma*I-A, 57–76) and E4 (*Eco*RI-C, 89.7–100) and, in addition, the following fragments from E1: *lane 4, Xba*I-B (3.8–28.8) plus *Hpa*I-E (0–4.5); *lane 5, Hpa*I-E; *lane 6, Xba*I-B; *lane 7,* no E1 fragment. In this and subsequent figures, the molecular weights (in kd) of major Ad2 structural polypeptides are indicated in the margin. Also marked is the position of the major AAV2 virion protein, VP3

quirement for E1 in the AAV assay. To test this we isolated a variety of fragments from the left end of the Ad2 genome. These fragments were tested for their ability to promote the AAV helper effect when coinjected with fragments spanning E2A and E4. As expected, fragments located entirely outside E1 did not promote helper activity. Unexpectedly, however, fragments containing either E1A or E1B alone could restore the helper effect. Results of a typical experiment are shown in Fig. 3, and the outcome of several experiments is summarized in Fig. 4. Possible cross-contamination of gel-purified DNA fragments is eliminated because cloned DNAs containing only E1A or E1B have also been successfully tried (RICHARDSON and WESTPHAL, in preparation). Gene products, rather than the genes them-

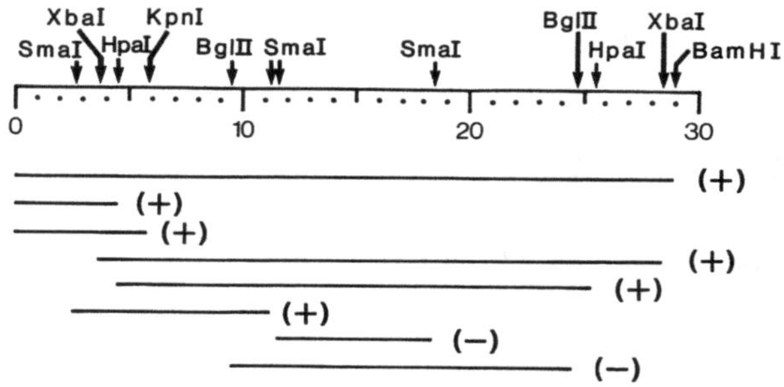

Fig. 4. AAV helper activity of DNA fragments from the left end of Ad2. The left 30% of the Ad2 genome is marked out in conventional map units (one map unit = 350 base pairs). The reader is referred to Fig. 2 for a map of transcripts from this section of the Ad2 genome and to Fig. 6 for a more detailed map of the region between 0 and 12 map units. Some cutting sites for restriction endonucleases are marked by *arrows*. DNA fragments generated by one or other of these enzymes from Ad2 virion DNA were purified on agarose gels and injected into AAV-infected cells together with fragments from E2A and E4 (see legend to Fig. 3). Fragments tested in this way are shown above; from *top to bottom* these are *Bam*HI-B, *Hpa*I-E, *Kpn*I-G, *Xba*I-B, *Hpa*I-C, *Sma*I-E, *Sma*I-F, and *Bgl*II-B. A + sign denotes that AAV antigen synthesis was detected

selves, are the active agents, since purified E1A- or E1B-specific mRNAs can substitute for their corresponding DNA fragments (Fig. 5). It therefore appears that E1A and E1B can participate in two independent regulatory pathways, both of which also involve E2A and E4. Consistent with this is the finding by LAUGHLIN et al. (1982) that deletion mutants of Ad5, mapping to either EIA or EIB are incapable of helping AAV. It should be emphasized at this point that care has been taken in all experiments involving injection of DNA to maintain a constant gene dose of approximately 100 copies of each fragment per cell. This is important, for we can overcome the requirement for any E1 gene product by increasing the number of E2A and E4 genes injected to 250 copies per cell (RICHARDSON and WESTPHAL, in preparation). This dramatic effect is consistent with experience using deletion and point mutants of E1A, which are only defective for growth at low multiplicities of infection (SHENK et al. 1979; NEVINS 1981).

3.3 Role of E1A

The detailed genetic arrangement of E1A is illustrated in Fig. 6. Two differently spliced mRNAs of 1.1 kb (13S) and 0.9 kb (12S) are produced early during infection, each of which gives rise to a different protein doublet when translated in vitro. We purified the 1.1-kb mRNA by hybridization to a DNA fragment consisting of sequences unique to this message. Injection of the 1.1-kb mRNA together with E2A and E4 DNA fragments sufficed to promote the helper effect (RICHARDSON and WESTPHAL, in preparation).

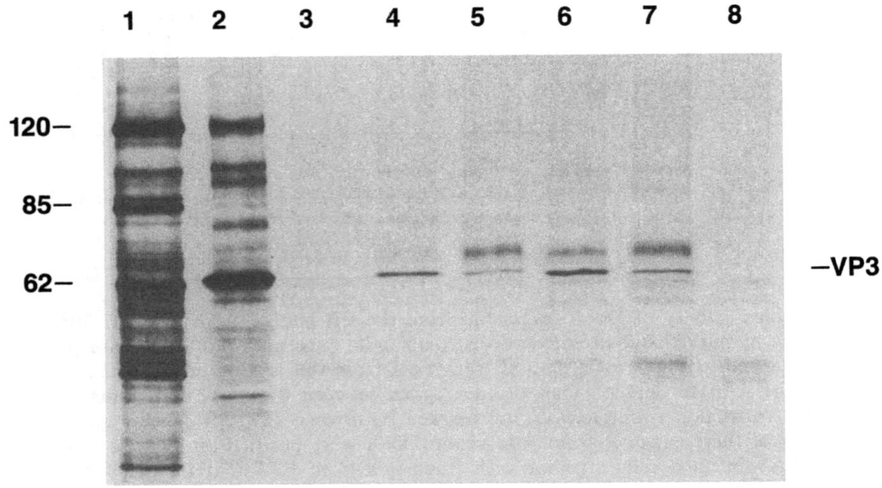

Fig. 5. AAV antigen synthesis in cells infected with AAV and injected with mRNA and DNA or coinfected with Ad2. Experimental details as for Fig. 3, except that mRNA was injected into the cytoplasm, DNA into the nucleus. *Lane 1,* Ad2 late protein markers; *lane 2,* Ad2 coinfection; *lane 3,* AAV alone. *Lanes 4–8* are all from cells injected with DNA fragments containing E2A and E4 (as in Fig. 3) and, in addition, the following mRNAs or DNA: *lane 4,* E1A and E1B mRNAs; *lane 5,* E1A mRNA; *lane 6,* E1B mRNA; *lane 7,* E1 DNA fragment (*Bam*HI-B, 0–29); *lane 8,* No E1 RNA or DNA

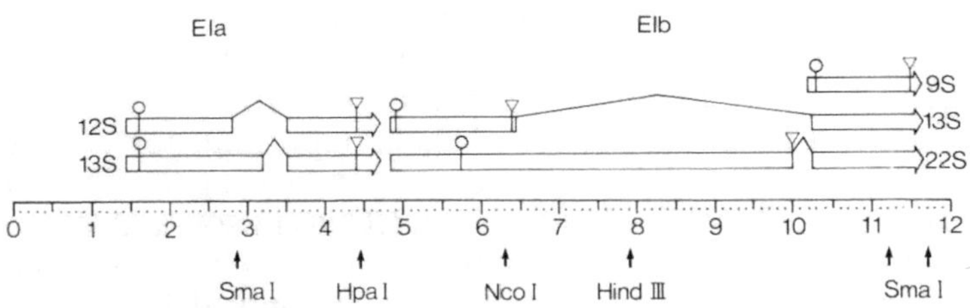

Fig. 6. Transcription map of Ad2 early region 1. The DNA sequence is marked off in conventional map units from the left end of the genome. One map unit represents 350 bp. Above this are the mRNAs found early after infection. Structural features of the mRNAs are denoted similarly in this and subsequent figures. Introns are marked as *caret* symbols linking the exons. An *arrowhead* marks the 3′ end of the molecule. *Vertical flag* markers delimit the open reading frames, a *circular flag* denoting the initiator AUG and a *triangular flag* the termination codon. E1A mRNAs share a common splice acceptor site and differ only in the size of their introns. Each E1A mRNA translates in vitro to give a different protein doublet in the range 40–55 kd (Halbert et al. 1979; Esche et al. 1980; Smart et al. 1981). Three mRNAs emanate from E1B. The unspliced message has its own promoter and specifies polypeptide IX, a virion component expressed at intermediate to late times after infection. The other mRNAs, a 13S and a 22S species, both specify the same 15-kd protein by initiation at the 5′ proximal AUG; in addition, the 22S mRNA specifies a 58-kd protein by initiation at an internal AUG (Bos et al. 1981). A minor, unmapped leftward transcript is also located within the E1 region (Katze et al. 1982). Also marked in the figure are some restriction sites mentioned in the text

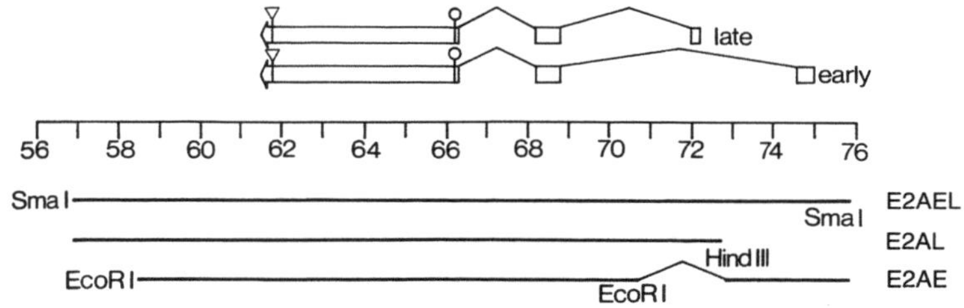

Fig. 7. Map of the E2A region of Ad2. General layout is the same as in Fig. 6. The *lower part* of the figure shows DNA fragments used for injection. The uppermost one (E2AEL) contains both early and late 1st leader segments and the middle one (E2AL) contains only the late leader. Both of these were purified from restriction digests of virion DNA. The lower one (E2AE) represents a recombinant plasmid in which the upstream segment was cloned into the *Hin*dIII site of pBR322 and the downstream segment into the *Eco*RI site of the same plasmid in the orientations shown. The two segments are therefore linked by ~30 bp of plasmid DNA, denoted by a *caret* symbol

Studies with mutants which do not produce one or other of the E1A mRNAs have shown (CARLOCK and JONES 1981; MONTELL et al. 1982) that the 1.1-kb message is responsible for the enhancement of early transcription, so correspondingly this very likely explains the role of E1A in the helper effect. In support of this conclusion we have found that sequences near the E2A mRNA start site are important for the action of E1A. Two E2A mRNA leader sequences are found during normal infection, one at map position 75 and the other at 72 (see Fig. 7). Use of these leaders is regulated during the course of infection, the one at 72 being introduced predominantly at late times (CHOW et al. 1979; GOLDENBERG et al. 1981). The leader near 72 is therefore described as "late," and the one near 75 as "early," although it is active both early and late. Figure 8 shows that E1A is able to activate the AAV helper effect when combined with E4 and an E2A gene (E2AEL) containing both early and late leaders, but not when only the late leader is present (E2AL). Both E2A DNA fragments can produce small, but detectable, amounts of 72-kd DNA-binding protein when microinjected alone at 100 copies per cell, and E2AEL, but not E2AL, is stimulated substantially when coinjected with E1A (RICHARDSON and WESTPHAL, in preparation). In contrast to E1A, E1B can complement both E2A fragments (Fig. 8).

3.4 Role of E1B

Since E1B could be substituted for E1A in some circumstances, it was possible that they both supplied similar functions, that is to activate E2A transcription. If so, it would appear from the experiment in Fig. 8 that E1B, unlike E1A, must be capable of activating the E2A late promoter. To test this further, we constructed a plasmid containing the E2A gene from which we specifically removed the late leader while leaving the early one intact

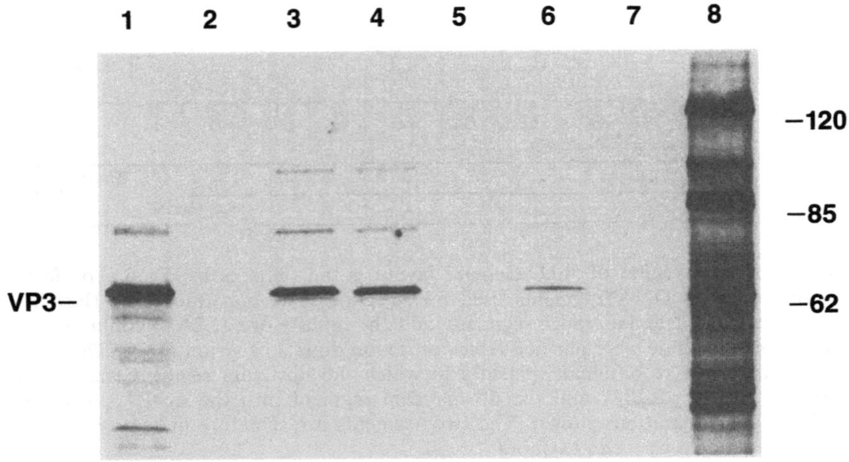

Fig. 8. AAV antigen synthesis in cells infected with AAV and injected with DNA fragments or coinfected with Ad2. Experimental details as for Fig 3. *Lane 8,* Ad2 late protein markers; *lane 1,* Ad2 coinfection; *lane 2,* AAV alone. *Lanes 3–7* are all from cells which received a DNA fragment containing E4 and, in addition, fragments from E1 and E2A as follows: *lane 3,* E1A, E1B, and E2AEL (see Fig. 7); *lane 4,* E1A and E2AEL; *lane 5,* E1A and E2AL; *lane 6,* E1B and E2AL; *lane 7,* E2AL

(see Fig. 7). This construct (with an E4 fragment) complements both E1A; and E1B for AAV helper activity (RICHARDSON and WESTPHAL, in preparation). It was still possible to explain these data by postulating that whereas E1A can activate only the early E2A promoter, E1B can activate both early and late. This idea was dispelled however, when we looked directly at production of radiolabeled 72-kd DNA-binding protein (DBP) by immunoprecipitation (RICHARDSON and WESTPHAL, in preparation). Although E1A clearly augments synthesis of the DBP, E1B has no detectable effect. This raised the possibility that E1B acts directly on E4 and caused us to reexamine some of our earlier experiments. As described earlier in this chapter, we had previously reported that injecting purified E2A mRNA followed by E4 DNA led to expression of the AAV helper effect. Our more recent experiments (RICHARDSON and WESTPHAL, in preparation) confirm this result, but in addition we now find that E1B mRNA can also complement E4 DNA. Indeed, E1A mRNA can also complement E4 DNA to a certain extent. We missed these effects previously because they are usually less marked than with E2A mRNA, and our E1 mRNA preparations are possibly more active than before. It is not easy to fit these findings into a simple pattern of regulation such as that proposed for E1A in the previous section. We discuss possible explanations in Sect. 4.

The E1B region of Ad2 gives rise to three major mRNAs and the same number of polypeptides (described in detail in Fig. 6). Which of these gene products is responsible for the effects described? Not all the available coding capacity of E1B is required for activity in the AAV assay. An extremely truncated DNA fragment, bounded by the *Hpa*I site at 4.4 and the *Nco*I site at 6.3 (see Fig. 6), is sufficient (RICHARDSON and WESTPHAL, in preparation). This fragment contains an intact promoter region and the coding information for all but 12 C-terminal amino acids of the 15-kd protein, but only a small fraction (61 of 495 amino acids) of the 58-kd protein. It seems improbable that such a severely truncated form of the 58-kd protein could retain biological activity, and therefore the 15-kd protein is very likely the important species, although we cannot exclude an additional role for the 58-kd moiety. The 15-kd protein is located in the plasma membrane during lytic infection (PERSSON et al. 1982). How it might succeed in modulating gene expression from this viewpoint is an interesting question.

Certain mutants of Ad12, which map to E1B (LAI FATT and MAK 1982), cause extensive degradation of both host and viral DNA (EZOE et al. 1981). This phenotype is recessive, being overcome by coinfection with a wild-type virus, and this led to the suggestion (EZOE et al. 1981) that an E1B product could protect DNA from degradation, perhaps by inactivating a nuclease. Such a role for E1B would have obvious repercussions on gene expression.

3.5 Role of E2A

The major product of the E2A gene is the 72-kd single-stranded DBP. This is the best characterized of the early proteins, due largely to the availability of a temperature-sensitive mutant of DBP, H5*ts*125 (ENSINGER and GINSBERG 1972). Apart from its role in DNA replication (see TAMANOI and STILLMAN, this volume, pp 75–87) DBP is involved in some aspects of mRNA metabolism, including transcription (CARTER and BLANTON 1978; NEVINS and JENSEN WINKLER 1980; HANDA et al. 1983), mRNA stability (BABICH and NEVINS 1981), and possibly translation (JAY et al. 1981; MCPHERSON et al. 1982). DBP also influences the host range of the virus (KLESSIG and GRODZICKER 1979) and has a function in virus assembly (NICOLAS et al. 1983). Which, if any, of these properties of DBP is important for the AAV helper effect is unclear. Experiments of JAY et al. (1981), using the mutant H5*ts*125, suggested that translation of AAV mRNAs was blocked at the nonpermissive temperature, but other workers found no effect of this mutation on AAV antigen production (HANDA et al. 1975; MCPHERSON et al. 1982) or DNA or RNA synthesis (STRAUSS et al. 1976). Nevertheless, there is general agreement on the requirement for a functional E2A gene in the helper effect (MYERS et al. 1980; JANIK et al. 1981; RICHARDSON and WESTPHAL 1981). Our own conclusion (RICHARDSON and WESTPHAL 1981) that E2A enhances expression of E4 seems at odds with the finding that DBP specifically represses transcription of E4 both in vivo (NEVINS and JENSEN WINKLER 1980) and in vitro (HANDA et al. 1983). This could mean that E2A influences a post-transcriptional event in a positive way, or that some minor product of E2A (RICHARDSON and WESTPHAL

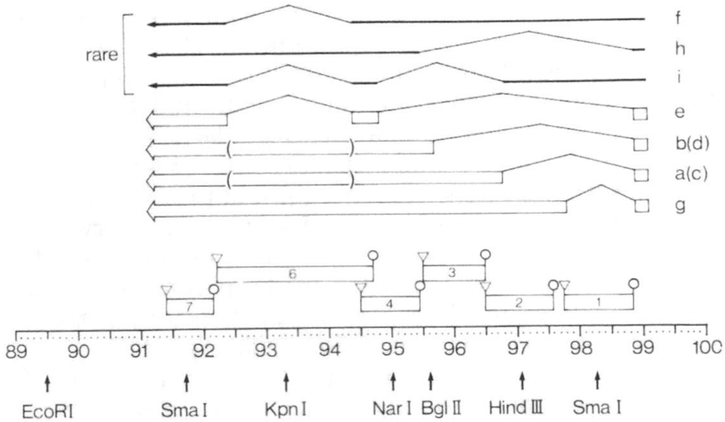

Fig. 9. Map of the E4 region of Ad2. Layout is the same as in Fig. 6, except that locations of open reading frames are shown separately from the mRNAs. The open reading regions are numbered according to HÉRISSÉ et al. (1981). The major mRNAs, a, b, c, d, e, and g, are drawn as *wide boxes*, while the remaining rare mRNAs are represented as *single lines*. mRNA c differs from mRNA a only in the removal of a second intron, marked in the figure by *parentheses*. mRNAs d and b are related in the same way. Also marked are some restriction sites mentioned in the text

1981; ASSELBERGS et al. 1983) is the active agent in our assay. Alternatively, there may be some other explanation for our results; we consider this possibility later (see Sect. 4).

3.6 Role of E4

In terms of its genetic organization, E4 is the most complex early region of adenovirus. The primary DNA sequence of the entire region has been determined (HÉRISSÉ et al. 1981; GINGERAS et al. 1982) and reveals several open reading regions capable of specifying polypeptides of greater than 10 kd. The locations of the open reading regions in E4 are shown in Fig. 9. The problem of identifying the origin of the AAV helper function is obvious. We have eliminated some of the coding regions by injecting DNA fragments progressively truncated at the 3′ end of the gene. The experiment shown in Fig. 10, for instance, shows that the terminal *Kpn*I fragment of Ad2 can supply the helper function, while the terminal *Hind*III fragment can not. We have in fact found (RICHARDSON and WESTPHAL, in preparation) that the right-most *Nar*I fragment (95.9–100) is active, but the *Bgl*II fragment (95.6–100) is not. Assuming that the only vital difference between these two fragments is loss of coding sequences between 95.0 and 95.6, this narrows the choice to protein coding regions 3 and 4 of HÉRISSÉ et al. (1981). Coding region 3 is the only one for which a product has been positively identified, this being an 11- to 14-kd polypeptide (SARNOW et al. 1982;

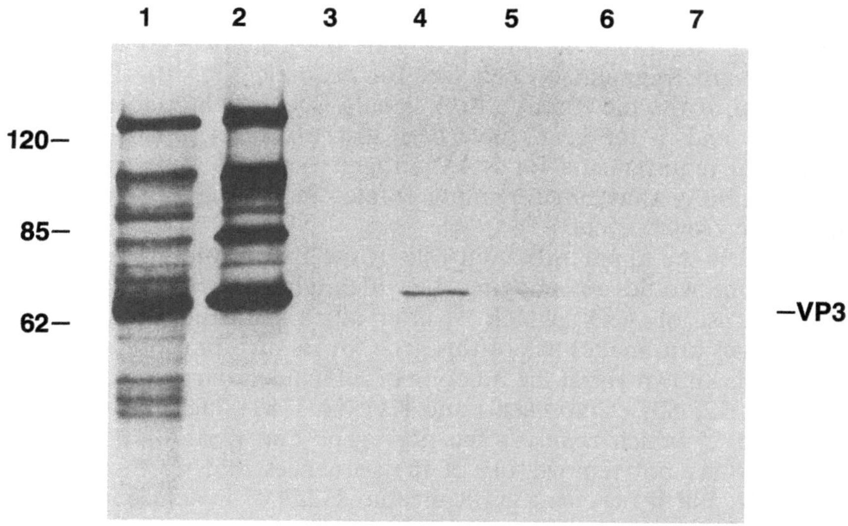

Fig. 10. AAV antigen synthesis in cells infected with AAV and injected with DNA fragments or coinfected with Ad2. Experimental protocol as in Fig. 3. *Lane 1,* Ad2 late protein markers; *lane 2,* Ad2 coinfection; *lane 3,* AAV alone. *Lanes 4–6* are all from cells which received DNA fragments from E1 (*Bam*HI-B, 0–29) and E2A (*Sma*I-A, 57–76) and in addition the following E4 fragments: *lane 4, Eco*RI-C (89.7–100); *lane 5, Kpn*I-F (93.5–100); *lane 6, Hind*III-K (97.1–100); *lane 7,* No E4 fragment

DOWNEY et al. 1983), preferentially associated with the nuclear matrix (SARNOW et al. 1982).

More mRNAs have been identified from E4 than any other early region (CHOW et al. 1979; KITCHINGMAN and WESTPHAL 1980; BERK and SHARP 1978). Figure 9 shows the structural relationships between open reading frames and mRNAs. In most cases the 5′ end of a mRNA body corresponds well with the beginning of a putative coding region (HÉRISSÉ et al. 1981). If we assume the general, but not invariant, rule that eukaryotic ribosomes initiate at the 5′ proximal AUG of an mRNA (KOZAK 1981), then mRNAs a and c will specify the product of coding region 3, while mRNAs b and d will specify the product of coding region 4 (see Fig. 9). We have attempted to distinguish between coding regions 3 and 4 by injecting subsets of E4 mRNAs selected by hybridization to various DNA fragments. RNA selected on the *Nar*I-*Sma*I fragment (95.0–98.3), which is homologous to mRNAs a, b, c, d, and g, can support AAV antigen production efficiently. However, RNA selected on a fragment (96.9–98.3) homologous to mRNAs a, c, and g appears not to help, even though it translates in vitro to give the 11-kd polypeptide from coding region 3 (RICHARDSON and WESTPHAL, in preparation). Thus, coding region 4 is probably the source of the AAV helper function. This reading frame is capable of encoding a 14-kd protein, which, however, has not yet been identified.

It should be mentioned that the level of AAV protein synthesis observed with E4 DNA fragments which truncate the gene is never as high as when a full-length fragment is used (see for example Fig. 10). This could be a reflection of the fact that control signals, such as the polyadenylation site at the 3′ end of the gene, have been lost. However, our assay detects the minimum requirements for AAV antigen synthesis, and we do not reject the possibility that other reading frames in E4 may also play a part in the AAV growth cycle.

What is the apparently vital role of the E4 product in the AAV growth cycle? One would not imagine that adenovirus carries around a gene for the sole use of AAV, which is, after all, a parasitic virus. Undoubtedly adenovirus also makes use of this gene for its own purposes. Unfortunately nothing is known about the function of E4 in adenovirus infection. A mutant of Ad2, H2*dl*807 (CHALLBERG and KETNER 1981), lacks sequences between 82.5 and 95 which removes the fiber gene and parts of E3 and E4. This deletion may not remove any of the sequences that we think are essential for AAV, but it may be significant that H2*dl*807 is deficient in production of some late proteins (other than fiber), although it can replicate its DNA apparently normally. This mutant helps AAV growth poorly, if at all (CARTER et al. 1983). The only other E4 mutants for which data have been published (SARNOW et al. 1982; DOWNEY et al. 1983) map in coding region 3 and grow as well as wild-type virus in tissue culture. It might be expected that the AAV helper function would be involved in adenovirus, as well as AAV, DNA replication. Nevertheless, E4 gene products do not seem to be required for replication of adenovirus DNA in vitro (LICHY et al. 1982). The in vitro assay measures initiation and elongation only on a double-stranded DNA template. It is conceivable that E4 is involved somehow with "filling in" the displaced single strand. Consistent with this is the fact that AAV itself has a single-stranded genome which must be converted to a double-stranded replicative form after infection (see SAMULSKI et al. 1982 for a model of AAV replication).

4 Conclusions

We believe that the effects we see using the microinjection technique accurately reflect those occurring during a regular infection. Our reasons for thinking this are severalfold. First, all the regions that we have identified as important for the AAV helper effect have also been implicated using mutants of adenovirus and DNA transfection experiments. Second, gene activation by the product of the E1A 1.1 kb mRNA, which is well established by more conventional genetics (BERK et al. 1979; JONES and SHENK 1979; NEVINS 1981; MONTELL et al. 1982), is also detected by our microinjection assays. Lastly, the critical importance of gene dose or multiplicity effects, also known from work with mutants (SHENK et al. 1979; NEVINS 1981),

Table 1. Helper activity of Ad2 DNA fragments

E1A	E1B	E2AEL	E2AE	E2AL	E4	AAV
100	.	100	.	.	100	+
.	100	100	.	.	100	+
100	.	.	100	.	100	+
.	100	.	100	.	100	+
100	.	.	.	100	100	–
.	100	.	.	100	100	+
.	.	250	.	.	250	+
.	250	.	.	.	250	–
.	.	.	.	.	500	–
.	.	500	.	.	.	–
.	500	.	.	.	.	–

The Table summarizes several experiments in which various DNA fragments were combined and tested for their ability to help AAV, as described in the text. The fragments used are listed in the column headings and their precise locations in the Ad2 genome can be found in the text. Each fragment was diluted to give the approximate gene dose written in the table, calculated assuming an injection volume of 10^{-11} ml (Graessmann et al. 1980). Fragments in each horizontal line were injected simultaneously. A + sign indicates that AAV protein synthesis was detected

is reemphasized in our experiments. Our microinjection assay may not, however, detect all the genes which have a bearing on AAV expression. For instance, Janik et al. (1981) found a requirement for RNA polymerase III transcript VA1. Also, since we inject purified genes, rather than the intact Ad2 genome, we might conceivably overlook *cis* regulatory effects of one region of the genome on another, if such effects exist.

Using the microinjection technique, we have identified four regions of the Ad2 genome, E1A, E1B, E2A, and E4, which are required under certain circumstances for expression of AAV antigens. Not all of these regions must necessarily be present simultaneously. Table 1 summarizes several experiments in which DNA fragments were injected into AAV-infected cells. We have never observed AAV expression in the absence of either E2A or E4. When injected at over 200 copies per cell, these two genes are themselves sufficient. At a dose of 100 copies per cell or less, however, a third gene becomes essential and this can be either E1A or E1B. The effect of E1A can easily be explained by its ability to enhance transcription from the other early genes, in this case E2A and E4 (see Sect. 3.3). The role of E1B is less clear. It does not simply duplicate the effect of E1A (see Sect. 3.4).

Injection of mRNA, rather than the genes themselves, may give some clues to the pattern of interactions among the Ad2 early genes and AAV. Table 2 summarizes several experiments involving injections of mRNA or combinations of mRNA and DNA. From the results tabulated, it appears that the crucial, direct-acting AAV helper function is located in E4, since injection of RNA from this region alone can activate AAV protein synthesis. Furthermore, we can overcome the requirement for E2A by injecting E1B

Table 2. Helper activity of Ad2 DNA and mRNA

E1A	E1B	E2A	E4	AAV
.	.	.	R	+
R	R	R	.	−
.	.	.	D	−
.	.	R	D	+
.	R	.	D	+
R	.	.	D	+/−
R	.	D	D	+

The Table summarizes experiments in which mRNA or combinations of mRNA and DNA were injected into the same cells and tested for AAV helper activity. DNA (*D*) was injected at 100 copies per cell into the nucleus and mRNA (*R*) into the cytoplasm. A + sign indicates that AAV protein synthesis was readily detected, and +/− that it was detected, but at a very low level

RNA with E4 DNA or, conversely, overcome the requirement for E1B by injecting E2A RNA with E4 DNA. If we assume that by injecting RNA rather than DNA we are merely bypassing the steps of mRNA biogenesis, these data imply that E1B or E2A gene products (and to a lesser degree E1A products) can activate expression of a direct-acting AAV helper function encoded in E4. However, we must be cautions about this interpretation, since it ignores effects of gene dose over which we have little control when injecting mRNA. We may therefore be supplying concentrations of gene product in excess of those prevailing in a cell infected with virus or injected with a controlled number of DNA fragments. An alternative explanation of our results should therefore be considered. E1B and E2A gene products, rather than interacting with E4, may interact directly with AAV. Their contributions to the AAV growth cycle may be such that E4 performs a central and indispensable role, while E1B and E2A augment AAV expression in ways that, although not strictly essential when adequate quantities of E4 product are around (e.g., when E4 mRNA is injected), nevertheless are critical in the presence of more physiological concentrations of E4 product (such as during infection or when DNA is injected). Such a scheme would be plausible if, for instance, net AAV DNA synthesis (and protein production) depended on competition between concomitant replication and degradation. To present just one of many possible scenarios, the E4 protein might stimulate DNA synthesis, the 72-kd DBP might bind to and protect progeny DNA molecules from nuclease attack, and the E1B product might inactivate a nuclease. In a scheme like this, small amounts of E1B and E2A products might cooperate to produce an effect similar to a larger amount of E1B or E2A alone. Until we can directly test the level of E4 proteins in microinjected cells, we cannot choose between these interpretations.

In any event, expression of the AAV helper effect is a complex process dependent on the synergistic interaction of several adenovirus early functions. Continued study of the helper effect is likely to lead to a better understanding of adenovirus early gene function, and gene regulation in general.

References

Asselbergs FAM, Smart JE, Mathews MB (1983) Analysis of expression of adenovirus DNA (fragments) by microinjection in *Xenopus* oocytes. Independent synthesis of minor early region 2 proteins. J Mol Biol 163:209–238

Babich A, Nevins JR (1981) The stability of early adenovirus mRNA is controlled by the viral 72K DNA-binding protein. Cell 26:371–379

Berget SM, Moore C, Sharp PA (1977) Spliced segments at the 5′ terminus of adenovirus 2 late mRNA. Proc Natl Acad Sci USA 74:3171–3175

Berk AJ, Sharp PA (1978) Structure of the adenovirus 2 early mRNAs. Cell 14:695–711

Berk AJ, Lee F, Harrison T, Williams J, Sharp PA (1979) Pre-early adenovirus 5 gene product regulates synthesis of early viral messenger RNAs. Cell 17:935–944

Bernards R, Houweling A, Schrier PI, Bos JL, van der Eb AJ (1982) Characterisation of cells transformed by Ad5/Ad12 hybrid early region 1 plasmids. Virology 120:422–432

Bos JL, ten Wolde-Kraamwinkel HC (1983) The E1B promoter of Ad12 in mouse L tk^- cells is activated by adenovirus region E1A. EMBO J 2:73–76

Bos JL, Polder LJ, Bernards R, Schrier PI, van den Elsen PJ, van der Eb AJ, van Ormondt H (1981) The 2.2 kb E1B mRNA of human Ad12 and Ad5 codes for 2 tumor antigens starting at different AUG triplets. Cell 27:121–131

Carlock LR, Jones NC (1981) Transformation-defective mutant of adenovirus type 5 containing a single altered E1A mRNA species. J Virol 40:657–664

Carter TH, Blanton RA (1978) Autoregulation of adenovirus type 5 early gene expression. II. Effect of temperature-sensitive early mutations on virus RNA accumulation. J Virol 28:450–456

Carter BJ, Marcus-Sekura CJ, Laughlin CA, Ketner G (1983) Properties of an adenovirus type 2 mutant, Ad2 *dl* 807, having a deletion near the right-hand genome terminus: Failure to help AAV replication. Virology 126:505–516

Challberg MD, Kelly TJ Jr (1979) Adenovirus DNA replication in vitro. Proc Natl Acad Sci USA 76:655–659

Challberg SS, Ketner G (1981) Deletion mutants of adenovirus 2: Isolation and initial characterisation of virus carrying mutations near the right end of the viral genome. Virology 114:196–209

Chow LT, Gelinas RE, Broker TR, Roberts RJ (1977) An amazing sequence arrangement at the 5′ ends of adenovirus 2 messenger RNA. Cell 12:1–8

Chow LT, Broker TR, Lewis JB (1979) Complex splicing patterns of RNAs from the early regions of adenovirus 2. J Mol Biol 134:265–303

Downey JF, Rowe DT, Bacchetti S, Graham FL, Bayley ST (1983) Mapping of a 14000-dalton antigen to early region 4 of the human adenovirus 5 genome. J Virol 45:514–523

Ensinger MJ, Ginsberg HS (1972) Selection and preliminary characterisation of temperature-sensitive mutants of type 5 adenovirus. J Virol 10:328–339

Esche H, Mathews MB, Lewis JB (1980) Proteins and messenger RNAs of the transforming region of wild-type and mutant adenoviruses. J Mol Biol 142:399–417

Ezoe H, Lai Fatt RB, Mak S (1981) Degradation of intracellular DNA in KB cells infected with *cyt* mutants of human adenovirus type 12. J Virol 40:20–27

Gingeras TR, Sciaky D, Gelinas RE, Bing-Dong J, Yen CE, Kelly MM, Bullock PA, Parsons

BL, O'Neill KE, Roberts RJ (1982) Nucleotide sequences from the adenovirus 2 genome. J Biol Chem 257:13475–13491

Ginsberg HS, Ensinger MJ, Kauffman RS, Mayer AJ, Lundholm U (1974) Cell transformation: A study of regulation with types 5 and 12 adenovirus temperature-sensitive mutants. Cold Spring Harbor Symp Quant Biol 39:419–426

Goldenberg CJ, Rosenthal R, Bhaduri S, Raskas H (1981) Coordinate regulation of two cytoplasmic RNA species transcribed from early region 2 of the adenovirus 2 genome. J Virol 38:932–939

Graessmann A, Graessmann M, Mueller C (1980) Microinjection of early SV40 DNA fragments and T antigen. Methods Enzymol 65:816–825

Graham L, van der Eb AJ (1973) Transformation of rat cells by DNA of human adenovirus 5. Virology 54:536–539

Halbert DN, Spector DJ, Raskas HJ (1979) In vitro translation products specified by the transforming region of adenovirus type 2. J Virol 31:621–629

Handa H, Shiroki K, Shimojo H (1975) Complementation of adeno-associated virus growth with temperature-sensitive mutants of human adenovirus types 12 and 5. J Gen Virol 29:239–242

Handa H, Kingston RE, Sharp PA (1983) Inhibition of adenovirus early region IV transcription in vitro by a purified viral DNA binding protein. Nature 302:545–547

Hérissé J, Rigolet M, Dupont de Dinechin S, Galibert F (1981) Nucleotide sequence of adenovirus 2 fragment encoding for the carboxylic region of the fiber protein and the entire E4 region. Nucleic Acids Res 9:4023–4042

Janik JE, Huston MM, Rose JA (1981) Locations of adenovirus genes required for the replication of adeno-associated virus. Proc Natl Acad Sci USA 78:1925–1929

Jay FT, Laughlin CA, Carter BJ (1981) Eukaryotic translational control: Adeno-associated virus protein synthesis is affected by a mutation in the adenovirus DNA-binding protein. Proc Natl Acad Sci USA 78:2927–2931

Jones N, Shenk T (1979) An adenovirus type 5 early gene function regulates expression of other early genes. Proc Natl Acad Sci USA 76:3665–3669

Katze MG, Persson H, Philipson L (1982) A novel mRNA and a low-molecular-weight polypeptide encoded in the transforming region of adenovirus DNA. EMBO J 7:783–789

Kitchingman GR, Westphal H (1980) The structure of adenovirus 2 early nuclear and cytoplasmic RNAs. J Mol Biol 137:23–48

Kitchingman GR, Lai S-P, Westphal H (1977) Loop structures in hybrids of early RNA and the separated strands of adenovirus DNA. Proc Natl Acad Sci USA 74:4392–4395

Klessig DF, Grodzicker T (1979) Mutations that allow human Ad2 and Ad5 to express late genes in monkey cells map in the viral gene encoding the 72K DNA-binding protein. Cell 17:957–966

Klessig DF, Quinlan MP, Grodzicker T (1982) Proteins containing only half of the coding information of early region 1B of adenovirus are functional in cells transformed with the herpes simplex thymidine kinase gene and adenovirus type 2 DNA. J Virol 41:423–434

Kozak M (1981) Mechanism of mRNA recognition by eukaryotic ribosomes during initiation of protein synthesis. Curr Top Microbiol Immunol 93:81–123

Lai Fatt RB, Mak S (1982) Mapping of an adenovirus function involved in the inhibition of DNA degradation. J Virol 42:969–977

Laughlin CA, Jones N, Carter BJ (1982) Effect of deletions in adenovirus early region 1 genes upon replication of adeno-associated virus. J Virol 41:868–876

Lichy JH, Field J, Horwitz MS, Hurwitz J (1982) Separation of the adenovirus terminal protein from its associated DNA polymerase: Role of both proteins in the initiation of adenovirus DNA replication. Proc Natl Acad Sci USA 79:5225–5229

McPherson RA, Ginsberg HS, Rose JA (1982) Adeno-associated virus helper activity of adenovirus DNA-binding protein. J Virol 44:666–673

Montell C, Fisher EF, Caruthers MH, Berk AJ (1982) Resolving the functions of overlapping viral genes by site-specific mutagenesis at a mRNA splice site. Nature 295:380–384

Myers MW, Laughlin CA, Jay FT, Carter BJ (1980) Adenovirus helper function for growth of adeno-associated virus: Effect of temperature-sensitive mutations in adenovirus early gene region 2. J Virol 35:65–75

Nevins JR (1981) Mechanism of activation of early viral transcription by the adenovirus E1A gene product. Cell 26:213–220

Nevins JR, Jensen Winkler J (1980) Regulation of early adenovirus transcription: A protein product of early region 2 specifically represses region 4 transcription. Proc Natl Acad Sci USA 77:1893–1897

Nicolas JC, Sarnow P, Girard M, Levine AJ (1983) Host range temperature-conditional mutants in the adenovirus DNA-binding protein are defective in the assembly of infectious virus. Virology 126:228–239

Ostrove JM, Berns KI (1980) Adenovirus early region 1B gene function required for rescue of latent adeno-associated virus. Virology 104:502–505

Persson H, Katze MG, Philipson L (1982) Purification of a native membrane-associated adenovirus tumor antigen. J Virol 42:905–917

Richardson WD, Westphal H (1981) A cascade of adenovirus early functions is required for expression of adeno-associated virus. Cell 27:133–141

Richardson WD, Carter BJ, Westphal H (1980) Vero cells injected with adenovirus type 2 mRNA produce authentic viral polypeptide patterns: Early mRNA promotes growth of adenovirus-associated virus. Proc Natl Acad Sci USA 77:931–935

Robinson AJ, Younghusband HB, Bellett AJD (1973) A circular DNA-protein complex from adenoviruses. Virology 56:54–69

Samulski RJ, Berns KI, Tan M, Muzyczka N (1982) Cloning of adeno-associated virus into pBR322: Rescue of intact virus from the recombinant plasmid in human cells. Proc Natl Acad Sci USA 79:2077–2081

Sarnow P, Hearing P, Anderson CW, Reich N, Levine AJ (1982) Identification and characterisation of an immunologically conserved adenovirus early region 4 11000 mr protein and its association with the nuclear matrix. J Mol Biol 162:565–583

Smart JE, Lewis JB, Mathews MB, Harter ML, Anderson CW (1981) Adenovirus type 2 early proteins: Assignment of the early region 1A proteins synthesised in vivo and in vitro to specific mRNAs. Virology 112:703–713

Shenk T, Jones N, Colby W, Fowlkes D (1979) Functional analysis of adenovirus type 5 host range deletion mutants defective for transformation of rat embryo cells. Cold Spring Harbor Symp Quant Biol 44:367–375

Stillman BW, Lewis JB, Chow LT, Mathews MB, Smart JE (1981) Identification of the gene and mRNA for the adenovirus terminal protein precursor. Cell 23:497–508

Straus SE, Ginsberg HS, Rose JA (1976) DNA minus temperature-sensitive mutants of adenovirus type 5 help adenovirus-associated virus replication. J Virol 17:140–148

Thimmappaya B, Weinberger C, Schneider RJ, Shenk T (1982) Adenovirus VAI RNA is required for efficient translation of viral mRNAs at late times after infection. Cell 31:543–551

Van den Elsen P, de Pater S, Houweling A, van der Veer J, van der Eb AJ (1982) The relationship between region E1A and E1B of human adenoviruses in cell transformation. Gene 18:175–185

Ward DC, Tattersall P (eds) (1978) The replication of mammalian parvoviruses. Cold Spring Harbor Laboratory, Cold Spring Harbor, New York

Weeks DL, Jones NC (1983) E1A control of gene expression is mediated by sequences 5′ to the transcriptional starts of the early viral genes. Mol Cell Biol 3:1222–1234

Westphal H, Lai S-P (1977) Quantitative electron microscopy of early adenovirus RNA. J Mol Biol 116:525–548

Wilkie NM, Ustacelebi S, Williams JF (1973) Characterisation of temperature-sensitive mutants of adenovirus type 5: Nucleic acid synthesis. Virology 51:499–503

Ziff EB (1980) Transcription and RNA processing by the DNA tumour viruses. Nature 287:491–499

Antibodies to Synthetic Peptides Targeted to the Transforming Genes of Human Adenoviruses: An Approach to Understanding Early Viral Gene Function

MAURICE GREEN, KARL H. BRACKMANN, LYNNE A. LUCHER, and JANEY S. SYMINGTON

1 Introduction

Adenoviruses (Ads) are DNA tumor viruses which contain double-stranded linear DNA genomes of molecular weight 20–25 million (GREEN et al. 1967). There are 31 well-defined human adenoviruses, many of which are ubiquitous in the human population. Adenoviruses commonly cause latent infections of lymphoid tissue and are mainly associated with respiratory disease, which can reach epidemic proportions in closed populations. By DNA homology measurements, human Ads 1–31 are classified into five groups, A through E, containing homologous transforming gene sequences (GREEN et al. 1979a; MACKEY et al. 1979a). Although probably all human Ads can transform cells and many can induce tumors in laboratory animals, there is no evidence that they play a significant role in human carcinogenesis.

Institute for Molecular Virology, St. Louis University Medical Center, 3681 Park Avenue, St. Louis, MO 63110, USA

Current Topics in Microbiology and Immunology, Vol. 109

An extensive search was made for the presence of DNA sequences representing each of the five human Ad groups in human tumors representing about 90% of the cancer incidence in the United States. Significant Ad genetic information was not detected under conditions that would have detected less than one transforming gene per tumor cell in most cases (Mackey et al. 1976; Green and Mackey 1977; Green et al 1979b; Mackey et al. 1979b; Wold et al. 1979; M. Green, unpublished data). Thus, although the human Ads are widespread and have oncogenic potential, their oncogenic potential does not appear to be manifested in their native human host species.

The human Ads, in particular group C Ad2 and closely related Ad5, possess several striking technical advantages as models to study the molecular biology of the human cell (Green 1978). Consequently, they have been developed into valuable experimental systems to study growth control, gene organization and regulation, transcription and RNA processing, DNA replication, and glycoprotein membrane assembly. The Ads have two life-styles – productive infection of permissive human cells and transformation of nonpermissive rodent cells (Green et al. 1970). The viral genome is expressed in two major stages during productive infection – an early stage, which precedes the initiation of viral DNA replication, and a late stage, which occurs subsequently. The viral genes that are expressed early during viral development are of great interest, because they encode over 20 early viral proteins that appear to function in several interesting biological processes. However, very little is known about the specific functions of the Ad early proteins. The identification, purification, and functional analysis of most of the early proteins is very difficult, mainly because they are synthesized at relatively low levels at early times after infection when host cell proteins are synthesized at high levels. Thus, there are numerous interesting questions concerning the Ad early proteins that are difficult to address with the current technology.

We wish to learn more about the Ad-coded early proteins – especially those encoded by the transforming genes contained in region E1. The Ad E1 proteins probably function in cell transformation, viral gene regulation, and possibly in viral DNA replication; therefore they are probes to understand growth control, cellular gene regulation, and cellular DNA replication. Monospecific antibodies directed against the early proteins could provide valuable tools for their further investigation. However, direct immunization of laboratory animals with purified proteins is not feasible, since only very small amounts of most Ad early proteins are present in infected and transformed cells. The synthetic peptide antibody approach provides a means of studing these proteins further (reviewed by Lerner 1982). The DNA sequence of most of the Ad2 and Ad5 genomes is now known. One can deduce the amino acid sequence of each early protein from the sequence of its genomic DNA and complementary DNA copies of viral mRNA. Peptides targeted to specific domains of each early protein molecule can be synthesized and used to generate antipeptide antibodies that recognize the native viral proteins. Antibodies raised against synthetic peptides targeted to early Ad proteins can be used as specific probes to study the genomic origin, structure, and function of each Ad early protein. Furthermore, antipeptide antibody provides a reagent to purify early proteins by antibody affinity chromatography.

In this manuscript, we first review briefly recent studies using antipeptide antibodies to investigate problems in virology. We then outline a general strategy using antipeptide antibodies to investigate the known Ad early proteins and to detect putative, as yet unidentified proteins whose existence is suggested by the presence of open translation reading frames as predicted from the Ad DNA sequence. We then describe the following studies, illustrated with data from our laboratory, applying the synthetic peptide antibody technology to study proteins that are encoded by the Ad transforming region: (a) the identification of Ad12 E1A-encoded and Ad2 E1B-encoded proteins that are synthesized in Ad-infected and -transformed mammalian cells and in *Escherichia coli* transformed by Ad cDNA-containing plasmids; (b) the direct identification of the gene boundaries and translation reading frames of the Ad2 E1B 19K and Ad2 E1B 53K proteins; (c) the tentative location of the coding region of the recently described major Ad2 E1B 20K protein that is related to the Ad2 E1B 53K protein (Green et al. 1982); and (d) the immunoprecipitation analysis of E1B proteins encoded by Ad2 cell transformation defective mutants isolated by Chinnadurai (1983) which contain lesions in the Ad2 E1B 19K protein.

2 Synthetic Peptide Antibody Technology: Applications to Virus Research

Although the possibility of using small peptides containing six to 16 amino acid residues for the generation of antibodies specific for whole protein molecules has been apparent for a number of years, the exploitation of this possibility has begun only recently. Studies since 1980 have emphasized the power of the synthetic peptide technology for investigating important problems in virology (reviewed by Lerner 1982) and have underscored the potential value of this approach for characterizing the structure and function of the numerous Ad early proteins. Below are summarized several of these studies which illustrate the following uses of antipeptide antibodies: (a) the detection of proteins predicted from the viral nucleotide sequence; (b) the study of viral protein structure and function; (c) the subcellular localization of viral-coded transforming proteins using antipeptide antibodies of predetermined specificity; (d) the identification of viral protein precursor molecules; and (e) the purification of viral proteins.

The synthetic peptide antibody technology has been especially valuable in studies on DNA and RNA tumor virus proteins that are implicated in cell transformation. The nucleic acid regions encoding the transforming genes of several tumor viruses have been cloned and sequenced. From the nucleotide sequence, the amino acid sequences of putative transforming proteins have been deduced and used to target synthetic peptides. Peptides were coupled to a carrier protein (usually bovine serum albumin or keyhole-limpet hemocyanin) and used to generate peptide-specific antibodies in rabbits. Antipeptide antibodies prepared in this manner have been used to identify by immunoprecipitation the proteins encoded by the transforming genes of SV40 (WALTER et al. 1980), polyoma virus (WALTER et al. 1981), Rous sarcoma virus (WONG and GOLDBERG 1981; SEFTON and WALTER 1982), feline sarcoma virus and Fujinami sarcoma virus (SEN et al. 1983), and the *ts*110 murine sarcoma virus (PAPKOFF and HUNTER 1983).

The structure-function relationships of viral-coded transforming proteins have been studied using antipeptide antibody. Many tumor viruses encode a tyrosine-specific protein kinase which appears to play a major role in cell transformation. Antipeptide antibody can help distinguish between enzyme activity that is an integral part of a viral protein and that which is associated with contaminating host proteins. Antibodies have been prepared against synthetic peptides, which encompass the tyrosine-acceptor amino acid residue in the viral protein kinase molecule that is self-phosphorylated by the Rous sarcoma virus $pp60^{src}$, and the polyoma middle T antigen, which also has protein kinase activity. The antipeptide antibodies were shown to inhibit the viral-associated protein kinase activity in vitro (WONG and GOLDBERG 1981; SCHAFFHAUSEN et al. 1982). Antipeptide antibody has also been shown to inhibit the tyrosine-specific protein kinase activity associated with the feline sarcoma virus and the Fujinami sarcoma virus (SEN et al. 1983).

Antipeptide antibodies have advantages over antitumor sera as reagents for immunocytochemical localization of viral proteins. For example, the specificity of cell staining by antitumor sera is difficult to assess and may reflect the interaction of the sera with transformation-induced cellular proteins, rather than with viral-coded transforming proteins. This is less of a problem with antibody against defined viral-coded peptides, especially when the antibody is purified by peptide affinity chromatography. Furthermore, the specificity of cell staining can be verified by peptide competition studies. By indirect immunofluorescence microscopy using antipeptide antibody, NIGG et al. (1982) have shown that $pp60^{src}$ is localized in transformation-induced rosette clusters at the ventral cell surface, in cell-cell contact areas, and in focal adhesion plaques. A codistribution of $pp60^{src}$ was observed with vinculin, a candidate substrate for the $pp60^{src}$ protein kinase. The domains of the SV40 large T antigen that are exposed on the cell surface were analyzed using antipeptide antibody. Evidence was provided that the C-terminal is exposed on the surface of SV40-transformed cells (DEPPERT and WALTER 1982).

The ability of antipeptide antibody to recognize the targeted native protein has suggested its use for purification of the protein by antibody affinity chromatography. The protein can be dissociated from the antibody-antigen complex by competition with peptide, conditions which are less denaturing than conventional methods that employ chaotropic agents. Using this approach, the middle T antigen of polyoma virus has been extensively purified by affinity chromatography using antipeptide antibody (WALTER et al. 1982).

Open reading frames that could encode exons of unknown viral-coded proteins have been revealed by the current application of rapid nucleic acid sequencing methods to viral genes. For example, 22 putative, as yet unidentified proteins have been identified by DNA sequence analysis of 43% of the Ad2 genome (GINGERAS et al. 1982). Antipeptide antibody targeted to these open reading frames may provide the means to identify new gene products as well as the immunological tools to further analyze their structure and function.

Precursors to the RNA genome-linked protein VPg encoded by the poliovirus genome were identified by the use of antipeptide antibodies (BARON and BALTIMORE 1982a; SEMLER et al. 1982). Antipeptide antibody targeted to the poliovirus p63 replicase protein was shown to block both replicase and polyuridylic acid polymerase activities, suggesting that both functions are the property of a single viral protein (BARON and BALTIMORE 1982b). Antipeptide antibody directed to the C-terminal of the glycosylated envelope polyprotein $Pr80^{env}$ of the Moloney leukemia virus was used to study membrane protein maturation (N. GREEN et al. 1981). Antibody targeted to peptides encoded by the influenza virus hemagglutinin (N. GREEN et al. 1982), by the surface antigen of hepatitis B virus (LERNER et al. 1981; NEURATH et al. 1982; DREESMAN et al. 1982), and by the VP1 protein of foot-and-mouth disease virus (BITTLE et al. 1982) have been shown to recognize the native viral protein. These studies suggest the exciting possibility that synthetic peptides targeted to specific domains of viral proteins may be useful as vaccines to protect against virus infection.

3 Genetic Organization of the Ad2 Genome: Potential Sites for Targeting Antipeptide Antibodies

The genetic organization of the Ad2 genome and the location of the major early and late mRNAs and early proteins are illustrated in Fig. 1. The viral genome of about 36000 base pairs in length is divided into 100 map units.

3.1 Early Viral Gene Expression

Six regions of the viral genome are transcribed during the early phase of Ad2 productive infection: early region 1A (E1A), E1B, E2 (A and B), E3, E4, and late region 1 (L1). Each early transcription unit generates a complex spectrum of spliced mRNAs from a single promoter. E1A, located at map

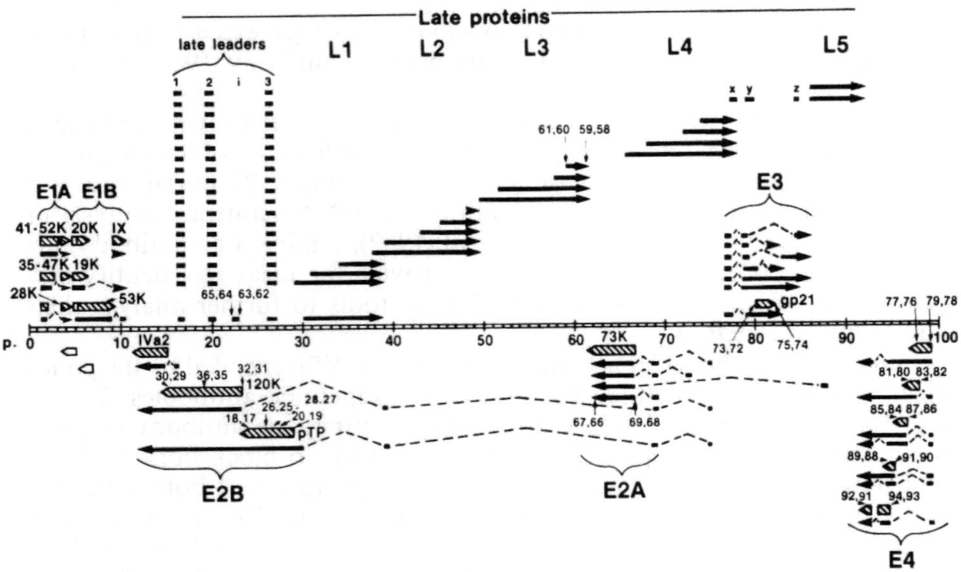

Fig. 1. Structure and gene products of the Ad2 genome and sites targeted for antipeptide antibodies. Viral mRNAs are shown as *dark arrows*. *Arrowheads* indicate 3′ ends and *dotted lines* show spliced-out introns. Early proteins are shown as *crosshatched arrows*. The locations of known early proteins are designated by the estimated molecular weight of the protein. *pTP*, the 80K terminal protein precursor. The *small numbers with arrows* pointing to viral proteins refer to synthetic peptides targeted to specific protein domains

position (mp) 1.3–4.5, and E1B, at mp 4.6–11.1, are of special interest because they encode all of the information necessary for Ad2-induced cell transformation. As many as nine early proteins have been identified in E1A and E1B. E1A is of particular interest because it appears to encode a factor or factors that regulate the expression of other Ad early genes (Jones and Shenk 1979a; Berk et al. 1979), as well as specific cellular genes (M.R. Green et al., to be published), possibly through the interaction with a cellular component (Katze et al. 1981). E1B is also of substantial interest because it encodes a protein or proteins whose functions appear to be necessary for the expression of a fully transformed cell (Houweling et al. 1980). The use of antipeptide antibodies to study the Ad-transforming gene products is discussed in a later section.

E2A and E2B use the same early promoter at mp 75 to express two sets of viral genes that function in viral DNA replication. E2A, whose coding region is located at mp 66–61, specifies a 73K DNA-binding phosphoprotein (DBP), which has been purified to homogeneity (Sugawara et al. 1977; Linne et al. 1977). DBP binds to single-stranded DNA (van der Vliet and Levine 1973) and to the terminals of double-stranded DNA (Fowlkes et al. 1979). DBP appears to function early during viral DNA chain elongation, but its precise role is not known. Antipeptide antibody targeted to different domains of DBP (e.g., peptides 66–69 in the E2A coding region shown in Fig. 1) could be useful for studies on the interaction of DBP with DNA and for the analysis of DBP function using soluble DNA replication systems that accurately initiate and elongate viral DNA molecules in vitro (reviewed by CHALLBERG and KELLY 1982). In addition, DBP has been recently shown to repress transcription in vitro from the E4 promoter (HANDA et al. 1983). Antipeptide antibody targeted to DBP domains may be useful in the further analysis of this interaction.

The large early transcription unit, E2B, encodes two proteins. The first is a primer protein, the 80K terminal protein precursor (pTP, see Fig. 1) (CHALLBERG et al. 1980, 1982), located at mp 28.9–23.5 (STILLMAN et al. 1981; GINGERAS et al. 1982; SMART and STILLMAN 1982; ALESTRÖM et al. 1982). The second is a novel DNA polymerase activity, a 140K protein (LICHY et al. 1982; STILLMAN et al. 1982; OSTROVE et al. 1983) coded by the large open reading frame (termed 120K in Fig. 1) located at mp 24.0–14.2 (GINGERAS et al. 1982; SMART and STILLMAN 1982; ALESTRÖM et al. 1982). The pTP and 140K proteins appear to function together in the initiation of Ad DNA replication. Antipeptide antibodies targeted to different domains of these proteins (see Fig. 1) would be invaluable for structure-function analyses using both in vitro DNA replication systems and reconstituted systems (NAGATA et al. 1982; OSTROVE et al. 1983).

E3, located at mp 76–86, expresses eight mRNAs using a common promoter at mp 76, as detected by electron microscopy (CHOW et al. 1979; KITCHINGMAN and WESTPHAL 1980). Six to eight RNA species have been identified in Ad2-infected and -transformed (T2C4 cell line) cells by RNA blot analysis (GREEN et al. 1981). From the DNA sequence of E3, 11 open reading frames have been identified that could encode polypeptides of 6.8K to 18.5K (HÉRISSÉ et al. 1980; HÉRISSÉ and GALIBERT 1981). However, only two E3-coded proteins are known, the E3 25K (referred to also as E3 19K) glycoprotein that is associated with the cell membrane (JENG et al. 1978; PERSSON et al. 1978, 1979; KORNFELD and WOLD 1981) and the E3 14K protein (PERSSON et al. 1978, 1979; WOLD and GREEN 1979). The functions of these proteins are unknown. The E3-coded glycoprotein is a useful model to study structure-function relationships of transmembrane glycoproteins. Antipeptide antibodies targeted to different protein domains (see Fig. 1) would be useful for such studies. Antipeptide antibodies directed against the other putative E3 proteins, as deduced from the DNA sequence of large open reading frames, would be useful for identifying and mapping the E3 proteins.

E4, located at mp 99.1 to 91.3, encodes a complex group of at least seven 5′ and 3′ coterminal mRNAs (BERK and SHARP 1978; CHOW et al. 1979; KITCHINGMAN and WESTPHAL 1980; GREEN et al. 1981; TIGGES and RASKAS 1982), which appear to be formed by differential splicing from a common nuclear precursor (CRAIG and RASKAS 1976). The expression of E4 genes is positively and negatively regulated by the gene products of E1 (BERK et al. 1979; JONES and SHENK 1979a) and E2 (NEVINS and WINKLER 1980), respectively. At least six (LEWIS et al. 1976; MATSUO et al. 1982), and as many as sixteen (TIGGES and RASKAS 1982), different E4 proteins have been observed by in vitro translation. Six of the in vitro synthesized proteins have been identified in vivo by immunoprecipitation, two-dimensional gel electrophoresis, and peptide mapping (BRACKMANN et al. 1980; MATSUO et al. 1982). The functions of the E4 proteins are unknown. The E4 11K protein has recently been shown to be associated with the nuclear matrix fraction (SARNOW et al. 1982). Antipeptide antibodies targeted to the six E4 open reading frames (HÉRISSÉ et al. 1981) (e.g., peptides 76 to 94 in Fig. 1) would be useful in identifying and characterizing the E4 proteins.

3.2 Late Viral Gene Expression

There is a shift to the late stages of Ad gene expression at the time of initiation of viral DNA replication, producing a major change in the pattern of viral gene transcription. Most late genes map in a region extending from the major late promoter at mp 16 to about mp 92 (Fig. 1). The late mRNAs that initiate at the major late promoter contain a tripartite leader segment derived from sequences at about mp 16, 19, and 26. A common, large nuclear precursor RNA molecule is processed to about 20 late mRNAs that comprise five families, L1 to L5 (Fig. 1). The mRNAs within each family contain different 5′ ends and coterminal 3′ ends. Late mRNAs and the mainly viral structural proteins they encode are produced in large quantities at late times after infection.

There are several novel sites for targeting antipeptide antibodies in genes expressed late after infection. For example, AKUSJÄRVI et al. (1981) have identified an open translational reading frame located at mp 60.6–62.7 following the hexon gene in the L3 coterminal family and have identified a late 14S mRNA which maps in this position. They have suggested that this open reading frame may code for a protease which is responsible for processing Ad proteins during virion maturation. The *ts*1 mutant, which is defective in virion maturation, maps in this region (HASSEL and WEBER 1978). In collaboration with U. PETTERSSON, we have synthesized peptides 58–61 targeted to the open reading frame located at mp 60.0–61.7 (Fig. 1). Antipeptide antibody that recognizes the putative protease would be useful for establishing an immunoassay to study virion assembly.

A second example is a putative agnoprotein that may be encoded by the Ad2 i-leader. The i-leader, located at mp 21–22.5, is spliced frequently between the second and third segments of the late mRNA tripartite leader. By DNA sequence analysis, VIRTANEN et al. (1982) have found that the i-leader contains an open translation reading frame that could code a protein of 16K. A protein of the predicted size was detected by in vitro translation of mRNA selected by hybridization to DNA containing i-leader sequences. By analogy with the SV40 late leader, which encodes a so-called SV40 agnoprotein (JAY et al. 1981), it is possible that an Ad agnoprotein exists.

In collaboration with U. PETTERSSON, we have synthesized peptides directed to the i-leader open reading frame (peptides 62–65 in Fig. 1). Antipeptide antibody directed against this protein would facilitate the identification and functional analysis of the putative Ad agnoprotein.

4 Adenovirus-Transforming Genes: Organization, Gene Products, and Functional Studies

The great promise of tumor virology is that understanding the functions of the viral-coded transforming gene products will provide insights into cellular growth control and malignancy. The Ad transforming region is located in early region E1 within the left 11–12% of the viral genome. This was demonstrated by transformation of cells with restriction fragments encoding the E1 region of group C Ad2 and Ad5 (GRAHAM et al. 1974; VAN DER EB et al. 1977), group A Ad12 (SHIROKI et al. 1977), and group B Ad7 (SEKIKAWA et al. 1978; DIJKEMA et al. 1979), and by analysis of the viral RNA sequences that are expressed in Ad-transformed cell lines (FLINT et al. 1976; FLINT 1977; YOSHIDA et al. 1979; YOSHIDA and FUJINAGA 1980; SAWADA and FUJINAGA 1980; GREEN et al. 1981).

Much is known about the Ad2 and Ad5 E1 transforming regions: the DNA encoding the transforming genes has been sequenced (VAN ORMONDT et al. 1978; BOS et al. 1981; GINGERAS et al. 1982) and the major E1-coded mRNAs and proteins have been identified. The discussion below will be limited mainly to the Ad2 and Ad5 transforming genes and gene products. Comparative analyses of Ad7 and Ad12 transforming genes (SUGISAKI et al. 1980; KIMURA et al. 1981; DIJKEMA et al. 1982) and gene products have shown that the transforming regions of groups A, B, and C possess strikingly similar molecular organizations. Unfortunately, not much is known about the mechanism of Ad-induced cell transformation. The central questions are which E1-coded proteins, if any, are required to initiate and/or to maintain cell transformation, and what are their functions. Studies relevant to these questions are briefly discussed below.

Virus mutants with lesions in E1 that are defective in cell transformation have been isolated (HARRISON et al. 1977; GRAHAM et al. 1978; JONES and SHENK 1979b; SOLNICK and ANDERSON 1982; HO et al. 1982). Two complementation groups of Ad5 host range mutants have been defined, one group (*hr*I) mapping in E1A and the second group (*hr*II) mapping in E1B (HARRISON et al. 1977; FROST and WILLIAMS 1978; GALOS et al. 1980; HO et al. 1982). Since both mutant groups are defective in transformation, two E1 genes appear to be involved (GRAHAM et al. 1978). One possibility is that E1A products are required to activate the E1B genes, which are the "true transforming" genes. However, this possibility appears to be ruled out by the finding that an Ad5 host range mutant (*hr*440) that is defective in the expression of E1A but is normal in the transcription of E1B does not transform baby rat kidney cells (SOLNICK and ANDERSON 1982). These results indicate that the expression of E1B alone is not sufficient for cell transformation and support the view that E1A expression may serve a specific function in the transformation process. More recently, cold-sensitive host range mutants of Ad5 were isolated (HO et al. 1982). Temperature shift experiments provided evidence that an E1A function is required for maintenance of transformation and that an E1B function may be involved in both initiation and maintenance of transformation. Most of the deletion mutants of JONES and SHENK (1979b) are defective in transformation. Since the deletions are quite large, correlating the deletion with specific proteins is not straightforward. Mutant *dl*311, which has a small deletion at the *Xba*I site at mp 3.85, is able to transform cells. The sequences at this *Xba*I site most likely are expressed in all E1A proteins; apparently, this domain is nonessential for transformation.

VAN DER EB and colleagues have attempted to define the transforming region by transforming cells with Ad5 restriction fragments that contain only portions of E1 and then analyzing the properties of the transformed cells. They found that cells transformed with Ad5 *Hsu*I-G (mp 0–8) (VAN DER EB et al. 1977) have essentially the same transformed phenotype as cells transformed by virions or by larger fragments. They also showed that cells could be transformed by *Hpa*I-E (mp 0–4.5), although these cells were only semitransformed, i.e., the only transformed cell properties that they possess are immortality and aneuploidy (HOUWELING et al. 1980). Incomplete transformation was also reported using Ad7 (DIJKEMA et al. 1979) and Ad12 (SHIROKI et al. 1979) restriction fragments encoding only E1A. These results suggest that although E1A proteins may confer immortality and aneuploidy, E1B proteins are responsible for inducing the other transformed cell properties (e.g., altered morphology, lack of contact inhibition, rapid growth rate, tumorigenicity). E1B 19K, but not E1B 53K, was synthesized by *Hsu*I-G-transformed cells, which suggests that E1B 19K is important in transformation (SCHRIER et al. 1979). Four well-characterized Ad2 transformed cell lines – F17, 8617, T2C4, and F4 – were all shown to contain E1A and E1B in an intact state and to synthesize the usual E1A- and E1B-specific mRNAs (GREEN et al. 1981). All of these cell lines synthesized E1B 19K (MATSUO et al. 1982), arguing for the importance of E1B 19K in cell transformation. Direct evidence for a major role for E1B 19K in cell transformation was provided by the isolation of a series of transformation-defective large-plaque (*lp*) mutants by CHINNADURAI (1983) that map by marker rescue experiments in E1B. The mutational defects in two of these mutants, *lp*3 and *lp*5, were shown to be in the N-terminal domain of E1B 19K (CHINNADURAI, 1983).

Our knowledge of the possible functions of Ad E1A and E1B proteins is very limited. There are several studies that imply a complex series of virus-coded controls which operate on the early genes of Ad2. Different early proteins begin to be synthesized and reach maximal levels at different times after infection (HARTER et al. 1976; PERSSON et al. 1978; ROSS et al. 1980). In addition, early genes are activated sequentially, possibly at the transcription level, through the specific activation of early promoters (NEVINS et al. 1979). Host range and deletion mutants in the E1A region interfere with the expression of other early regions (BERK et al. 1979; JONES and SHENK 1979a), suggesting that an Ad early gene function specified by E1A regulates the expression of other early genes. Protein inhibitor studies by different laboratories have suggested that virus-coded products encoded in E1A may control early viral gene expression, but the level of control appears to be controversial. In one laboratory, evidence was obtained for control by an E1A and/or E1B product on the accumulation and the in vivo translation of Ad early mRNAs (PERSSON et al. 1981; KATZE et al. 1981, 1983), leading to the proposal that an E1A or E1B gene product regulates the accumulation of viral mRNA by inactivating a cellular factor responsible for the turnover of mRNA. NEVINS (1981) found evidence for the control of early gene function by an E1A product at the level of transcription and proposed that the role of the E1A product was to inactivate a cellular repressor of early viral promoters.

Late gene expression is tied, in an obligatory manner, to early gene expression either indirectly, as a result of viral DNA replication, or directly, through the action of one of the early proteins. Studies by THOMAS and MATHEWS (1980) have provided evidence that the early-to-late switch in Ad gene expression requires the replication of the viral DNA template and that the accumulation of early gene products by itself is insufficient. During the replication of SV40 and T4 bacteriophage, both template replication *and* an early protein appear to be necessary for late gene expression. Thus it is conceivable that an E1 protein is required for the early-to-late switch.

It is possible that some of the Ad early proteins autoregulate their own synthesis. For example, the two E1A 35–47K and 41–52K protein genes are not amplified during late stages of infection, whereas the E1A 28K protein gene is amplified late; all three genes most likely utilize the same major promoter. Similarly, the E1B 53K gene is not amplified late, whereas the E1B 19K gene apparently is amplified; the amplification, or lack of it, could be mediated through the action of early proteins, perhaps E1 proteins, with the transcription apparatus of these early genes.

It is clear that an understanding of Ad-induced cell transformation will require a detailed knowledge of the functions of the E1 proteins. Ad E1 proteins provide a unique opportunity to study eukaryotic gene regulation. An important approach to establishing the functions of E1 genes involves biological and biochemical studies with purified E1-coded proteins. Antibody to peptides encoded by the individual E1-coded proteins would provide the means to purify E1 proteins by antibody affinity chromatography. In addition, antipeptide antibodies would provide specific probes to study E1 proteins. Below are discussed the results of studies on the development of antipeptide antibodies targeted to E1 proteins.

5 Molecular Characterization of Proteins Encoded by Adenovirus-Transforming Genes Using Antipeptide Antibodies

5.1 Organization of Ad2 E1 Region

Figure 2 illustrates our current knowledge of the organization and expression of the Ad2 E1 region. Cell-free translation studies have detected five proteins that are coded by E1A: two proteins of 41–52K translated from 13S mRNA, two proteins of 35–47K from 12S mRNA, and a protein of 28K from 9S mRNA. Three proteins are coded by E1B: a protein of 53K translated from 22S mRNA, a protein of 19K from both 22S and 13S mRNA, and a protein of 20K (LEWIS et al. 1976; HARTER and LEWIS 1978; HALBERT et al. 1979; ESCHE et al. 1980; BOS et al. 1981; GREEN et al. 1982; MATSUO et al. 1982). The E1B 20K polypeptide, which shares sequences with E1B 53K (GREEN et al. 1982), must be translated in the same reading frame as E1B 53K, but its mRNA must have a spliced structure that allows it to initiate or terminate differently from 53K. If the E1A and E1B proteins are required to maintain cell transformation, then transformed cells should retain the genes for these proteins in an intact state and should express their genes as mRNA and protein. The E1A (mp 1.3–4.5) and E1B (mp 4.6–11.1) sequences of four Ad2-transformed cell lines – F17, 8617, F4, and T2C4 – are present in an intact (uninterrupted) state and are expressed at the RNA (FLINT 1977; FLINT et al. 1976; SAMBROOK et al. 1979; GREEN et al. 1981) and protein (MATSUO et al. 1982; data presented in this paper) levels.

Using Ad tumor sera, Ad2- or Ad5-specific proteins have been immunoprecipitated from Ad2- or Ad5-infected cells (JOHANSSON et al. 1978; LEVINSON and LEVINE 1977; GILEAD et al. 1976; GREEN et al. 1979c; WOLD and GREEN 1979) and from Ad2- or Ad5-transformed cells (SCHRIER et al. 1979; ROSS et al. 1980; MATSUO et al. 1982). These proteins are of sizes similar to those produced by cell-free translation. The identity of in vitro translated and in vivo immunoprecipitated E1 proteins was demonstrated by peptide map analysis (MATSUO et al. 1982; GREEN et al. 1982). The E1B 19K protein was unrelated by peptide mapping to the E1B 53K and E1B 20K proteins (BRACKMANN et al. 1980; GREEN et al. 1982; MATSUO et al. 1982). Polypeptides of 28K and 14K–15K (three species), which are unrelated to each other or to any other E1 proteins, can also be immunoprecipitated using antisera against F17 cells; it is not known whether these polypeptides are coded by E1 (GREEN et al. 1979c).

In order to study the Ad2 E1A and E1B proteins further, our laboratory has synthesized peptides targeted to various domains of the Ad2 E1A and E1B proteins. These are illustrated by the numbers above the protein coding regions in Fig. 2. Of additional interest, the E1A 9S mRNA encoding the E1A 28K protein is expressed only late after infection, although it apparently uses the same promoter as the E1A 13S and 12S mRNAs, which are synthe-

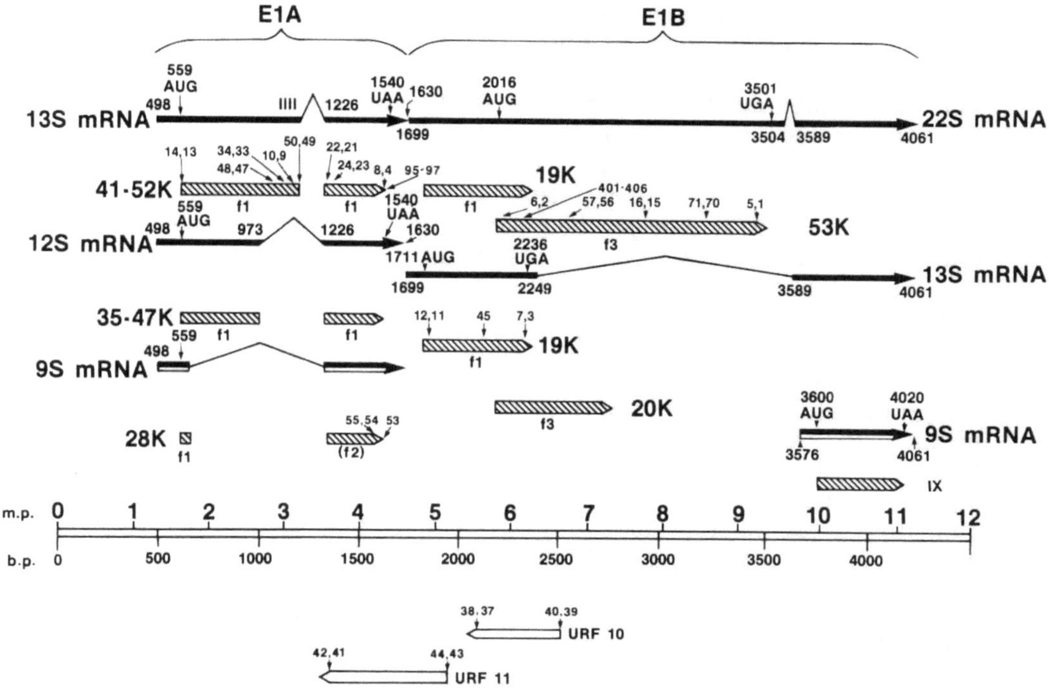

Fig. 2. Structure and gene products of the Ad2 transforming region, E1A and E1B, and sites targeted for antipeptide antibodies. Viral mRNAs are shown as *dark arrows* with designated S values. *Arrowheads* indicate 3′ ends and *dotted lines* show spliced-out introns. E1 proteins are shown as *crosshatched arrows* with molecular weights. *Nucleotide numbers* refer to transcription initiation and termination sites, splice donor and acceptor sites, and translation initiation and termination sites. *f1, f2, f3,* translation reading frames; *URF10, URF11* open reading frames that code for putative E1-coded proteins (GINGERAS et al. 1982). *Smaller numbers with arrows* pointing to E1 proteins refer to synthetic peptides targeted to specific domains of the protein

sized as early as 1 h postinfection. The E1A 13S, 12S, and 9S mRNAs were thought to be translated completely in reading frame 1. Recent studies (A. VIRTANEN, U. PETTERSSON, personal communication) have provided evidence for a frameshift generated by a different splice for the 9S mRNA, which would generate a different carboxyl portion for the late E1A 28K protein. This may explain some of the unusual properties of the 9S gene. In collaboration with U. PETTERSSON, we are synthesizing peptides (Fig. 2, peptides 53, 54, and 55) encoded in reading frame 2 in order to confirm this possibility.

There are two unidentified open reading frames, URF10 and URF11, that are encoded in the Ad2 E1 region on the l-strand (Fig. 2). We are preparing antibodies to peptides 37–42 directed to sequences in these open reading frames (Fig. 2) to provide probes for the detection and analysis of these putative gene products.

5.2 Studies Using Antipeptide Antibodies Targeted to the Ad2 E1B 53K, 19K, and 20K Proteins

In order to study the Ad2 E1B proteins further, our laboratory has synthesized six peptides of eight to 16 amino acid residues (see Fig. 3), targeted to the N- and C-terminals of E1B 19K or E1B 53K, based on the amino acid sequence of these proteins deduced from the Ad2 DNA sequence (GINGERAS et al. 1982). The peptides were coupled to keyhole-limpet hemocyanin or bovine serum albumin, either through the free N-terminal by reaction with glutaraldehyde (KAGAN and GLICK 1979) or through the sulfhydryl group of the cysteine residue placed at the C-terminal of each peptide (N. GREEN et al. 1982). The coupled peptide-protein preparations were used to generate antibodies in rabbits. Our initial objectives were to use the antipeptide antibodies to establish by direct analyses the coding boundaries of the E1B 19K and E1B 53K genes, the reading frames of E1B 19K and E1B 53K, and the location of the E1B 20K gene. Previous studies have shown that E1B 19K does not share amino acid sequences with E1B 53K and, thus, must be translated in a different reading frame (BRACKMANN et al. 1980; GREEN et al. 1982; MATSUO et al. 1982) and that E1B 20K shares most of its sequences with E1B 53K, but is unrelated to E1B 19K. Our further objectives are to use monospecific antibodies to purify and functionally analyze the Ad2 E1B proteins.

The results of our studies, which are described below, are consistent with the genomic location and expression of the E1B T antigens predicted by the Ad2 DNA sequence, as described by GINGERAS et al. (1982). Transcription of both the E1B 22S and 13S mRNAs is initiated at nucleotide 1699 (see Fig. 3). There is an open reading frame (frame 1) from the first ATG at nucleotide 1711 to a TGA at nucleotide 2236 in both the 22S and 13S mRNAs which could encode a protein of 20.5K. The 15K protein translated from E1B selected mRNA in vitro by LEWIS et al. (1976), which is probably the same as the E1B 19K protein described in our studies (GREEN et al. 1982), is known to initiate translation in vitro at nucleotide 1711 (ANDERSON and LEWIS 1980) and therefore is probably coded by this open reading frame. The E1B 19K T antigen should terminate at nucleotide 2235, before the splice donor site of its mRNA. We have synthesized and prepared antibodies against Peptide 12 (15 amino acids), coded close to the N-terminal of the 19K gene at nucleotide 1729, and peptide 7 (16 amino acids), coded at the C-terminal of the 19K gene at nucleotide 2235 (Fig. 3).

There is a second open reading frame (frame 3) between the second ATG of the E1B-22S mRNA at nucleotide 2016 and a TGA at nucleotide 3501 (see Fig. 3) which could encode a 54.5K protein. The large E1B T antigen (E1B-53K) could be coded by this open reading frame. Alternatively, translation of the large E1B T antigen could initiate in reading frame 3 at the third ATG at nucleotide 2202, or the fourth ATG at nucleotide 2235, which would yield a protein of about M_r 49000. As described below, studies with anti-peptide antibodies prove that translation of the E1B 53K T antigen initiates at the second ATG at nucleotide 2016. Transla-

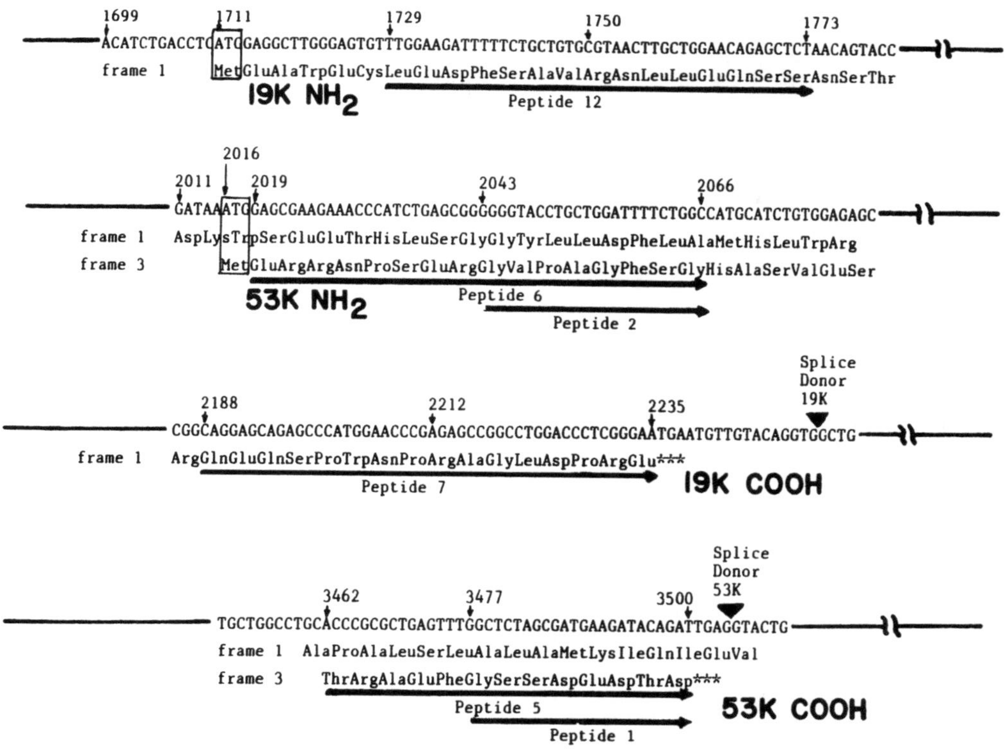

Fig. 3. Genomic location, DNA sequence, and amino acid sequence of the coding regions of the Ad2 E1B 19K and E1B 53K T antigens to which synthetic peptides are directed. The DNA sequence and amino acid sequence are taken from GINGERAS et al. (1982). Peptide 12 contains a 15 amino acid sequence close to the N-terminal and peptide 7 contains the 16 amino acid sequence at the C-terminal of the E1B 19K T antigen. Peptide 6 contains the 16 amino acid sequence at the N-terminal of the E1B 53K T antigen; Peptide 2 contains the 8 amino acid sequence at the C-terminal of Peptide 6. Peptide 5 contains the 13 amino acid sequence at the C-terminal of the E1B 53K T antigen; Peptide 1 contains the 8 amino acid sequence at the C-terminal of Peptide 5. See text for discussion of viral gene location and regulatory signals

tion of the E1B 53K T antigen should terminate at nucleotide 3501 before the splice donor site of the 22S mRNA. The N-terminal of the E1B 53K in frame 3 overlaps the C-terminal of the E1B 19K in frame 1 by about 216 nucleotides, or about 70 amino acids. The E1B 19K and E1B 53K T antigens should therefore not share amino acid sequences if they are translated in different reading frames; in accord with this prediction, the peptide maps of E1B 19K and E1B 53K indicate that they do not share sequences (Bos et al. 1981; BRACKMANN et al. 1980; GREEN et al. 1982). We have synthesized two peptides encoded at the N-terminal of the E1B 53K at nucleotide 2019, peptide 2 (eight amino acids) and peptide 6 (16 amino acids) (Fig. 3). We have also synthesized two peptides encoded at the C-terminal of the E1B 53K gene at nucleotide 3500, peptide 1 (eight amino acids) and peptide 5 (13 amino acids).

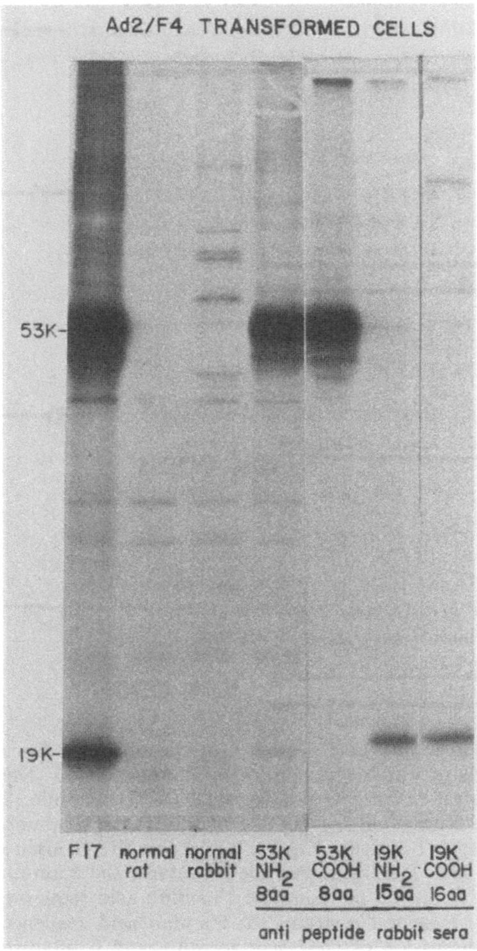

Fig. 4. Immunoprecipitation of Ad2 E1B 53K and E1B 19K proteins from extracts of [^{35}S]methionine-labeled F4 cells by antipeptide antibody targeted against the N-terminal (peptide 2) and C-terminal (peptide 1) of E1B 53K and the N-terminal (peptide 12) and C-terminal (peptide 7) of E1B 19K. (Data from GREEN, BRACKMANN, LUCHER, SYMINGTON, and KRAMER (to be published)

Figure 4 illustrates the efficacy and specificity of antipeptide antibody in the immunoprecipitation of the Ad2 E1B 53K and E1B 19K proteins from the Ad2-transformed cell line F4 (GREEN, BRACKMANN, LUCHER, SYMINGTON, and KRAMER, to be published). F4 cells contain multiple copies of the right 5% of the viral genome fused to the left 68% of the genome. F4 cells synthesize 24S and 19S hybrid mRNAs that contain both E4 and E1A sequences, as well as the normal E1A and E1B mRNAs (SAMBROOK et al. 1979; GREEN et al. 1981). Antibodies directed against peptide 2 (Gly-Val-Pro-Ala-Gly-Phe-Ser-Gly) and peptide 1 (Gly-Ser-Ser-Asp-Glu-Asp-Thr-Asp), encoded at the N- and C-terminal respectively of the E1B 53K protein (Fig. 3), immunoprecipitated specifically the E1B 53K protein, as shown in Fig. 4. The identity of this protein with the Ad2 E1B 53K was confirmed by peptide mapping the bands of immunoprecipitated protein.

Antibodies directed against peptide 12 (Leu-Glu-Asp-Phe-Ser-Ala-Val-Arg-Asn-Leu-Leu-Glu-Gln-Ser-Ser) and peptide 7 (Gln-Glu-Gln-Ser-Pro-Trp-Asn-Pro-Arg-Ala-Gly-Leu-Asp-Pro-Arg-Glu), encoded at the N- and C-terminal respectively of the E1B 19K protein (Fig. 3), immunoprecipitated a 19K protein (Fig. 4), which was shown to be authentic E1B 19K by peptide map analysis. Of interest, both hydrophilic and hydrophobic peptides were effective in inducing antibodies that recognized the native Ad2 E1B proteins. The homologous peptides, at a concentration of 0.5 µg/ml, completely blocked the immunoprecipitation of the E1B 19K or E1B 53K, thus establishing the specificity of the antipeptide antibody reaction. These results provide direct evidence at the protein level that the E1B 19K and E1B 53K proteins overlap in sequence and are translated in frames 1 and 3 respectively, as shown in Fig. 3, thus confirming the results of peptide mapping (BRACKMANN et al. 1980) and DNA sequence analysis (GINGERAS et al. 1982). Similar conclusions were reported for the structural organization of the E1B region of Ad5 and Ad12, based on DNA sequence analyses, S1 nuclease mapping, and peptide mapping (BOS et al. 1981).

Data relevant to the genomic location of the E1B 20K protein (GREEN et al. 1982) was obtained using antibody to peptide 6, the 16-amino-acid peptide at the N-terminal of the E1B 53K protein (see Fig. 3). Antipeptide antibody immunoprecipitated the E1B 20K protein, in addition to the E1B 53K protein (LUCHER, BRACKMANN, SYMINGTON, and GREEN, to be published). This is illustrated in Fig. 5, which concerns the analysis of the Ad2 *lp* mutants described by CHINNADURAI (1983), as discussed below. Peptide mapping studies proved that the immunoprecipitated 20K band is identical to the E1B 20K described previously by GREEN et al. (1982). These results provide preliminary evidence that the E1B 20K protein initiates close to the N-terminal of the 53K protein, probably utilizing the second ATG at nucleotide 2016 in frame 3, as does the 53K protein (Fig. 3).

5.3 Use of Antipeptide Antibodies to Analyze the Ad2 Large Plaque Mutants that Are Defective in Cell Transformation

CHINNADURAI (1983) has isolated and characterized a series of Ad2 *lp* mutants that produce large, clear plaques on human KB cells and are defective in inducing transformation of the established rat embryo cell line 3Y1. He has mapped two of the mutants, *lp*3 and *lp*5, within the E1B transforming gene region. In collaboration with G. CHINNADURAI, we have used antipeptide antibody preparations targeted to the E1B 53K and E1B 19K proteins to identify the proteins synthesized in KB cells infected with the *lp* mutants.

As shown in Fig. 5, antipeptide antibody directed against the 16-amino-acid peptide encoded at the E1B 53K N-terminal immunoprecipitated both the E1B 53K and the E1B 20K protein from wild-type Ad2-infected cells. Antipeptide antibody directed against the E1B 53K C-terminal peptide im-

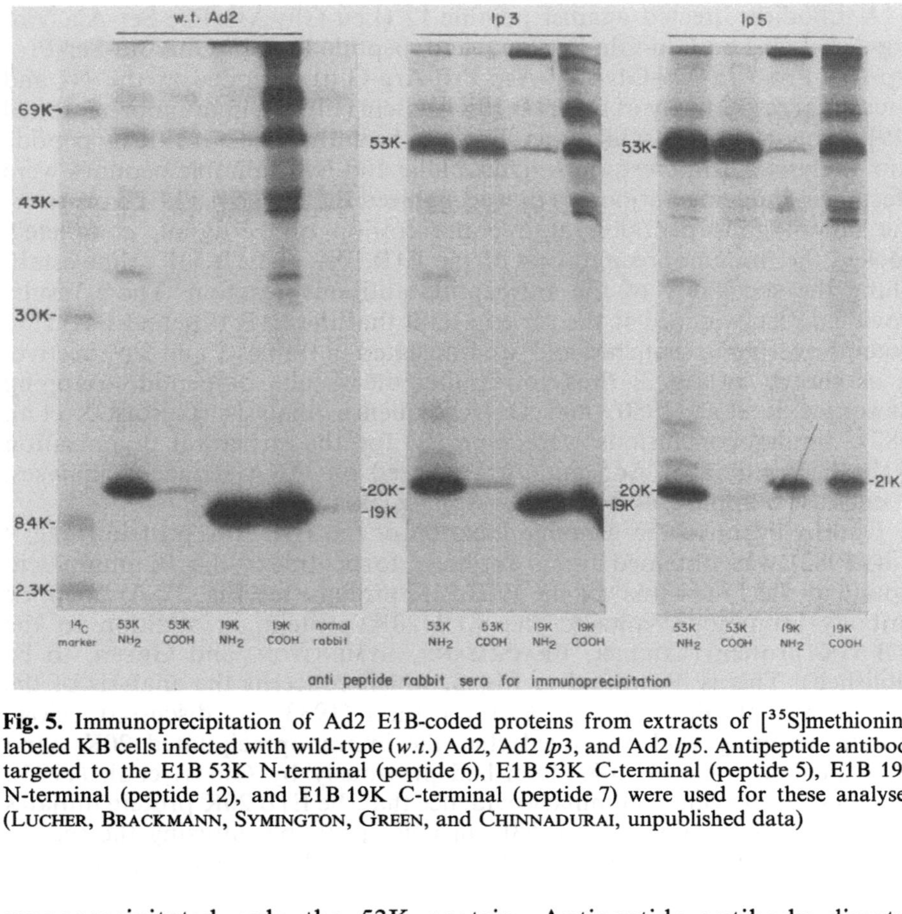

Fig. 5. Immunoprecipitation of Ad2 E1B-coded proteins from extracts of [^{35}S]methionine-labeled KB cells infected with wild-type (*w.t.*) Ad2, Ad2 *lp*3, and Ad2 *lp*5. Antipeptide antibody targeted to the E1B 53K N-terminal (peptide 6), E1B 53K C-terminal (peptide 5), E1B 19K N-terminal (peptide 12), and E1B 19K C-terminal (peptide 7) were used for these analyses. (LUCHER, BRACKMANN, SYMINGTON, GREEN, and CHINNADURAI, unpublished data)

munoprecipitated only the 53K protein. Antipeptide antibody directed against the E1B 19K C- and N-terminals immunoprecipitated the expected E1B 19K protein. Cells infected with *lp*3 produced normal amounts of E1B 53K, E1B 20K, and E1B 19K. In contrast, *lp*5-infected cells did *not* synthesize the E1B 19K, but instead synthesized a new protein of 21K.

The antipeptide antibody data shown in Fig. 5 confirm the results of DNA sequence analysis, immunoprecipitation, and cell-free translation reported recently by CHINNADURAI (1983). Mutant *lp*3 has a single base pair change at the N-terminal of the E1B 19K protein, resulting in the substitution of valine for alanine at nucleotide 1717 (CHINNADURAI, 1983). Antipeptide antibody analysis did *not* detect any difference in the size of the E1B 19K protein in *lp*3 and wild-type infected cells, consistent with the results of DNA sequence analysis. There are two mutational changes in *lp*5. One results in the substitution of tyrosine for aspartic acid near the N-terminal region at nucleotide 1954, the other converts the termination codon at nucleotide 2236 to a leucine residue, resulting in an increased size of the 19K protein when translated either on the 22S or 13S mRNA (CHINNADURAI,

1983). As shown by analysis with antipeptide antibody in Fig. 5, a 21K protein is synthesized instead of a 19K protein, consistent with DNA sequence analysis.

The findings of CHINNADURAI (to be published) provide direct genetic evidence for an essential role for the E1B 19K protein in Ad2-induced cell transformation. In addition, these data indicate that the N-terminal portion of the 19K transforming protein is an essential functional domain for the induction of cell transformation.

5.4 Use of Antipeptide Antibodies to Identify Ad2 and Ad12 E1A-Encoded Proteins in Infected and Transformed Mammalian Cells and in *E. coli* Cells Transformed by E1A cDNA-Containing Plasmids

Very recently, several laboratories have successfully prepared antipeptide antibodies that recognize the Ad E1A proteins. FELDMAN and NEVINS (1983) prepared antibody against the 13-amino-acid peptide, His-Tyr-His-Arg-Arg-Asn-Thr-Gly-Asp-Pro-Asp-Ile-Met, which is specific for the unique region of Ad5 E1A 13S mRNA. The antipeptide antibody failed to immunoprecipitate the protein product of the Ad5 E1A 13S mRNA, presumably because of the low levels of these proteins in infected cells. However, a major 46K protein and a minor 44K protein were detected by protein blot analysis (FELDMAN and NEVINS 1983). Cell fractionation studies and indirect immunofluorescence microscopy using antipeptide antibody suggested that the proteins encoded by the Ad5 E1A 13S mRNA are present in both cytoplasm and nucleus, the nuclear form being associated primarily with the nuclear matrix fraction, i.e., a nuclear fraction that is resistant to solubilization by detergent, DNase, and high salt concentration. This is of interest because Ad E1A proteins appear to act as positive regulators of gene expression, and the nuclear matrix fraction has been associated with transcriptionally active genes. However, further studies are needed to provide direct evidence that the E1A proteins function as transcriptional regulatory proteins.

In contrast to the results with antipeptide antibodies directed to the unique region of the Ad5 E1A 13S mRNA (FELDMAN and NEVINS 1983), antipeptide antibodies directed against the common C-terminal peptide of Ad5 E1A 13S and E1A 12S mRNA, Tyr-Gly-Lys-Arg-Pro-Arg-Pro, immunoprecipitated four major and two minor proteins from early infected cells. The proteins had apparent values of 52K, 50K, 48.5K, 45K, 37.5K, and 35K (P. BRANTON, personal communication). Three species were coded by the E1A 13S mRNA, and three by the E1A 12S mRNA, as indicated by immunoprecipitation analysis of cells infected by several Ad5 mutants (P. BRANTON, personal communication).

Antipeptide antibody has been prepared against the synthetic peptide Arg-Glu-Gln-Thr-Val-Pro-Val-Asp-Leu-Ser-Val-Lys-Arg-Pro-Arg, encoded at the common C-terminal of the Ad12 E1A 13S and 12S mRNAs. The

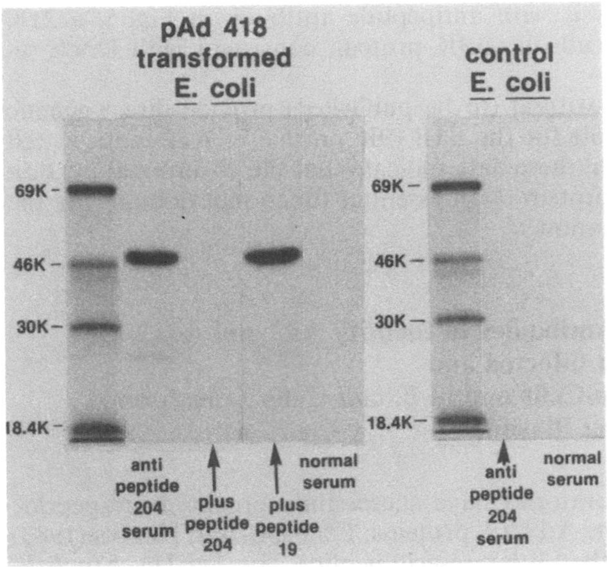

Fig. 6. Immunoprecipitation of an Ad12 E1A-specific protein of 47–48K from extracts of [^{35}S]methionine-labeled *E. coli* transformed by pAd418 (an expression plasmid containing a cDNA copy of the Ad12 E1A 13S mRNA) and induced with isopropyl thiogalactoside. Antibody against a synthetic peptide encoded at the common C-terminal of the Ad12 E1A 13S and 12S mRNAs was used for immunoprecipitation analysis. (LUCHER, KIMELMAN, BRACKMANN, SYMINGTON, and GREEN, manuscript in preparation)

antipeptide antibody immunoprecipitated two proteins 47K and 45K from extracts of [^{35}S]methionine-labeled Ad12-infected KB cells treated with arabinosyl cytosine, as detected by one-dimensional electrophoresis on 10% polyacrylamide gels (LUCHER, BRACKMANN, KIMELMAN, SYMINGTON, and GREEN, manuscript in preparation). The Ad12 E1A-specific antipeptide antibody was used to analyze *E. coli* W3110i^q clones that were transformed by two plasmids, pAd416 and pAd418. pAd416 and pAd418 contain cDNA copies of Ad12 E1A 12S and 13S mRNA, respectively, under the control of the tac (trp-lac) promoter (KIMELMAN, LUCHER, BRACKMANN, PTASHNE, and GREEN, manuscript in preparation). After induction with 1 m*M* isopropyl thiogalactoside for 2 h, followed by labeling with [^{35}S]methionine, extracts of these clones contained high levels of either a 47K or 48K protein which was not present in uninduced cells or in *E. coli* cells without plasmid. The Ad12 E1A antipeptide antibody immunoprecipitated both the 47K and 48K proteins, as well as the minor 27K protein; the 27K protein may be a proteolytic breakdown product or a protein translated by initiation at a second AUG in Ad12 E1A mRNAs. Figure 6 shows the immunoprecipitation analysis of pAd418 transformed cells. When the homologous synthetic peptide (peptide 204) was included in the immunoprecipitation assay, immunoprecipitation of the 47K–48K protein was completely inhibited (Fig. 6). When a nonhomologous peptide was included (peptide 19), no inhibition was observed. The Ad12 E1A antipeptide antibody should prove useful for the purification of the Ad12 E1A proteins produced in *E. coli* transformed by pAd416 and pAd418.

6 Projections on Adenovirus-Transforming Proteins

Establishing the functions of the adenovirus early proteins, and in particular the early region E1-encoded transforming proteins, will be a challenging and exciting task. Knowledge of the functions of the E1 proteins is very limited. The absence of purified E1 proteins and of specific reagents to measure E1 proteins has handicapped progress in this area. The availability of antipeptide antibodies of known specificity for different domains of each E1-encoded protein should soon remedy this deficiency. One can optimistically look forward to the availability of microgram quantities of purified E1 proteins for further studies and to an expanded knowledge of the properties and functions of E1 proteins.

Acknowledgments. Research support was provided by Grant 1419 from The Council For Tobacco Research, U.S.A., Inc. and Grants CA21824, CA29561, and CA09222 from the National Cancer Institute. We thank C.E. Mulhall for editorial assistance and G. Chinnadurai and W.S.M. Wold for reading the manuscript.

References

Akusjärvi G, Zabielski J, Perricaudet M, Pettersson U (1981) The sequence of the 3′ non-coding region of the hexon mRNA discloses a novel adenovirus gene. Nucleic Acids Res 9:1–17

Aleström P, Akusjärvi G, Pettersson M, Pettersson U (1982) DNA sequence analysis of the region encoding the terminal protein and the hypothetical N-gene product of adenovirus type 2. J Biol Chem 257:13492–13498

Anderson CW, Lewis JB (1980) Amino-terminal sequence of adenovirus type 2 proteins: hexon, fiber, component IX, and early protein 1B–15K. Virology 104:27–41

Baron MH, Baltimore D (1982a) Antibodies against the chemically synthesized genome-linked protein of poliovirus react with native virus-specific proteins. Cell 28:395–404

Baron MH, Baltimore D (1982b) Antibodies against a synthetic peptide of the poliovirus replicase protein: reaction with native, virus-encoded proteins and inhibition of virus-specific polymerase activities in vitro. J Virol 43:969–978

Berk AJ, Sharp PA (1978) Structure of the adenovirus 2 early mRNAs. Cell 14:695–711

Berk AJ, Lee F, Harrison T, Williams J, Sharp PA (1979) Pre-early adenovirus 5 gene product regulates synthesis of early viral messenger RNAs. Cell 17:935–944

Bittle JL, Houghten RA, Alexander H, Shinnick TM, Sutcliffe JG, Lerner RA (1982) Protection against foot-and-mouth disease by immunization with a chemically synthesized peptide predicted from the viral nucleotide sequence. Nature 298:30–33

Bos JL, Polder LJ, Bernards R, Schrier PI, van den Elsen PJ, van der Eb AJ, van Ormondt J (1981) The 2.2 kb E1B mRNA of human Ad12 and Ad5 codes for two tumor antigens starting at different AUG triplets. Cell 27:121–131

Brackmann KH, Green M, Wold WSM, Cartas M, Matsuo T, Hashimoto S (1980) Identification and peptide mapping of human adenovirus type 2-induced early polypeptides isolated by two-dimensional gel electrophoresis and immunoprecipitation. J Biol Chem 255:6772–6779

Challberg MD, Kelly TJ Jr (1982) Eukaryotic DNA replication: viral and plasmid model systems. Annu Rev Biochem 51:901–934

Challberg MD, Desiderio SV, Kelly TJ Jr (1980) Adenovirus DNA replication in vitro: Characterization of a protein covalently linked to nascent DNA strands. Proc Natl Acad Sci USA 77:5105–5109

Challberg MD, Ostrove JM, Kelly TJ Jr (1982) Initiation of adenovirus DNA replication: detection of covalent complexes between nucleotide and the 80-kilodalton terminal protein. J Virol 41:265–270
Chinnadurai G (1983) Adenovirus 2 lp^+ locus codes for a 19K tumor antigen that plays an essential role in cell transformation. Cell 33:759–766
Chow LT, Broker TR, Lewis JB (1979) Complex splicing patterns of RNAs from the early regions of adenovirus 2. J Mol Biol 134:265–303
Craig EA, Raskas HJ (1976) Nuclear transcripts larger than the cytoplasmic mRNAs are specified by segments of the adenovirus genome coding for early functions. Cell 8:205–213
Deppert W, Walter G (1982) Domains of simian virus 40 large T-antigen exposed on the cell surface. Virology 122:56–70
Dijkema R, Dekker BMM, van der Feltz MJM, van der Eb AJ (1979) Transformation of primary rat kidney cells by DNA fragments of weakly oncogenic adenoviruses. J Virol 32:943–950
Dijkema R, Dekker BMM, van Ormondt, H. (1982) Gene organization of the transforming region of adenovirus type 7 DNA. Gene 18:143–156
Dreesman GR, Sanchez Y, Ionescu-Matiu I, Sparrow JT, Six HR, Peterson DL, Hollinger FB, Melnick JL (1982) Antibody to hepatitis B surface antigen after a single inoculation of uncoupled synthetic HBsAg peptides. Nature 295:158–160
Esche H, Mathews MB, Lewis JB (1980) Proteins and messenger RNAs of the transforming region of wild-type and mutant adenoviruses. J Mol Biol 142:399–417
Feldman LT, Nevins JR (1983) Localization of the adenovirus $E1A_a$ protein, a positive-acting transcriptional factor, in infected cells. Mol Cell Biol 3:829–838
Flint SJ (1977) Two "early" mRNA species in adenovirus type 2-transformed rat cells. J Virol 23:44–52
Flint SJ, Sambrook J, Williams JF, Sharp PA (1976) Viral nucleic acid sequences in transformed cells. IV. A study of the sequences of adenovirus 5 DNA and RNA in four lines of adenovirus 5-transformed rodent cells using specific fragments of the viral genome. Virology 72:456–470
Fowlkes DM, Lord ST, Linném T, Pettersson U, Phillipson L (1979) Interaction between the adenovirus DNA-binding protein and double-stranded DNA. J Mol Biol 132:163–180
Frost E, Williams J (1978) Mapping temperature-sensitive and host-range mutations of adenovirus type 5 by marker rescue. Virology 91:39–50
Galos R, Williams J, Shenk T, Jones N (1980) Physical location of host-range mutants of adenovirus type 5; deletion and marker-rescue mapping. Virology 104:510–513
Gilead Z, Jeng Y-H, Wold WSM, Sugawara K, Rho KM, Harter ML, and Green M (1976) Immunological identification of two adenovirus 2-induced early proteins possibly involved in cell transformation. Nature (London) 264:263–266
Gingeras TR, Sciaky D, Gelinas RE, Bing-Dong J, Yen CE, Kelly MM, Bullock PA, Parsons BL, O'Neill KE, Roberts RJ (1982) Nucleotide sequences from the adenovirus 2 genome. J Biol Chem 257:13475–13491
Graham FL, Anderson PJ, Mulder C, Heijneker HL, Warnaar SO, de Vries FAJ, Fiers W, van der Eb AJ (1974) Studies on in vitro transformation by DNA and RNA fragments of human adenoviruses and simian virus 40. Cold Spring Harbor Symp Quant Biol 39:637–650
Graham FL, Harrison T, Williams J (1978) Defective transforming capacity of adenovirus type 5 host-range mutants. Virology 86:10–21
Green M (1978) Adenoviruses: Model systems of virus replication, human cell molecular biology, and neoplastic transformation. Perspect Biol Med 21:373–397
Green M, Mackey JK (1977) Are oncogenic human adenoviruses associated with human cancer? Analysis of human tumors for adenovirus transforming gene sequences. Cold Spring Harbor Conf Cell Prolif 4:1027–1042
Green M, Piña M, Kimes RC, Wensink PC, MacHattie LA, Thomas CA (1967) Adenovirus DNA. I. Molecular weight and conformation. Proc Natl Acad Sci USA 57:1302–1309
Green M, Parsons JT, Piña M, Fujinaga K, Caffier H, Landgraf-Leurs I (1970) Transcription of adenovirus genes in productively infected and in transformed cells. Cold Spring Harbor Symp Quant Biol 35:803–818

Green M, Mackey JK, Wold WSM, Rigden P (1979a) Thirty-one human adenovirus serotypes (Ad1-31) form five groups (A–E) based upon DNA genome homologies. Virology 93:481–492

Green M, Wold WSM, Mackey JK, Rigden P (1979b) Analysis of human tonsil and cancer DNAs and RNAs for DNA sequences of group C (serotypes 1, 2, 5, 6) human adenoviruses. Proc Natl Acad Sci USA 76:6606–6610

Green M, Wold WSM, Brackmann KH, Cartas MA (1979c) Identification of families of overlapping polypeptides coded by early "transforming" gene region 1 of human adenovirus type 2. Virology 97:275–286

Green M, Wold WSM, Büttner W (1981) Integration and transcription of group C human adenovirus sequences in the DNA of five lines of transformed rat cells. J Mol Biol 151:337–366

Green M, Brackmann KH, Cartas MA, Matsuo T (1982) Identification and purification of a protein encoded by the human adenovirus type 2 transforming region. J Virol 42:30–41

Green M, Brackmann KH, Lucher LA, Symington JS, Kramer TA (to be published) Human adenovirus 2 E1B–19K and E1B–53K tumor antigens: Antipeptide antibodies targeted to the NH_2 and COOH termini. J Virol

Green MR, Treisman R, Maniatis T (to be published) Transcription activation of cloned human β-globin genes by viral immediate-early gene products. Cell

Green N, Shinnick TM, Witte O, Ponticelli A, Sutcliffe JG, Lerner RA (1981) Sequence-specific antibodies show that maturation of Moloney leukemia virus envelope polyprotein involves remival of a COOH-terminal peptide. Proc Natl Acad Sci USA 78:6023–6027

Green N, Alexander H, Olson A, Alexander S, Shinnick TM, Sutcliffe JG, Lerner RA (1982) Immunogenic structure of the influenza virus hemagglutinin. Cell 28:477–487

Halbert DN, Spector DJ, Raskas HJ (1979) In vitro translation products specified by the transforming region of adenovirus type 2. J Virol 31:621–629

Handa H, Kingston RE, Sharp PA (1982) Inhibition of adenovirus early region IV transcription in vitro by a purified viral DNA binding protein. Nature 302:545–547

Harrison T, Graham F, Williams J (1977) Host-range mutants of adenovirus type 5 defective for growth in HeLa cells. Virology 77:319–329

Harter ML, Lewis JB (1978) Adenovirus type 2 early proteins synthesized in vitro and in vivo: Identification in infected cells of 38000- to 50000-molecular-weight protein encoded by the left end of the adenovirus type 2 genome. J Virol 26:736–749

Harter ML, Shanmugam G, Wold WSM, Green M (1976) Detection of adenovirus type 2-induced early polypeptides using cycloheximide pre-treatment to enhance viral protein synthesis. J Virol 19:232–242

Hassel JA, Weber J (1978) Genetic analysis of adenovirus type 2. VIII. Physical locations of temperature-sensitive mutations. J Virol 28:671–678

Hérissé J, Galibert F (1981) Nucleotide sequence of the *Eco*RI E fragment of adenovirus 2 genome. Nucleic Acids Res 9:1229–1240

Hérissé J, Courtois G, Galibert F (1980) Nucleotide sequence of the *Eco*RI D fragment of adenovirus 2 genome. Nucleic Acids Res 8:2173–2192

Ho Y-S, Galos R, Williams J (1982) Isolation of type 5 adenovirus mutants with a cold-sensitive host-range phenotype: genetic evidence of an adenovirus transformation maintenance function. Virology 122:109–124

Houweling A, van den Elsen PJ, van der Eb AJ (1980) Partial transformation of primary rat cells by the left-most 4.5% of adenovirus 5 DNA. Virology 105:537–550

Jay G, Nomura S, Anderson CW, Khoury G (1981) Identification of the SV40 agnogene product: A DNA binding protein. Nature 291:346–349

Jeng Y-H, Wold WSM, Green M (1978) Evidence for an adenovirus type 2-coded early glycoprotein. J Virol 28:314–323

Johansson K, Persson H, Lewis AM, Pettersson U, Tibbetts C, Philipson L (1978) Viral DNA sequences and gene products in hamster cells transformed by adenovirus type 2. J Virol 27:628–639

Jones N, Shenk T (1979a) An adenovirus type 5 early gene function regulates expression of other early genes. Proc Natl Acad Sci USA 76:3665–3669

Jones N, Shenk T (1979b) Isolation of adenovirus type 5 host range deletion mutants defective for transformation of rat embryo cells. Cell 17:683–689

Kagan A, Glick SM (1974) In: Jaffe and Behrman (eds) Methods of Hormone Radioimmunoassay. "Oxytocin." Academic Press, New York. First Edition. pp 173–185

Katze MG, Persson H, Philipson L (1981) Control of adenovirus early gene expression: Post-transcriptional control mediated by both viral and cellular gene products. Mol Cell Biol 1:807–813

Katze MG, Persson H, Johansson B-M, Philipson L (1983) Control of adenovirus gene expression: cellular gene products restrict expression of adenovirus host range mutants in non-permissive cells. J Virol 46:50–59

Kimura T, Sawada Y, Shinawawa M, Shimizu Y, Shiroki K, Shimojo H, Sugisaki H, Takanami M, Uemizu Y, Fujinaga K (1981) Nucleotide sequence of the transforming early region E1B of adenovirus type 12 DNA: structure and gene organization, and comparison with those of adenovirus type 5 DNA. Nucleic Acids Res 9:6571–6589

Kitchingman GR, Westphal H (1980) The structure of adenovirus 2 early nuclear and cytoplasmic RNAs. J Mol Biol 137:23–48

Kornfeld R, Wold WSM (1981) Structures of the oligosaccharides of the glycoprotein coded by early region E3 of adenovirus 2. J Virol 40:440–449

Lerner RA (1982) Tapping the immunological repertoire to produce antibodies of predetermined specificity. Nature 299:592–596

Lerner RA, Green N, Alexander H, Liu F-T, Sutcliffe JG, Shinnick TM (1981) Chemically synthesized peptides predicted from the nucleotide sequence of the hepatitis B virus genome elicit antibodies reactive with the native envelope protein of Dane particles. Proc Natl Acad Sci USA 78:3403–3407

Levinson A, Levine AJ (1977) The isolation and identification of the adenovirus group C tumor antigens. Virology 76:1–11

Lewis JB, Atkins JF, Baum PR, Solem R, Gesteland RF, Anderson CW (1976) Location and identification of the genes for adenovirus type 2 early polypeptides. Cell 7:141–151

Lichy JH, Field J, Horwitz MS, Hurwitz J (1982) Separation of the adenovirus terminal protein precursor from its associated DNA polymerase: Role of both proteins in the initiation of adenovirus DNA replication. Proc Natl Acad Sci USA 79:5225–5229

Linné T, Jörnvall H, Philipson L (1977) Purification and characterization of the phosphorylated DNA-binding protein from adenovirus type 2-infected cells. Eur J Biochem 76:481–490

Lucher LA, Brackmann KH, Symington JS, Green M (to be published) Antibody directed to a synthetic peptide encoding the NH_2 terminal 16 amino acids of the adenovirus type 2 E1B–53K tumor antigen recognizes the E1B–20K tumor antigen. Virology

Mackey JK, Rigden PM, Green M (1976) Do highly oncogenic group A human adenoviruses cause human cancer? Analysis of human tumors for adenovirus 12 transforming DNA sequences. Proc Natl Acad Sci USA 73:4657–4661

Mackey JK, Wold WSM, Rigden P, Green M (1979a) Transforming regions of group A, B, and C adenoviruses: DNA homology studies with twenty-nine human adenovirus serotypes. J Virol 29:1056–1064

Mackey JK, Green M, Wold WSM, Rigden P (1979b) Analysis of human cancer DNA for DNA sequences of human adenovirus type 4. J Natl Cancer Inst 62:23–26

Matsuo T, Wold WSM, Hashimoto S, Rankin AR, Symington J, Green M (1982) Polypeptides encoded by early transforming region E1B of human adenovirus 2: immunoprecipitation from transformed and infected cells and cell free translation of E1B-specific mRNA. Virology 118:456–465

Nagata K, Guggenheimer RA, Enomoto T, Lichy JH, Hurwitz J (1982) Adenovirus DNA replication in vitro: identification of a host factor that stimulates synthesis of the preterminal protein-dCMP complex. Proc Natl Acad Sci USA 79:6438–6442

Neurath AR, Kent SBH, Strick N (1982) Specificity of antibodies elicited by a synthetic peptide having a sequence in common with a fragment of a virus protein, the hepatitis B surface antigen. Proc Natl Acad Sci USA 79:7871–7875

Nevins JR (1981) Mechanism of activation of early viral transcription by the adenovirus E1A gene product. Cell 26:213–220

Nevins JR, Winkler JJ (1980) Regulation of early adenovirus transcription: A protein product of early region 2 specifically represses region 4 transcription. Proc Natl Acad Sci USA 77:1893–1897

Nevins JR, Ginsberg HS, Blanchard J-M, Wilson MC, Darnell JE Jr (1979) Regulation of the primary expression of the early adenovirus transcription units. J Virol 32:727–733

Nigg EA, Sefton BM, Hunter T, Walter G, Singer SJ (1982) Immunofluorescent localization of the transforming protein of Rous sarcoma virus with antibodies against a synthetic *src* peptide. Proc Natl Acad Sci USA 79:5322–5326

Ostrove JM, Rosenfeld P, Williams J, Kelly TJ Jr (1983) In vitro complementation as an assay for purification of adenovirus DNA replication proteins. Proc Natl Acad Sci USA 80:935–939

Papkoff J, Hunter T (1983) Detection of an 85000-dalton phosphoprotein in *ts*110 murine sarcoma virus-infected cells with antiserum against a v-*mos* peptide. J Virol 45:1177–1182

Persson H, Oberg B, Philipson L (1978) Purification and characterization of an early protein (E14K) from adenovirus type 2-infected cells. J Virol 28:67–79

Persson H, Signäs C, Philipson L (1979) Purification and characterization of an early glycoprotein from adenovirus type 2-infected cells. J Virol 29:938–948

Persson H, Katze MG, and Philipson L (1981) Control of adenovirus early gene expression: accumulation of viral mRNA after infection of transformed cells. J Virol 40:358–366

Ross SR, Flint SJ, Levine AJ (1980) Identification of the adenovirus early proteins and their genomic map positions. Virology 100:419–432

Sambrook J, Greene R, Stringer J, Mitchison T, Hu S-L, Botchan M (1979) Analysis of the sites of integration of viral DNA sequences in rat cells transformed by adenovirus 2 or SV40. Cold Spring Harbor Symp Quant Biol 44:569–584

Sarnow P, Hearing P, Anderson CW, Reich N, Levine AJ (1982) Identification and characterization of an immunologically conserved adenovirus early region 11000 M_r protein and its association with the nuclear matrix. J Mol Biol 162:565–583

Sawada Y, Fujinaga K (1980) Mapping of adenovirus 12 mRNAs transcribed from the transforming region. J Virol 36:639–651

Schaffhausen B, Benjamin TL, Pike L, Casnellie J, Krebs E (1982) Antibody to the nonapeptide Glu-Glu-Glu-Glu-Tyr-Met-Pro-Met-Glu is specific for polyoma middle T antigen and inhibits in vitro kinase activity. J Biol Chem 257:12467–12470

Schrier PI, van den Elsen PJ, Hertoghs JJL, van der Eb AJ (1979) Characterization of tumor antigens in cells transformed by fragments of adenovirus type 5 DNA. Virology 99:372–385

Sefton BM, Walter G (1982) Antiserum specific for the carboxy terminus of the transforming protein of Rous sarcoma virus. J Virol 44:467–474

Sekikawa K, Shiroki K, Shimojo H, Ojima S, Fujinaga K (1978) Transformation of a rat cell line by an adenovirus 7 DNA fragment. Virology 88:1–7

Semler BL, Anderson CW, Hanecak R, Dorner LF, Wimmer E (1982) A membrane-associated precursor to poliovirus VPg identified by immunoprecipitation with antibodies directed against a synthetic hepapeptide. Cell 28:405–412

Sen S, Houghten RA, Sherr CJ, Sen A (1983) Antibodies of predetermined specificity detect two retroviral oncogene products and inhibit their kinase activities. Proc Natl Acad Sci USA 80:1246–1250

Shiroki K, Handa H, Shimojo H, Yano S, Ojima S, Fujinaga K (1977) Establishment and characterization of rat cell lines transformed by restriction endonuclease fragments of adenovirus 12 DNA. Virology 82:462–471

Shiroki K, Shimojo H, Sawada Y, Uemizu K, Fujinaga K (1979) Incomplete transformation of rat cells by a small fragment of adenovirus 12 DNA. Virology 95:127–136

Smart JE, Stillman BW (1982) Adenovirus terminal protein precursor. J Biol Chem 257:13499–13506

Solnick D, Anderson MA (1982) Transformation-deficient adenovirus mutant defective in expression of region 1A but not region 1B. J Virol 42:106–113

Stillman BW, Lewis JB, Chow LT, Mathews MB, Smart JE (1981) Identification of the gene and mRNA for the adenovirus terminal protein precursor. Cell 23:497–508

Stillman BW, Tamanoi F, Mathews MB (1982) Purification of an adenovirus-coded DNA polymerase that is required for initiation of DNA replication. Cell 31:613–623

Sugawara K, Gilead Z, Green M (1977) Purification and molecular characterization of adenovirus type 2 DNA-binding protein. J Virol 21:338–346

Sugisaki H, Sugimoto K, Takanami M, Shiroki K, Saito I, Shimojo H, Sawada Y, Uemizu

Y, Uesugi S-I, and Fujinaga K (1980) Structure and gene organization in the transforming *Hind*III-G fragment of Ad12. Cell 20:777–786

Thomas GP, Mathews MB (1980) DNA replication and the early to late transition in adenovirus infection. Cell 22:523–533

Tigges MA, Raskas HJ (1982) Expression of adenovirus 2 early region E4: assignment of the early region 4 polypeptides to their respective mRNAs, using in vitro translation. J Virol 44:907–921

Van der Eb AJ, Mulder C, Graham FL, Houweling A (1977) Transformation with specific fragments of adenovirus DNAs. I. Isolation of specific fragments with transforming activity of adenovirus 2 and 5 DNA. Gene 2:115–132

Van der Vliet PC, Levine AJ (1973) DNA-binding proteins specific for cells infected by adenovirus. Nature New Biol 246:170–174

Van Ormondt H, Maat J, de Ward A, van der Eb A (1978) The nucleotide sequence of the transforming *Hpa*I-E fragment of adenovirus type 5 DNA. Gene 4:309–328

Virtanen A, Aleström P, Persson H, Katze MG, Pettersson U (1982) An adenovirus agnogene. Nucleic Acids Res 10:2539–2548

Walter G, Scheidtmann K-H, Carbone A, Laudano AP, Doolittle RF (1980) Antibodies specific for the carboxy- and amino-terminal regions of simian virus 40 large tumor antigen. Proc Natl Acad Sci USA 77:5197–5200

Walter G, Hutchinson MA, Hunter T, Eckhart W (1981) Antibodies specific for the polyoma virus middle-size tumor antigen. Proc Natl Acad Sci USA 78:4882–4886

Walter G, Hutchinson MA, Hunter T, Eckhart W (1982) Purification of polyoma virus medium-size tumor antigen by immunoaffinity chromatography. Proc Natl Acad Sci USA 79:4025–4029

Wold WSM, Green M (1979) Adenovirus type 2 early polypeptides immunoprecipitated by antisera to five lines of adenovirus-transformed rat cells. J Virol 30:297–310

Wold WSM, Mackey JK, Rigden P, Green M (1979) Analysis of human cancer DNAs for DNA sequences of human adenovirus serotypes 3, 7, 11, 14, 16, and 21 in group B. Cancer Res 39:3479–3484

Wong TW, Goldberg AR (1981) Synthetic peptide fragment of *src* gene product inhibits the *src* protein kinase and crossreacts immunologically with avian *onc* kinases and cellular phosphoproteins. Proc Natl Acad Sci USA 78:7412–7416

Yoshida K, Fujinaga K (1980) Unique species of mRNA from adenovirus type 7 early region 1 in cells transformed by adenovirus type 7 DNA fragment. J Virol 36:337–352

Yoshida K, Sekikawa K, Fujinaga K (1979) Mappings of adenovirus type 7 cytoplasmic RNA species synthesized early in lytically infected cells and synthesized in transformed cells. J Virol 32:339–344

On the Mechanism of Recombination Between Adenoviral and Cellular DNAs: The Structure of Junction Sites

W. Doerfler, R. Gahlmann, S. Stabel, R. Deuring, U. Lichtenberg, M. Schulz, D. Eick, and R. Leisten

1 Introduction

The integration of adenovirus DNA into host cell DNA was studied in considerable detail in abortively and productively infected cells, as well as in transformed and tumor cells induced by human adenoviruses (for a review

Institute of Genetics, University of Cologne, Weyertal 121, D-5000 Cologne 41

Current Topics in Microbiology and Immunology, Vol. 109

see Doerfler 1982). More than 70 different adenovirus-transformed cell lines and tumor cell lines from adenovirus-induced tumors exhibited nonidentical patterns of viral DNA insertion into the host genome. In this context hamster, mouse, and rat systems were investigated. From these studies, there was no evidence for highly specific sites of viral DNA insertion into the cellular genome as judged by the results of Southern blotting analyses. Of course, integration patterns in established transformed or tumor cell lines could have been modified by postintegrational alterations, such as rearrangements, amplifications, or partial deletions, and thus possible specificities could have been obscured. On the other hand, in several instances it could be demonstrated that the cellular nucleotide sequences at the sites of junction were not altered (Gahlmann et al. 1982; Stabel and Doerfler 1982; Gahlmann and Doerfler 1983). Obvious sequence specificities at the sites of insertion have so far not been found in the adenovirus, SV40, or polyoma virus system.

Thus, the problem of specificity could be approached meaningfully only by analyzing the sites of insertion at the nucleotide level (Deuring et al. 1981b; Visser et al. 1982; Westin et al. 1982; Gahlmann et al. 1982; Stabel and Doerfler 1982). Seven different sites of junction between adenoviral and cellular DNA derived from different cell lines have been sequenced so far in our laboratory. Adenovirus DNA insertion into mammalian cell DNA was used as a model to investigate the mechanism of foreign DNA insertion, in general. Foreign DNA insertion might represent a hallmark in evolution and tumor biology and is considered a precondition for gene therapy. Current research has mainly been focused on the following topics: (a) What characteristics do the sites of foreign DNA insertion exhibit, and at what sequence(s) does foreign DNA recombine with cellular DNA? Is insertion random? (b) What is the mechanism of foreign DNA insertion? (c) What effects do inserted genes have on the expression of host genes abutting the inserted DNA or of genes located at more remote sites? At present, these questions can only partly be answered.

We have chosen the adenovirus system in these investigations for two reasons. Firstly, the molecular biology of this virus system has been unraveled in considerable detail and nucleotide sequence information on adenovirus DNA is available. Secondly, the entire gamut of virus-host cell interactions is realized with adenoviruses. One could therefore hope to follow the insertion of foreign DNA under a number of biologically different, but defined conditions.

In this review we shall summarize the main features of inserted adenoviral genomes and then describe in detail the structure of the sites of junction that have been sequenced so far. We will also try to deduce more general conclusions about the insertion of foreign DNA into mammalian cell DNA. At present, the mechanism of foreign DNA insertion in mammalian cells is not understood at the molecular level. It is possible that adenovirus DNA insertion represents a special case, in that viral proteins, e.g., the terminal protein, could play an important role in mediating the recombination event.

2 Résumé of Main Previous Findings

The mode of insertion of the adenoviral genome is influenced by the permissivity of the virus-cell system. As pointed out previously, the adenovirus type 12 (Ad12) genome is usually inserted in the hamster or mouse genome without internal deletions (FANNING and DOERFLER 1976; SUTTER et al. 1978; STABEL et al. 1980; KUHLMANN and DOERFLER 1982; STARZINSKI-POWITZ et al. 1982), although in a few Ad12-induced tumors, deletions of parts of the Ad12 genome have been observed (KUHLMANN et al. 1982). Ad12 DNA is completely abortive in hamster cells (DOERFLER 1968, 1969; FANNING and DOERFLER 1976; ORTIN et al. 1976; ESCHE et al. 1979) and in mouse cells (STARZINSKI-POWITZ et al. 1982). In contrast, in the patterns of integration of the DNAs of adenovirus type 2 (Ad2) and adenovirus type 5 (Ad5), which are both permissive or at least semipermissive in hamster cells, extensive deletions of the viral genome do regularly occur (SAMBROOK et al. 1974; VISSER et al. 1979; VARDIMON and DOERFLER 1981). It is therefore reasonable to postulate that Ad12 DNA can persist intact or nearly intact in hamster cells, because the nonpermissive disposition of hamster cells toward Ad12 does not allow Ad12 DNA to replicate, and thus cells containing the entire Ad12 genome are not selected against. In Ad2- or Ad5-infected hamster cells, complete free viral genomes will replicate and eventually destroy the host cells. Transformants with intact integrated Ad2 or Ad5 genomes are not likely to be found, but in most cases the persisting viral genomes carry deletions. Of course, there may exist additional, more complex reasons for the differences in integration patterns found in Ad2- and Ad12-transformed cells.

The patterns of integration of the highly oncogenic simian adenovirus type 7 (SA7) in virus-induced hamster tumor cells are very complicated and imply that deletions and amplifications of the SA7 genome have occurred [TIMME, SOBOLL, NEUMANN and DOERFLER, unpublished results; compare also TIKCHONENKO (1984)].

In general, multiple copies of adenoviral DNA were inserted in the host genome, usually at the same or at a very limited number of sites. The number of viral genome equivalents per cell ranged from one to two to over 30, both in cells that were transformed in culture and in Ad12-induced tumor cells isolated from tumor-bearing hamsters or mice. From several lines of evidence it appeared plausible to argue that one or a limited number of genomes were integrated initially and that viral DNA sequences, probably together with abutting cellular sequences, were postintegrationally amplified. Adenoviral DNA was found to be inserted into unique or repetitive cellular DNA sequences in different cell lines or tumors (DEURING et al. 1981 b; GAHLMANN et al. 1982; STABEL and DOERFLER 1982). In a few Ad12-transformed cell lines, terminal sequences of the integrated adenoviral genomes were disproportionately amplified (STABEL et al. 1980). Multiple inserted adenoviral genomes were usually not arranged in true tandem, with one entire genome being followed by another (STABEL et al. 1980; KUHL-

MANN et al. 1982; STARZINSKI-POWITZ et al. 1982). In the cases analyzed in this respect, it was found that individual genomes were separated by sequences not identical to authentic viral DNA. These sequences could have constituted cellular, or possibly rearranged viral, DNA. Deletions and rearrangements at the right termini of most of the 20–22 Ad12 genomes integrated in cell line T637 have been demonstrated (EICK and DOERFLER 1982).

From the analysis of over 70 different adenovirus-transformed and tumor cell lines investigated in our laboratory and from the results of several other laboratories, no evidence emerged for the notion that adenoviral DNA would insert at identical or similar sites in the host genome. Similar conclusions were derived from studies on the insertion of SV40 or polyoma DNA (KETNER and KELLY 1976; BOTCHAN et al. 1976; SAMBROOK et al. 1979; STRINGER 1981, 1982; HAYDAY et al. 1982; MENDELSOHN et al. 1982). Absence of specificity was documented by the results of Southern blot analyses and of determinations of nucleotide sequences across the sites of junctions in a few cases.

From several Ad12-transformed or tumor cell lines, morphological revertants were isolated which had arisen spontaneously and had lost part or all of the integrated adenoviral genome copies. Cell line T637, which carried 20–22 genome equivalents of Ad12 DNA, lost all of these copies or retained one-half or one genome equivalent or one genome and a fraction of a second one per revertant cell. The revertants were different from the parent line (GRONEBERG et al. 1978; GRONEBERG and DOERFLER 1979; EICK et al. 1980). From an Ad12-induced hamster tumor line, morphological variants were isolated that had gradually lost all detectable traces of viral DNA, but preserved their oncogenic phenotype, the absence of viral DNA notwithstanding. It was therefore concluded that – at least in these variants – persistence of adenoviral DNA was not an essential precondition for expression of the oncogenic phenotype (KUHLMANN et al. 1982). It could not be ruled out that a viral DNA fragment of extremely short length (10–100 nucleotides) might have persisted in these cells. The occurrence of the morphological revertants and variants simultaneously proved that foreign DNA could not only be fixed in cellular DNA but could also be excised from it more or less completely.

By reassociation kinetics measurements (FANNING and DOERFLER 1976) as well as by SOUTHERN (1975) blot analysis (STABEL et al. 1980), scattered fragments of viral DNA could be detected in an integrated form in Ad12-transformed cell lines in addition to the intact or nearly intact DNA molecules. Apparently, in addition to nearly intact integrated viral DNA molecules some cell lines also contained fragments of viral DNA which might have been inserted independently of the intact viral DNA molecules. It is not known how stably these scattered fragments of viral DNA are inserted in the cellular genome.

Analyses of nucleotide sequences at several sites of integration of adenoviral DNA in hamster, mouse, or rat DNA revealed that terminal viral nucleotides were deleted at the sites of junction (DEURING et al. 1981b; VISSER et al. 1982; GAHLMANN et al. 1982; STABEL and DOERFLER 1982;

GAHLMANN and DOERFLER 1983, see also Table 1). In the Ad2-transformed hamster cell line HE5, cellular nucleotides were not deleted at the site of insertion (GAHLMANN and DOERFLER 1983). A detailed discussion of the structure and the peculiarities of each site of junction analyzed will be presented below.

The mechanism of foreign (viral) DNA insertion in mammalian cells is at present not understood. It is conceivable that patchy homologies between viral and cellular DNA sequences might have played a role in some recombination events (GUTAI and NATHANS 1978; DEURING et al. 1981b; STRINGER 1982; GAHLMANN et al. 1982; STABEL and DOERFLER 1982). Such short homologies, particularly certain combinations of nucleotide patches, might stabilize the recombination complex. Moreover, patch homologies between adenoviral DNA and cellular DNA are abundant and might help to explain the apparent lack of specificity in insertion sites. Depending on the combinations of patch homologies selected, insertion of foreign DNA might be facilitated at a very large number of sites. When nonessential sites in unique DNA or in repetitive DNA are hit by the insertion event, the cell might survive and even gain selective advantages from the added genetic information. Depending on the site of insertion, viral DNA sequences may be linked to cellular enhancers or vice versa and cells with viral DNA inserted at such sites may exhibit the transformed phenotype due to more efficient expression of viral or cellular DNA.

It will also be interesting to elucidate in what conformations the adenoviral genome can recombine with host DNA. Circular viral DNA – perhaps not a covalently linked circle, but one stabilized by the terminal viral protein (ROBINSON et al. 1973) – has sometimes been implicated in that precursory function (ROBINSON et al. 1973; SAMBROOK et al. 1974; VARDIMON and DOERFLER 1981). Evidence for the presence of covalently closed circular viral DNA molecules in abortively or productively infected cells has been difficult to obtain (DOERFLER et al. 1972, 1973). Recently, evidence has been published for adenoviral molecules which are joined end to end, but it is not yet certain that these molecules actually represent circular structures (RUBEN et al. 1983).

The expression of integrated viral genomes in transformed or tumor cell lines will not be dealt with in this chapter. Suffice it to say that, in general, early viral genes are expressed and late genes are shut off (SHARP et al. 1974; ORTIN et al. 1976; SAWADA and FUJINAGA 1980; SCHIRM and DOERFLER 1981; ESCHE and SIEGMANN 1982), although in most Ad12-transformed hamster lines the late regions of the Ad12 genome persist in a perfectly preserved form. In general, these late regions are extensively methylated (SUTTER et al. 1978; SUTTER and DOERFLER 1979, 1980; KRUCZEK and DOERFLER 1982). In Ad12-induced rat tumor cell lines, some of the late segments of the integrated Ad12 genomes can be expressed (IBELGAUFTS et al. 1980; SCHIRM and DOERFLER 1981).

3 Methods Used in the Analyses of Inserted Viral Genomes

The problem of viral DNA insertion, in particular with respect to the possible specificity of integration sites and the mechanism of integration, could be approached best by cloning sites of junctions between viral and cellular DNA from a number of transformed and tumor cell lines and by determining the nucleotide sequences at these sites. In general, the following standard procedures were followed:

1. The sizes of the junction fragments were determined as off-size fragments, usually after cleavage of transformed cell or tumor cell DNA with different restriction endonucleases, blotting, and hybridization to ^{32}P-labeled, cloned adenoviral DNA fragments comprising the termini of viral DNA. Off-size fragments with the host-virus DNA junction were identified in comparison with the cleavage pattern of authentic viral DNA. This approach has been described in detail (SUTTER et al. 1978; STABEL et al. 1980), and the integration maps of many different adenovirus-transformed and tumor cell lines have been published (for a review see DOERFLER 1982). All these cell or tumor lines were obtained by transforming cells with Ad 2 or Ad 12 or inducing tumors in animals with Ad12. Cells analyzed in our laboratory were not transformed with viral DNA fragments.

2. DNA fragments with a length corresponding to the off-size fragments to be cloned were selected by gel electrophoresis or by zone sedimentation on sucrose density gradients. The latter method was usually preferred. Gradient fractions were characterized by subsequent gel electrophoresis. Adenovirus DNA cut with the same restriction endonuclease was used as size marker and off-size positions were determined in this way.

3. The selected size-class of *Eco*RI-cut DNA was ligated with the "arms" of λgtWES·λB DNA (TIEMEIER et al. 1976). The arms of this or other λ vector DNAs, i.e., the terminal *Eco*RI fragments of the vector, were purified by zone sedimentation on sucrose density gradients. The ligated DNA was then packaged in vitro in λ phage heads (HOHN and MURRAY 1977; HOHN 1979). Phage plaques containing the junction fragment to be cloned were identified by the method of BENTON and DAVIS (1977), using cloned terminal viral DNA fragments as hybridization probes. The scheme followed in some of these cloning experiments is summarized in Fig. 1. It proved advantageous to perform the initial cloning of junction sites in a λ vector, since with that system large enough numbers of plaques per petri dish could be produced and screened. Comparable colony numbers could not have been attained using plasmids, since, upon insertion of foreign DNA into plasmids, transformation efficiencies sometimes dropped precipitously. A comparable drop in packaging and plaquing efficiencies was not observed with phage λ DNA. Usually, the desired fragments were detected in about 0.5×10^6 λ plaques, equivalent to about 40–50 petri dishes. In some instances, much higher numbers of plaques were screened without detecting a cloned junction fragment. With DNA from the Ad12-transformed cell line T637, several millions of plaques were screened without isolating a

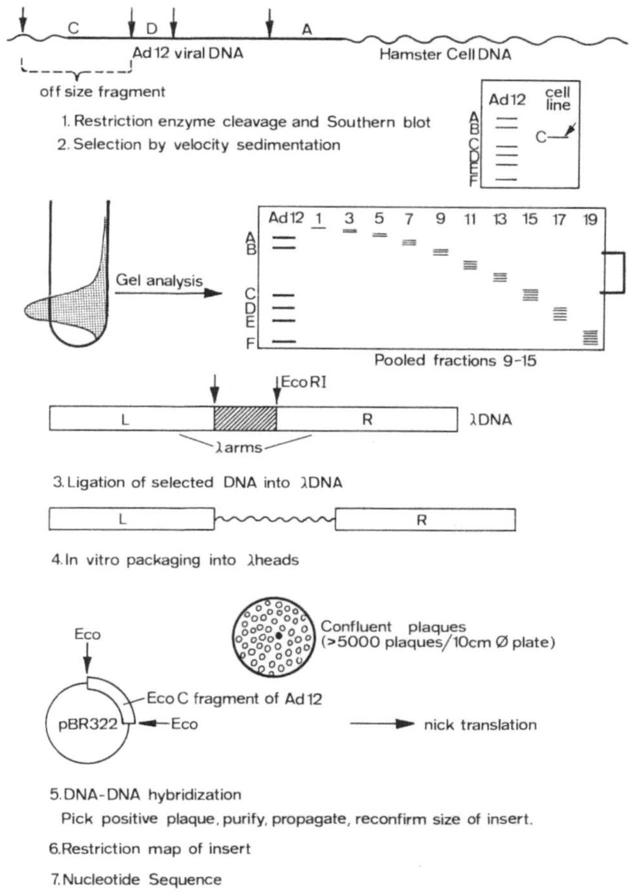

Fig. 1. Schematic presentation of major steps in the cloning of junction sequences. Details of the cloning procedure have been described in the text

single junction fragment. Internal fragments of adenoviral DNA could be readily cloned from T637 DNA. It has been shown that in this cell line peculiar DNA structures at the sites of junction between Ad12 and cell DNA and between Ad 12 DNA copies probably forestall the cloning of these sites (EICK et al. 1983, see Sect. 5.12).

4. The cloned insert was characterized further by restriction analyses using restriction endonucleases that cut adenovirus DNA frequently. Junction fragments were again identified by Southern blotting and by comparison with the authentic restriction pattern of adenovirus DNA. These junction fragments assumed off-size positions relative to the known virion DNA fragments. In most instances, the off-size fragments or parts thereof were subcloned in a suitable vector, usually in the plasmid pBR322, either directly or using appropriate linkers when necessary.

Table 1. Cloned and sequenced junction sites between viral and cellular DNA from adenovirus-

Number	Cell line	Species	Transforming agent	Type of line	Approximate copy number
1	F4	Rat	Ad2	Transformed rat embryo brain	16
2	CLAC3	Hamster	Ad12	Tumor line[a]	4–5
3	HE5	Hamster	UV irradiated Ad2	Transformed embryo cells	2–3
4	CLAC1	Hamster	Ad12	Tumor line[a]	10–13
5	BHK268-C31	Hamster	Ad5 (fragmented DNA)	DNA transformed primary baby hamster kidney cells	2
6	5RK20	Rat	UV irradiated Ad5	Primary rat kidney	1
7	SYREC, viral recombinant	Human	Ad12	Virion-encapsidated viral recombinant	–
8	Ad7-Hinf recombinant	Human	Ad7	Virion-encapsidated	–
9	T1111(2)	Hamster	Ad12	Tumor	10–11
10	CBA12-1-T	Mouse	Ad12	Tumor	20–30
11	HA12/7	Hamster	Ad12	Transformed primary hamster embryo	3–4

[a] Established from different tumors induced by Ad12 in newborn hamsters

5. Lastly, a detailed restriction map of the subcloned junction fragment was established. It was also ascertained that the subcloned fragment still contained cellular DNA by hybridizing the clone back to cellular DNA from transformed or untransformed cells. Moreover, it could also be demonstrated that cloning and subcloning of the junction fragment did usually not lead to deletions or rearrangements of the DNA insert.

6. Employing the Maxam and Gilbert sequencing technique (Maxam and Gilbert 1977, 1980) and terminal labeling at the 5′ or 3′ ends of distinct fragments, the nucleotide sequence across the sites of junction was determined. In most instances it was considered necessary to sequence several hundred nucleotides into the cellular segment, as it was conceivable that decisive recombinatorial signals might have been situated remote from the actual site of junction.

7. A special approach was designed to clone certain fragments that did not fit into available λ phage vectors or to avoid the use of nucleotide linkers (R. Gahlmann and W. Doerfler, unpublished results). The DNA from transformed or tumor cells was cut with the restriction endonuclease

transformed cell lines, from adenovirus-induced tumors, and from viral recombinants

Terminus sequenced	Type of cellular sequence	Nucleotides deleted	References
Right	Unknown	2	SAMBROOK et al. (1979)
Left	Unique	45	DEURING et al. (1981 b)
Left and right and internal link between remaining viral DNA	Unique	10 (l) 8 (r)	GAHLMANN et al. (1982), GAHLMANN and DOERFLER (1983)
Left	Repetitive	174	STABEL and DOERFLER (1982)
(Left)	Unknown	(572)	WESTIN et al. (1982)
Linkage of adjacent viral molecules	–	62 (l) 107 (r)	VISSER et al. (1982)
Left	Repetitive		DEURING et al. (1981 a); DEURING and DOERFLER (1983)
–	Repetitive (human Hinf family)	–	SHIMIZU et al. (1983)
Left		64	U. LICHTENBERG, W. DOERFLER, in preparation
Left	Unique	9	M. SCHULZ, W. DOERFLER, in preparation
Left			R. JESSBERGER, S. STABEL, and W. DOERFLER, unpublished

*Pst*I, *Hind*III, or *Bam*HI; then the off-size fragment to be cloned was size-selected by zone sedimentation on sucrose density gradients and ligated into the appropriate site of plasmid pBR322 DNA. The resealed plasmid carrying the insert was subsequently cleaved with *Eco*RI and ligated to the arms of phage λgtWES·λB DNA. Upon in vitro packaging of λ DNA, appropriate plaques containing the junction fragment were identified by the Benton-Davis technique using terminal virion DNA fragments as probes.

4 Cell Lines and Junction Sites Analyzed

Adenovirus-transformed cell lines and adenovirus-induced tumors or cell lines established from them constitute a source of clonal recombinants between viral and cellular DNA. As much as it would be desirable to investigate such recombinants shortly after their inception, sites of insertion could not easily be studied at that stage, as insertion might occur at many different sites. Thus, it was more realistic to concentrate on the study of clonal lines. Eventually, we will have to return to investigations on cells shortly after infection with virions or subviral particles or after transfection with viral DNA or with cloned DNA fragments. Such experiments have been initiated.

In Table 1, the sites of junction between adenovirus DNA and cell DNA that have been cloned and sequenced so far have been summarized. Different adenoviruses and cells from different species were used in these investigations. In symmetric recombinant (SYREC2) DNA (No. 7, Table 1), which was found to be encapsidated into adenovirions, linkage between Ad12 DNA and human KB cell DNA was shown by sequence analysis (Deuring and Doerfler 1983). SYREC DNA was generated in productively infected human cells (Deuring et al. 1981 a), and its existence constituted proof for the occurrence of recombination between the adenoviral and host genomes in productively infected cells. In cell line 5RK20 (No. 6, Table 1), the link between the adjacent right and left termini of integrated Ad5 genomes was sequenced (Visser et al. 1982). An internal link between the right and left termini of Ad2 DNA carrying a deletion was also sequenced in the Ad2-transformed cell line HE5 (Gahlmann and Doerfler 1983).

The sequence data from all the transformed and tumor cell lines represented unequivocal proof for the covalent linkage between adenoviral and cellular DNA. The findings that adenovirus DNA in infected and transformed cells was associated with high molecular weight DNA (Doerfler 1968, 1970; Groneberg et al. 1977) and that adenovirus DNA was linked to unique (Deuring et al. 1981 b; Gahlmann et al. 1982) or repetitive DNA sequences (Stabel and Doerfler 1982) argue for the chromosomal location of inserted viral DNA.

In situ DNA hybridization analyses have been performed in a CELO virus (avian adenovirus)-induced rat tumor cell line. The tumor cells carry about 33% of the viral DNA molecule linked to cellular DNA. This sequence has been repeated 160 times and is distributed on only a few chromosomes per hypotetraploid tumor cell (Yasue and Ishibashi 1982).

In the following section, the nucleotide sequences at individual integration sites will be presented and their analyses discussed. These sequence data did not reveal evidence for specific integration sites based on nucleotide sequence, in the sense that adenoviral DNA integrated at a few highly specific sequences in all cell lines. Certain regularities could, however, be observed; e.g., in all lines investigated in this respect, the viral DNA was linked via its termini to cellular DNA and viral nucleotides were deleted at the site of junction. Only in exceptional cases was Ad2 DNA linked to cell DNA via internal fragments (Vardimon and Doerfler 1981). As pointed out previously, we consider it likely that the terminal protein of adenoviral DNA can play a role in integration. It has also to be considered that signals of possible specificity for the integration event might not reside in nucleotide sequence, but rather in the structure of chromatin or of DNA-protein complexes. At the present stage of technical developments, the possibilities to unravel such structural signals are still very limited. Hence, any

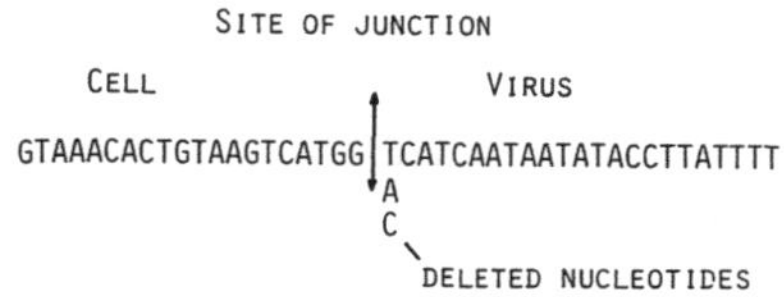

Fig. 2. Nucleotide sequence at site of junction between Ad2 DNA and rat cell DNA in cell line F4. (SAMBROOK et al. 1980)

statements with respect to specific sites of viral-cellular recombinations, or the lack thereof, will have to be phrased with caution at the present time. Thus, the concept of specificity of integration sites has to be viewed in the light of highly complex interactions between viral and cellular genomes.

5 Nucleotide Sequences at Individual Sites of Junction

5.1 Ad2-Transformed Rat Embryo Brain Line F4
(GALLIMORE 1974)

The right terminal junction sequence was determined (SAMBROOK et al. 1979). The cellular DNA sequence identified comprised only 20 nucleotides, and very short patch homologies between viral and cellular DNA could be detected (Fig. 2). No information was available on the nature or location of this cellular sequence.

5.2 Ad12-Induced Hamster Tumor Line CLAC3
(STABEL et al. 1980)

The tumor cell line CLAC3 carried four to five Ad12 genome equivalents per diploid genome. Using the λgtWES$\cdot\lambda$B vector DNA as cloning vehicle, a left terminal site of junction between Ad12 DNA and hamster cell DNA was cloned and sequenced (DEURING et al. 1981b). The colinearly inserted Ad12 genome exhibited the authentic virion DNA sequence starting with base pair 46 (Fig. 3a). The first 82 base pairs of the cloned fragment were not viral and contained scattered patchlike octa- to undecanucleotide pair homologies to in part remote sequences within the left terminal 2722 base pairs of Ad12 DNA. The phenomenon of patch homologies and its significance will be dealt with below. No homologies were observed between the deleted string of the 45 left terminal nucleotides of viral DNA and the cellular sequence replacing it. At the site of junction, a stemmed loop could be constructed, based on extensive regions of dyad symmetry (Fig. 3b).

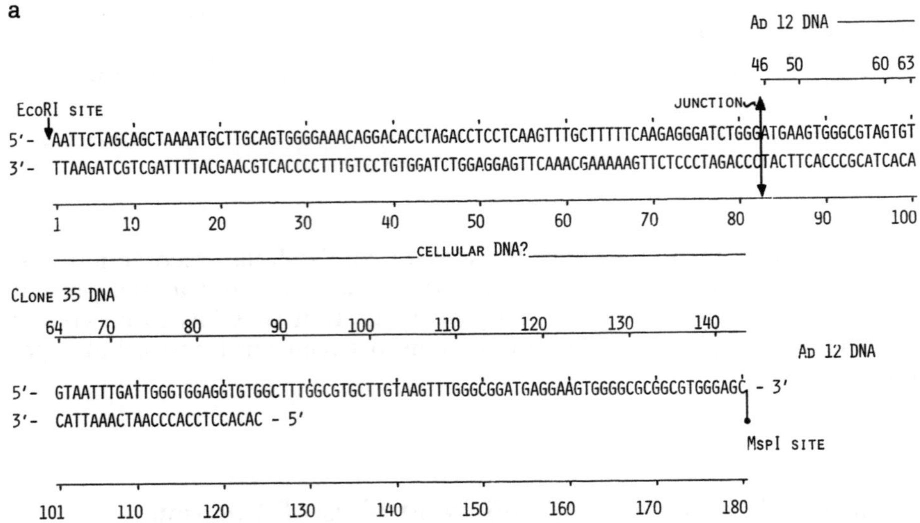

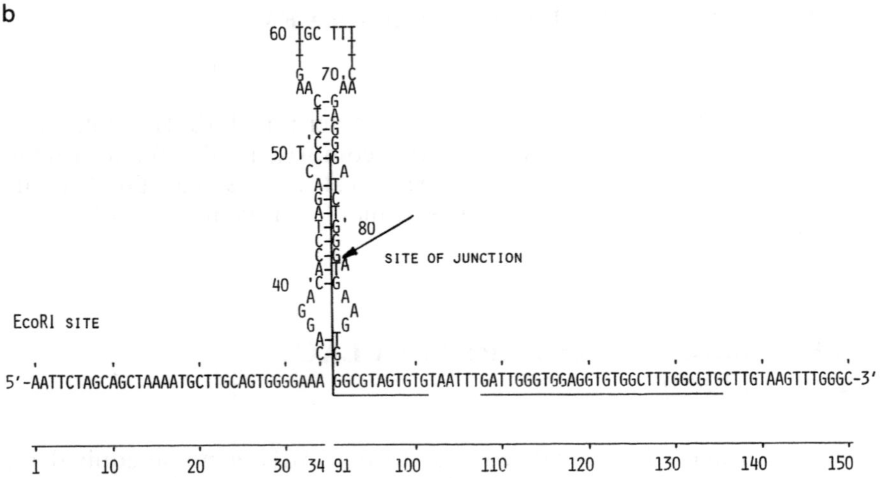

Fig. 3a, b. Junction site between Ad12 DNA and hamster cell DNA in cell line CLAC3 (Deuring et al. 1981b). **a** Nucleotide sequence at site of linkage. **b** Region of dyad symmetry and possible stemmed loop at site of junction. There is at present no proven way to ascertain the actual occurrence of such structures in vivo

Across the site of junction and close by in the viral DNA sequence, there were stretches of 27 base pairs that exhibited 70% homology (underlined sequences in Fig. 3b). The biological significance of these peculiarities is not understood.

5.3 Ad2-Transformed Hamster Cell Line HE5
(JOHANSSON et al. 1978; VARDIMON and DOERFLER 1981)

The cell line HE5 was generated by transformation of primary hamster embryo cells with UV-inactivated Ad2 (COOK and LEWIS 1979). Each cell contained approximately two copies of Ad2 DNA, and the viral DNA exhibited a major internal deletion (VARDIMON and DOERFLER 1981) extending from about 35 to 82 map units of Ad2 DNA (GAHLMANN and DOERFLER 1983). The junction sites of both termini of Ad2 DNA with hamster cell DNA, as well as the site of linkage of the two Ad2 DNA fragments at the internal joint, were cloned and sequenced (GAHLMANN et al. 1982; GAHLMANN and DOERFLER 1983). The unoccupied site of cellular DNA from cell line HE5 corresponding to the insertion site was also cloned and sequenced (GAHLMANN et al. 1982). This site was unique or occurred only a few times per cell. The results indicated that there was an almost perfect insert of the Ad2 DNA molecule via the termini. Not a single cellular nucleotide pair was lost or altered in the integration event, while eight and ten nucleotide pairs of Ad2 DNA were deleted at the right and left termini respectively. Figure 4a presents a survey of nucleotide sequences at the site of insertion; in Fig. 4b the cellular nucleotide sequence at the site of insertion is reproduced. The cellular DNA sequence and the abutting viral sequences at either end exhibited short patch homologies. Patches exceeding octannucleotides in length detected at the right terminus have been mapped as shown in Fig. 5.

Patch homologies ranging in length from dodeca- to octanucleotides were detected by computer analyses at sites also quite remote from the points of linkage. The nucleotide sequence at the right arm of the Ad2 genome was compared with randomly selected sequences of 401 nucleotides in length from vertebrate or prokaryotic DNA. In all cases, similar numbers of patch homologies were observed, indicating that patch homologies of up to 12 nucleotides long were abundant between Ad2 DNA and randomly chosen DNA sequences (Table 2). It was conceivable that such homologies could have played a role in the insertion event, perhaps by stabilizing different DNA molecules at certain stages in the recombination process. The abundance of patch homologies, which were quite different in sequence in each case, was consistent with the apparent lack of specificity with respect to integration sites of adenovirus DNA (cf. DOERFLER 1982), since many different sites could have been chosen in this way.

The internal junction between the right and left arms of Ad2 DNA remaining after the deletion was also cloned and sequenced (Fig. 6). Comparison of the sequence data with those of authentic Ad2 DNA (GINGERAS et al. 1982; ROBERTS et al. 1982) revealed that map position 35% from the left end was linked to map position 82% from the right end. A dinucleotide, GT, of unknown derivation was interspersed between the two arms of Ad2 DNA (Fig. 6). Intercalation of nucleotides between recombining molecules has also been described in somatic CV1 monkey cells, rejoining SV40 and pBR322 DNA molecules (WILSON et al. 1982). It is also interesting to note

a

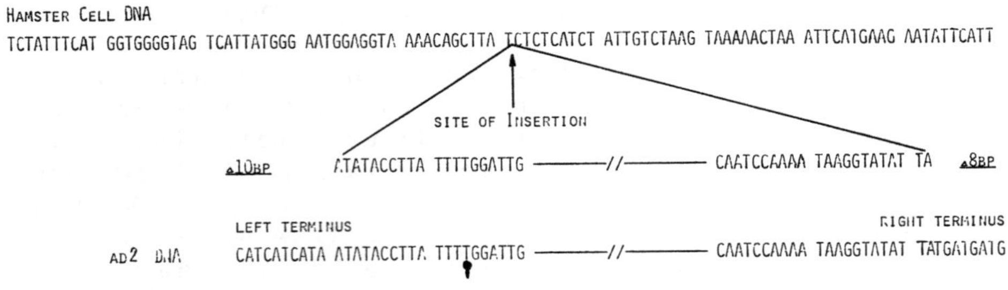

b

5′ ACCTTCAACA AATGAACAGC ACAGATTAAG CATAATGCTG CCTGACCATC

ATTTTATTAC TACTAAAATC CCCTTTGCTC TCTATTTCAT GGTGGGGTAG

SITE OF INSERTION OF AD2 DNA

TCATTATGGG AATGGAGGTA AAACAGCTTA T↕CTCTCATCT ATTGTCTAAG

TAAAAACTAA ATTCATGAAG AATATTCATT TTTAAGAGCA TAGATTTCTG

AATTAGAAAA AAGTTGTTTT TGTTCTGTTT TGGATAAAAT CTTGCTACAT 3′

Fig. 4a, b. Insertion of Ad2 DNA into cellular DNA in cell line HE5 (GAHLMANN and DOERFLER 1983). **a** Insertion without loss of cellular nucleotides. *Δ* Indicates the deletion of 10 left terminal and 8 right terminal adenoviral nucleotides at the site of junction. The viral genome is schematically presented by its terminal sequences and an *interrupted line* designating the remainder of the genome with an internal deletion between map units 35 and 82. **b** Unique cellular sequence around the site of insertion (↕). This sequence was also recloned from the unoccupied site in cell line HE5

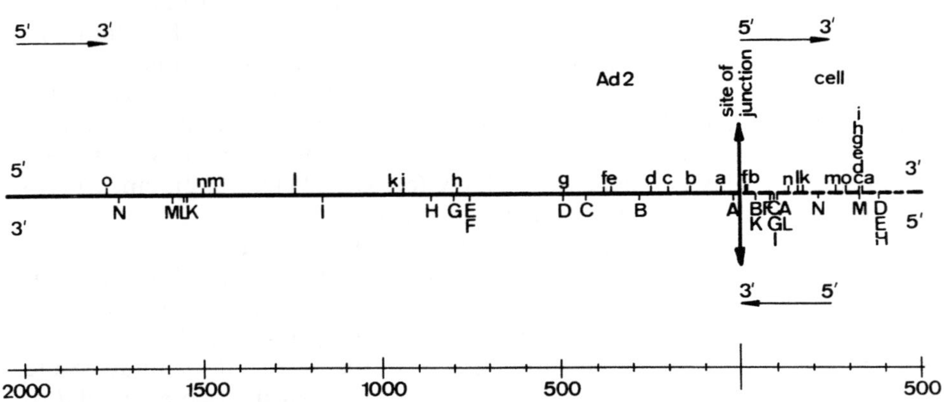

Fig. 5. Map of patches of homology between the right terminus of Ad2 DNA and the abutting hamster cell DNA sequence in cell line HE5 (GAHLMANN et al. 1982). Scale at the bottom indicates number of nucleotides from junction site. *Letters* refer to individual nucleotide patches (not shown)

Table 2. Frequency of oligonucleotides common between the right end of Ad2 DNA (*Eco*RI fragments F, D, E, and C, (l-strand)) and randomly selected prokaryotic and eukaryotic DNA sequences[a]

DNA sequence from	Length of oligonucleotide				
	8	9	10	11	12
a) Prokaryotic organisms					
E. coli, lac y	45	9	2	–	–
E. coli, rec A	36	10	2	1	–
Phage λ, 12	39	13	3	1	–
Phage λ, rex	41	12	2	1	–
b) Eukaryotic organisms					
HE5 hamster line, cell DNA in clone,					
l-strand	50	11	2	1	1
r-strand	54	11	2	2	–
Human preproinsulin	52	11	–	–	–
Human interferon 1 B	47	12	10	–	1
Human ε globin	43	10	1	–	–
Human Ig, kappa chain	58	13	2	1	–
Murine β globin	38	14	1	2	1
Murine IgG	46	13	3	1	–
Chicken ovalbumin	39	11	2	–	1

[a] The prokaryotic and eukaryotic sequences screened were 401 nucleotides long

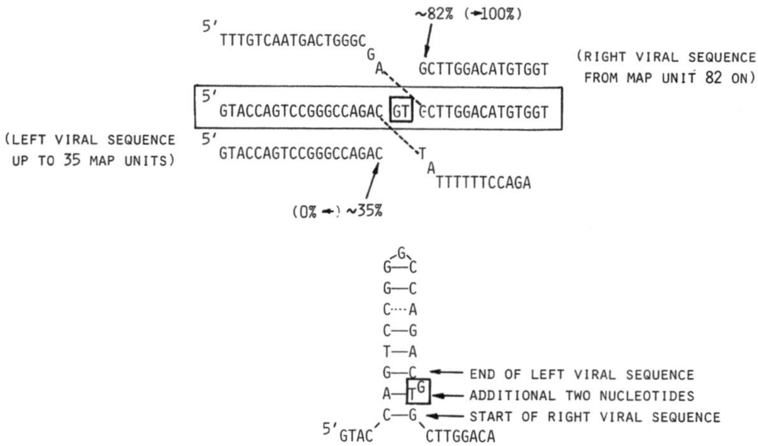

Fig. 6. Internal viral junction site in cell line HE5 (GAHLMANN and DOERFLER 1983). Sequence at the junction site is *boxed*. The additional GT dinucleotide is also indicated. Sequences outside the box and connected by *broken lines* are the immediately adjacent viral sequences that have been deleted. A region of dyad symmetry is apparent at the site of linkage (*bottom*)

that at the site of junction between the two arms a short inverted repeat can give rise to a cruciform structure (Fig. 6).

The apparent mode of insertion of adenovirus DNA preserved in the continuity of cellular DNA sequence in this cell line and involving linkage

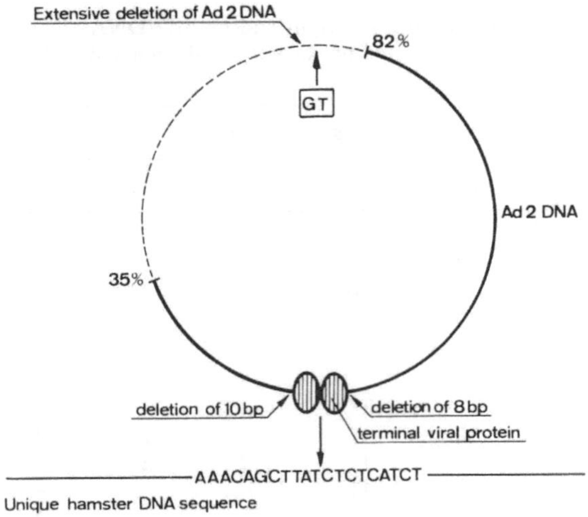

Fig. 7. Hypothetical model of Ad2 insertion. A circular intermediate has not been proven. In fact, a linear intermediate is equally likely. This model ascribes a guiding function in recombination to the terminal viral protein. Other models cannot be ruled out

to cellular DNA by the termini of viral DNA suggests a model of foreign DNA insertion in which both termini of Ad2 DNA would have to interact with adjacent nucleotides in cellular DNA either simultaneously or in succession. One possible, but unproven, intermediate structure of viral DNA participating in this event would be a circular adenovirus DNA molecule in which the termini were linked together by the terminal viral protein (GAHLMANN and DOERFLER 1983). It is not clear at what stage the deletion was introduced into Ad2 DNA. Such a deletion could have occurred before, during, or after the integration event. In the transformation experiment that led to cell line HE5, Ad2 virions irradiated with ultraviolet light had been used. This procedure might predispose the genome to deletion events. A possible model of the integration event is presented in Fig. 7.

Preliminary results (GAHLMANN and DOERFLER 1983) indicate that the cellular DNA into which Ad2 DNA has been inserted is at least partly expressed as cytoplasmic RNA in cell line HE5 and in primary LSH hamster embryo cells from which cell line HE5 was derived.

5.4 Ad12-Induced Tumor Cell Line CLAC1
(STABEL et al. 1980; STABEL and DOERFLER 1982)

This cell line was established from an Ad12-induced hamster tumor, and each cell of this line contained 10–12 copies of colinearly integrated viral DNA, which seemed to be inserted at a limited number of sites. A left terminus of Ad12 DNA linked to cellular DNA was molecularly cloned and sequenced (Figs. 8, 9). The first 174 nucleotides of Ad12 DNA at the site of linkage were deleted (Fig. 9a, b). Within 43 nucleotides of cellular DNA starting from the linkage point, there were one hepta-, one tri-, two

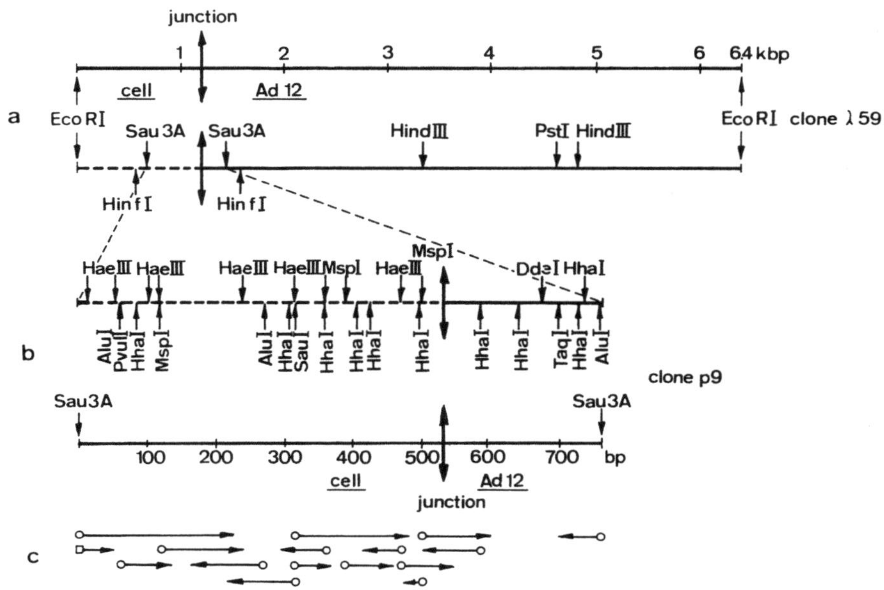

Fig. 8a–c. Restriction maps (**a, b**) and sequencing strategy (**c**) at site of junction between Ad12 DNA and repetitive hamster cell DNA in cell line CLAC1 (STABEL and DOERFLER 1982). The site of junction is designated by a double-headed vertical arrow

tetra-, and one pentanucleotide which were identical and arranged in the same order as in the 174 deleted viral nucleotides and the cellular sequence replacing them (underlined sequences in Fig. 9b). In addition, there were patch homologies, 22 octa-, twelve nona-, and one decanucleotides between the left terminal 2320 bp of Ad12 DNA and the sequenced 529 cellular nucleotides at the site of junction (cf. Fig. 16). Two types of patch homologies have been noted: One between the deleted terminal viral DNA sequence and the cellular sequence replacing it (STABEL and DOERFLER 1982), another one between the persisting viral DNA and the adjacent cellular DNA sequence. In the cellular sequence, an internal undecanucleotide repeat can be detected (brackets in Fig. 9a). The sequence GCCC is repeated six times in succession (underlined sequence in Fig. 9a), and the nucleotide sequences GCC, GCCC, and GCCCC occur 25, 12, and two times respectively and comprise 25% of the entire cellular sequence of 529 base pairs. The cellular DNA sequence corresponding to the fragment spanning the junction site was also cloned from BHK21 (B3) hamster cells and was determined by the Maxam-Gilbert technique. This cellular sequence was represented in the hamster genome several hundred times as determined by Southern blot analysis. Up to the linkage site with viral DNA, this cellular sequence was almost identical to the equivalent sequence from CLAC1 hamster cells (Fig. 10). Based on the results gleaned from this analysis, a model can be suggested, in which the insertion of Ad12 DNA into cellular DNA in cell

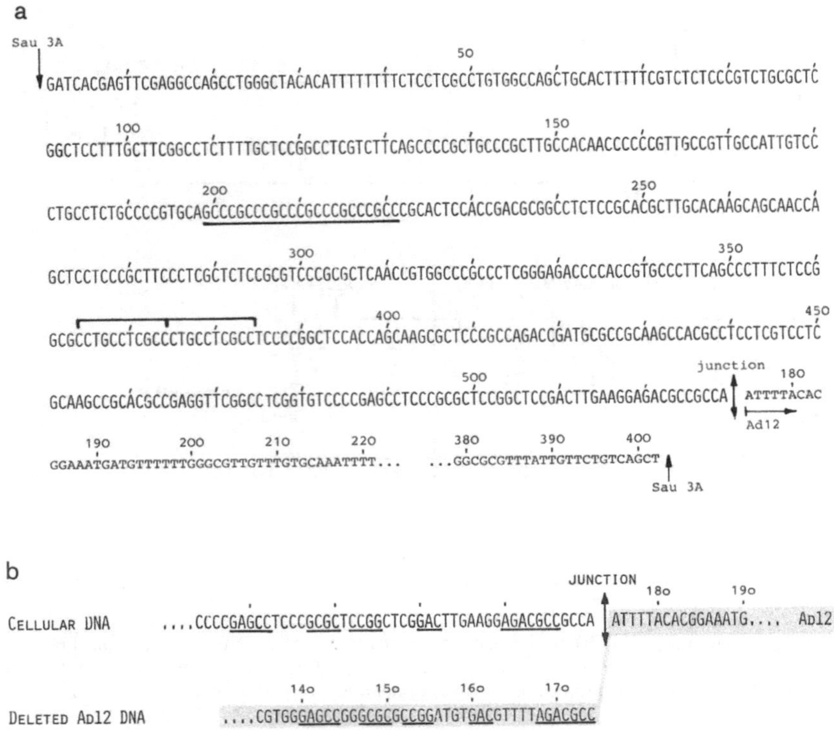

Fig. 9a, b. Nucleotide sequence at site of junction between the left terminus of Ad12 DNA and hamster cell DNA (STABEL and DOERFLER 1982). **a** Nucleotide sequence. **b** Junction site with patch homologies between deleted viral DNA sequence and cellular sequence replacing it. Patch homologies are underlined. Ad12 sequence is shaded

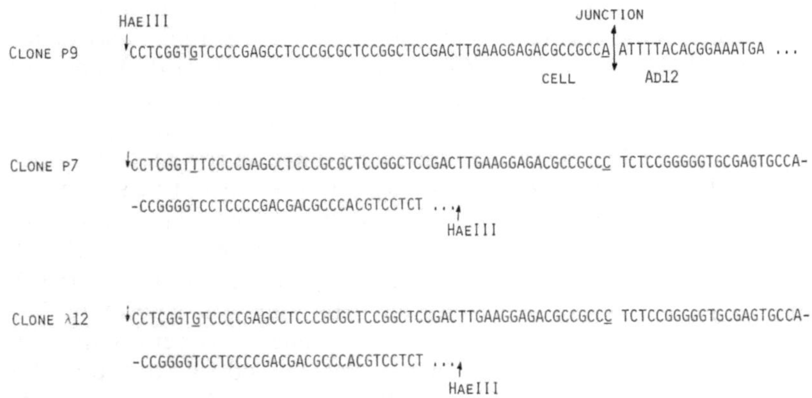

Fig. 10. The recloned cellular sequence from nonvirus-transformed hamster cells is identical to the cellular junction sequence in cell line CLAC1 (STABEL and DOERFLER 1982). Clone p9 represents the junction sequence from cell line CLAC1. Clones p7 and λ12 are different clones from BHK21 (B3) hamster cells

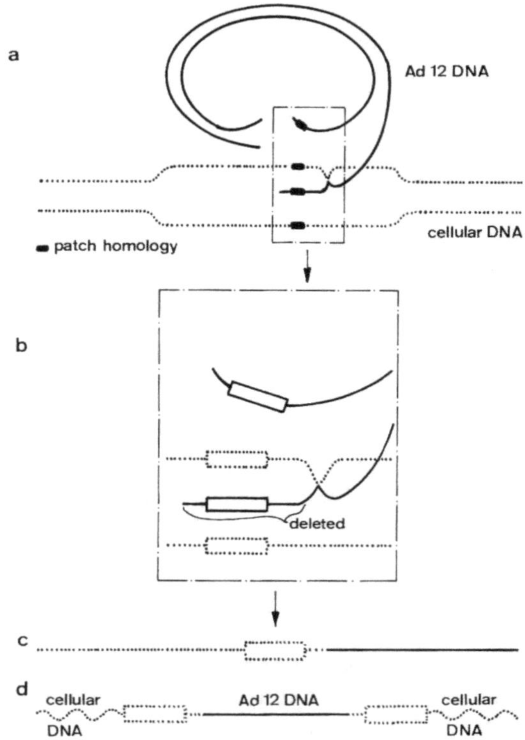

Fig. 11. a Hypothetical model of adenovirus DNA integration (STABEL and DOERFLER 1982). The terminus of one strand of adenovirus DNA invades double-stranded cellular DNA. The intermediate in recombination could be circular or linear. Site of recombination could be determined by patterns of patch homologies. Recombination occurred outside the realm of these patches (**b**). The model explains the deletion of terminal viral nucleotides. It is, of course, possible that the circular structure forms only at the time of insertion. Inserted viral DNA, **c** and **d**

line CLAC1 was mediated by multiple, short sequence homologies (Fig. 11) in an invasive recombination event. Multiple sets of short patch homologies might be recognized as patterns in independent integration events. As is apparent from Fig. 11, this model also accounts for the loss of terminal viral DNA sequences in line CLAC1. Such losses have proved a general phenomenon in the insertion of foreign (adenoviral) DNA by this mechanism. The model proposed was not intended to describe all aspects of the insertion mechanism. The model refers to the situation at the site of recombination at a certain moment in one particular recombination event that led to cell line CLAC1 and does not predict whether linear or circularized viral DNA molecules are intermediates in the reaction. As mentioned above, it is conceivable that circularized adenoviral DNA is required for recombination.

Another molecular feature of this model predicts (Fig. 12) that depending on the combination of patch homologies, that may happen to interact,

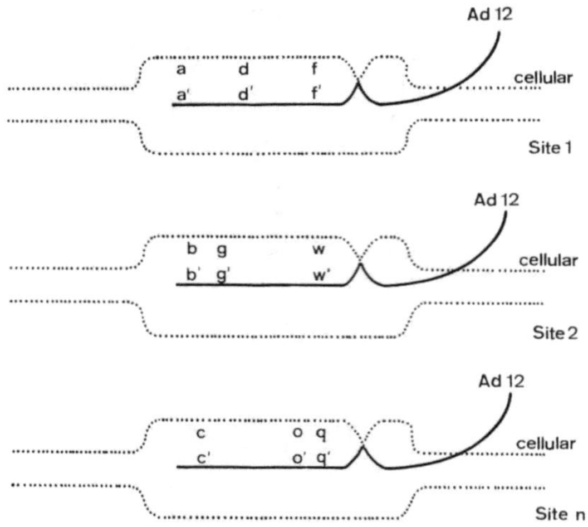

Fig. 12. Patterns of patch homologies determine site of insertion (STABEL and DOERFLER 1982). Letters *a, d, f; b, g, w;* and *c, o, q* represent combinations of patch homologies

different sites of recombination could be selected. Thus, a certain combination of patchy homologies may be decisive and would also account for the large number of possible sites. It should be added, however, that distinct patches, like at the junction site analyzed in cell line CLAC1, cannot always be found and that the insertion mechanism may not be entirely dependent on this feature.

5.5 The Ad5-DNA Transformed Hamster Cell Line BHK268-C31 (WESTIN et al. 1982)

The cell line BHK268-C31 carried two Ad5 DNA inserts, which were derived from the left terminus of Ad5 DNA. One of the viral inserts was cloned in λ Charon 4a DNA and consisted of a 4.4-kb fragment of Ad5 DNA which proved colinear with the authentic Ad5 genome between map positions 1.5 and 14.2. The junctions between viral and flanking cellular DNA were sequenced. The junction sequences exhibited no specific features, except patch homologies of octa- and nonanucleotides between viral and cellular DNA. The junction at the left terminus involved nucleotide 573 of Ad5 DNA; the first 572 nucleotides were deleted. Since cell line BHK268-C31 was generated by transformation of primary baby hamster kidney cells with fragmented Ad5 DNA, the occurrence of such deletions was not surprising. It was of interest that salmon sperm DNA, which had been used as carrier in the transfection experiment, was integrated adjacent to the right terminus of Ad5 DNA. This finding again raises the question to what extent fixation of the carrier DNA used in most transfection experiments contributes to the cellular transformation event.

5.6 Linkage Between Right and Left Termini of Ad5 DNA in Transformed Rat Line 5RK20
(VISSER et al. 1982)

In this particular rat cell line, a site of linkage between the right and left termini of Ad5 DNA was sequenced. This analysis revealed perfect congruence with the authentic Ad5 DNA sequence and a direct linkage of base pair 63 of the left-hand end of the viral genome to base pair 108 of the right terminus. Cellular sequences, rearranged viral DNA, or base pairs of unknown origin, as found at the internal viral junction in cell line HE5, were not detected at this internal junction. It was concluded that joining of the viral termini might have anteceded the integration event (VISSER et al. 1982). Obviously, it will be difficult from the data available to deduce the possible mechanism of this insertion event.

5.7 Linkage Between Ad12 DNA and Human KB Cell DNA in SYREC2 Which Is Encapsidated into Virus Particles
(DEURING et al. 1981a)

We have previously described a symmetric recombinant SYREC between the left terminus of Ad12 DNA and human KB cell DNA (DEURING et al. 1981a). The human cell DNA contained, at least in part, repetitive sequences recurring several hundred times in cellular DNA. SYREC DNA was encapsidated into viral particles and exhibited the same or nearly the same length as authentic Ad12 DNA. It was likely that SYREC DNAs constituted a collection of similarly but not identically constructed molecules. Accordingly, the length of the left terminal Ad12 DNA segment involved in the SYREC structure might vary somewhat from recombinant to recombinant. In any event, the presence of the left terminal sequences was important, as it was demonstrated that the left-most 400 nucleotides and specifically a sequence between bp 290 and 390 of adenoviral DNA were essential in packaging viral DNA (HAMMARSKJÖLD and WINBERG 1980, 1983). On denaturation and renaturation, the recombinant molecules were converted to molecules half the length of Ad12 DNA. Thus, SYREC represented a symmetrically duplicated inverted repeat of the type ABCDD′C′B′A′ with the left terminus of Ad12 DNA flanking the molecule on either end. The occurrence of SYREC molecules provided proof for the generation of virus-host DNA recombinants in productively infected cells as described earlier (BURGER and DOERFLER 1974; SCHICK et al. 1976; TYNDALL et al. 1978). It was therefore mandatory to confirm the recombinant structure of SYREC DNA by determining the nucleotide sequence across the point of fusion between Ad12 DNA and human KB cellular DNA. A 6.4-kb *Bam*HI fragment was cloned from SYREC2 DNA into the plasmid pBR322. The imme-

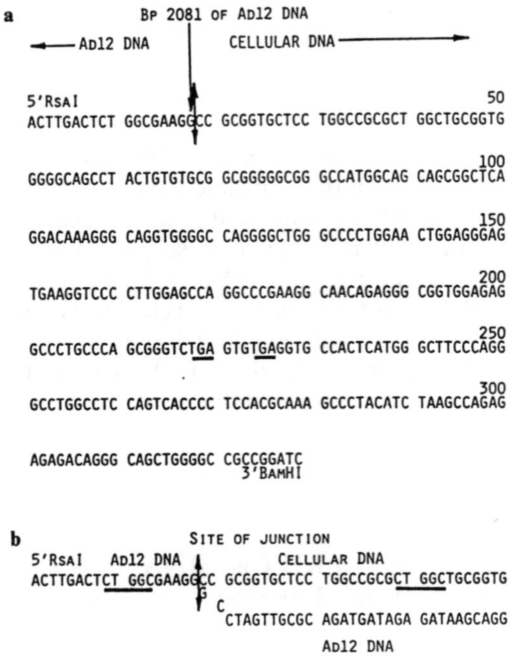

Fig. 13a, b. Site of junction between the left terminal 2081 nucleotides of Ad12 DNA and KB cellular DNA in SYREC2 DNA (Deuring et al. 1981a; Deuring and Doerfler 1983). **a** Nucleotide sequence. The Ad12 Elb protein starts at base pair 1846 of Ad12 DNA and continues in an open reading frame beyond the site of junction for another 66 codons. Two termination codons TGA have been indicated. **b** Comparison between the deleted viral nucleotide sequence and the cellular sequence replacing it. *CTGGC* is a common pentanucleotide between the remaining viral and the adjoining cellular DNA sequences

diate junction site was recloned as a 1.4-kb *Msp*I fragment, again in the plasmid pBR322. The results in Fig. 13 present the nucleotide sequence across the joint between the left terminus of Ad12 DNA and KB cell DNA (Fig. 13a). A comparison with the nucleotide sequence at the left terminus of authentic Ad12 DNA (Sugisaki et al. 1980; Bos et al. 1981; Kimura et al. 1981) revealed that the SYREC2 DNA analyzed contained 2081 nucleotides from the left end of Ad12 DNA covalently linked to KB cellular DNA (Deuring and Doerfler 1983). The 6.4 kb *Bam*HI fragment hybridized to Southern blots of KB DNA. Thus, the DNA fragment adjacent to the 2081 nucleotides of Ad12 DNA was indeed cellular. Sequence comparisons between the KB cellular sequence and the deleted Ad12 sequence from nucleotide 2082 on to the right showed no obvious homologies that might have been instrumental in recombination (Fig. 13b).

It was also interesting to note that the SYREC1 DNA molecule isolated in 1980 had only about 700–1150 nucleotides of Ad12 DNA derived from the left terminus (Deuring et al. 1981a). The recombinant SYREC2 described here was isolated in 1977 and carried the first 2081 nucleotides from the left end of Ad12 DNA as determined by sequence analysis (Deuring and Doerfler 1983). These data suggest that SYREC DNA molecules may undergo alterations during continued passage, e.g., selective deletions of viral nucleotides. It is, however, also conceivable that at different times, different SYREC populations became the preponderant ones or that different recombinants were chosen for the study at different times.

5.8 Insertion of a Human Hinf DNA Family into the Adenovirus Type 7 Genome
(SHIMIZU et al. 1983)

A variant adenovirus genome has been identified that carries a new family of human repetitive sequences, a Hinf DNA family, which is tandemly arranged in the mutant adenovirus type 7 genome (SHIMIZU et al. 1983). The repeats consist of a 319 bp unit which occurs as a dimerized unit of two related but distinct subunits of 172 bp and 147 bp. The Hinf insertion in adenovirus type 7 DNA resembles the sequence of eukaryotic transposable elements.

5.9 Linkage Between the Left Terminus of Ad12 DNA and Hamster Cell DNA in the Ad12-Induced Tumor T1111(2)
(U. LICHTENBERG and W. DOERFLER, manuscript in preparation)

The tumor T1111(2) was induced by injecting Ad12 into newborn Syrian hamsters; each cell contained 10–11 viral genome equivalents (KUHLMANN et al. 1982). A left terminal off-size fragment was cloned in λgtWES·λB DNA, and a *Hha*I fragment carrying the site of junction was subcloned in pBR322. The off-size band proved interesting, as it disappeared upon subsequent passage of the T1111(2) cells in culture (KUHLMANN et al. 1982). The nucleotide sequence across the site of junction was determined (Fig. 14a) (U. LICHTENBERG and W. DOERFLER, manuscript in preparation). The data revealed that the terminal 64 nucleotides had been deleted at the site of junction. Thus, deletions of terminal viral nucleotides again proved a general principle in this type of recombination process. Unlike the findings in tumor cell line CLAC1, comparison of the deleted sequence of 64 nucleotides of Ad12 DNA with the hamster cellular sequence replacing it did not reveal patchy homologies (Fig. 14b). Thus, the insertion event was apparently not directed by a mechanism similar to the one proposed for cell line CLAC1 in Fig. 11 or alternatively, cellular DNA had been deleted.

An interesting peculiarity of this junction site was uncovered by computer analysis of the nucleotide sequence (Fig. 14a). The left terminal nucleotide number 65 of Ad12 DNA was linked to cellular DNA of unique type or of low abundancy in hamster cell DNA. At 135 nucleotides to the left from this junction site, 64 nucleotides of Ad12 DNA from the left viral DNA terminus (nucleotides number 1297 to 1361) were inserted in the opposite direction into cellular DNA. It is unknown at what stage or in what way this insertion or partial duplication of viral sequences had occurred. This finding indicates that complicated rearrangements of sequences can occur occasionally before, during, or after the integration event.

a

```
T1111(2)

5' GGAGGCCGCA AAAGAGAGGA GTGTGTGAAG

CELLULAR
DNA   AD12 DNA
GTGATGCGTT CACTGACAGA TCAACAGGCA
     BP 1361

CTGTTTGTTC TCTTTCTTCC TCTTGAATCA

AD12 DNA      CELLULAR DNA
AATCCAAAAA TGCTTTCATT CTTATATTTT
       BP 1297

TGGTTGTGCC ATCACCACCC AGCTTGATTC

TGAATATGAC ATGGTTATTG CCTACATCAA

CTACCAAAAC CTGCGGTTGC CTGCAAAAGA
                          SITE OF JUNCTION
          CELLULAR DNA          AD12
TGAAGCCAGA TAAATTTCTT GCACTAATTT
                         BP 65

GATTGGGTGG AGGTGTGGCT TTGGCGTGCT

TGTAAGTTTG GGCGGATGAG GAAGTGG 3'
```

b

```
T1111(2)

SITE OF JUNCTION
CELLULAR DNA        AD12 DNA

CCTACATCAA CTACCAAAAC CTGCGGTTGC CTGCAAAAGA TGAAGCCAGA TAAATTTCTT GCACTAATTT GATTGGGTGG AGGTGTGGCT TTGGCGTGCT
                                                                     G
                                                                    T
CTATATATAT AATATACCTT ATACTGGACT AGTGCCAATA TTAAAATGAA GTGGGCGTAG TG
BP 1                                         AD12 DNA                BP 65
```

Fig. 14a, b. Site of junction between the left terminal Ad12 DNA and hamster cell DNA in the tumor T1111(2) (U. LICHTENBERG and W. DOERFLER, manuscript in preparation). **a** Nucleotide sequence. Note the inverted insert of viral nucleotides into cellular DNA. **b** Sequence comparison between 64 deleted adenoviral nucleotides and the hamster cell sequence replacing them

5.10 Linkage Between the Left Terminus of Ad12 DNA and Mouse Cell DNA in the Ad12-Induced Tumor CBA-12-1-T (M. SCHULZ and W. DOERFLER, manuscript in preparation)

The tumor CBA-12-1-T was induced by injecting Ad12 into newborn CBA/J mice (STARZINSKI-POWITZ et al. 1982), and the integration patterns of the approximately 20–30 viral genomes persisting were determined by standard

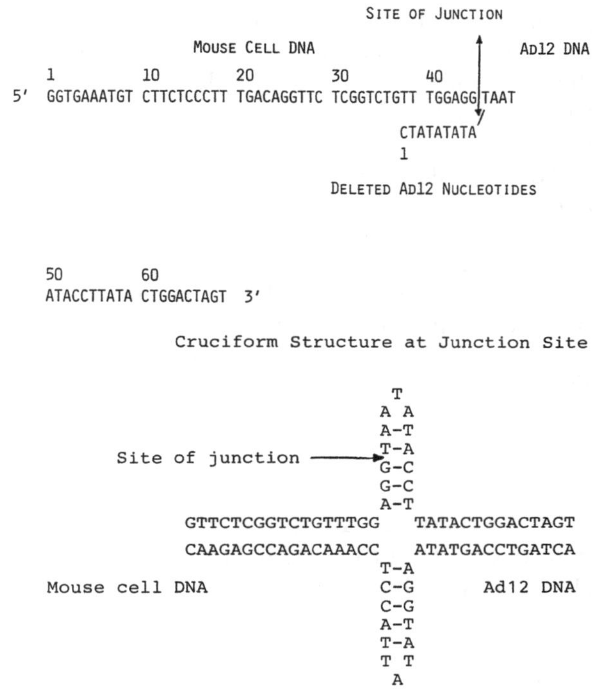

Fig. 15. Nucleotide sequence at the junction site between the left terminus of Ad12 DNA and mouse DNA in tumor cell line CBA-12-1-T (M. SCHULZ and W. DOERFLER, manuscript in preparation). At site of junction, a cruciform structure is detectable

methods. The left terminal *Eco*RI junction between Ad12 DNA and mouse cell DNA was cloned in λgtWES·λB DNA, and the *Pst*I junction fragment was subcloned in pBR322 and sequenced (Fig. 15). At the left terminus the first nine nucleotides of Ad12 DNA are deleted. Nucleotide 10 has been linked to cellular DNA apparently of the unique type. The deleted viral nucleotide sequence has no homology to cellular DNA replacing it. At the site of junction, there is a short palindromic sequence detectable (Fig. 15) perhaps similar to the one in line CLAC3 (Fig. 3b).

5.11 Left Terminal Junction Site of Ad12 DNA to Hamster Cell DNA or to Ad12 DNA in Cell Line HA12/7
(STABEL et al. 1980)

The hamster cell line HA12/7 was derived from primary hamster embryo cells which were transformed in culture by Ad12. Each cell carried about three to four Ad12 DNA equivalents, and the integration patterns were described in detail (STABEL et al. 1980). These patterns were characterized by a peculiar selective amplification of the left terminal Ad12 DNA sequences (STABEL et al. 1980). Moreover, there is evidence that the E1 region of integrated Ad12 DNA in cell line HA12/7 is expressed abundantly

(Schirm and Doerfler 1981; Esche and Siegmann 1982). It was decided to clone and sequence the left terminal junction site of viral DNA from this line, since it was conceivable that the terminal amplifications were relevant to explain the levels of expression of the E1 region. A *Pst*I off-size fragment of about 4 kb (see Stabel et al. 1980, Fig. 9c) was cloned into pBR322, and the *Eco*RI-cut plasmid vector was subsequently cloned into λgtWES·λB DNA to facilitate detection of clones containing the junction fragment. This junction fragment is presently being analyzed.

In propagating the λ phage clones containing the 4-kb insert, we observed an interesting instability, in that in some of the clones specific deletions were introduced (S. Stabel, R. Jessberger, and W. Doerfler, unpublished experiments). This instability may suggest the existence of peculiar DNA structures at or around the site of junction (cf. Sect. 5.12).

5.12 Excision of Amplified Viral DNA at Palindromic Sequences from the Ad12-Transformed Hamster Cell Line T637 (Eick et al. 1983)

In the Ad12-transformed cell line T637, 20–22 copies of viral DNA are integrated into cellular DNA. Morphological revertants of cell line T637 (Groneberg et al. 1978) that had lost all but about one or one-half of a viral DNA copy were considered useful tools in elucidating the mechanisms of insertion and excision of the Ad12 DNA molecules. It was shown that specific sites of linkage between viral genomes were lost in many of the morphological revertants. The same sites proved highly sensitive to endogenous nucleases, as shown by autoincubation of T637 nuclei. The failure of all attempts to molecularly clone the sites of junction in cell line T637 suggested that palindromic sequences might exist at these sites and prompted experiments in which T637 DNA was denatured, briefly renatured, and subsequently digested with S1 nuclease. The fold-back structures generated in this way were of distinct sizes and could be localized to the termini of the integrated Ad12 genome. Evidence was adduced that identical DNA sequences at these termini generated the fold-back structures and at the same time were highly sensitive to autodigestion by endogenous nucleases. One of these sites appeared to be less sensitive to nuclease digestion than the other one. The results of experiments in which T637 DNA was treated with endonuclease VII (Kemper and Garabett 1981), an enzyme presumably recognizing Holliday structures (Holliday 1964) in DNA (Mizuuchi et al. 1982), strengthened the notion that palindromic DNA sequences existed at the sites of linkage of Ad12 genomes in cell line T637. It appeared likely that only part of these sequences assumed the cruciform configuration at a given time. Thus, there was evidence of special sequence arrangements at the sites of viral-viral or viral-cellular DNA junctions (Eick et al. 1983).

	HE5 (r) v-c	c-v	CLAC3 v-c	c-v	CLAC1 v-c	c-v	SYREC 2 v-c	c-v	T1111(2) v-c	c-v
PBR322 TOTAL SEQUENCE	10:1 9:1 8:1	11:1 10:1 8:10	8:4	9:1 8:3	12:1 10:2 9:9 8:25	11:1 10:3 9:11 8:25	10:2 9:7 8:19	10:1 9:2 8:24	9:2 8:7	9:4 8:7
AD12 LEFT TERMINAL 2320 NUCLEOTIDES	9:5 8:10	11:1 9:2 8:11	11:2 9:1	9:2 8:4	9:5 8:16	10:1 9:7 8:6	9:1 8:5	10:1 9:1 8:8	9:2 8:4	8:2
AD2 LEFT 38 %	12:1 10:3 9:10 8:40	9:13 8:36	10:1 9:3 8:12	9:3 8:8	12:1 11:2 10:8 9:41 8:152	11:1 10:7 9:24 8:102	14:1 11:2 10:6 9:15 8:61	11:3 10:3 9:6 8:64	11:1 10:1 9:2 8:25	9:2 8:17
AD2 EcoRI F, D, E, C FRAGMENTS	12:1 11:1 10:2 9:11 8:50	11:2 10:2 9:11 8:52	10:1 9:3 8:12	8:12	12:1 11:3 10:8 9:16 8:62	12:1 11:2 10:9 9:17 8:84	11:1 10:4 9:10 8:33	12:1 10:1 9:7 8:35	9:3 8:14	11:2 10:3 9:6 8:19

Fig. 16. Computer-aided analysis of cellular junction sequences analyzed. The nucleotide sequences of cellular junction sequences read in both directions (5′ – 3′), viral-cellular (v-c) or cellular-viral (c-v), were compared, with the aid of a CDC76 computer, to the entire sequence of the *E. coli* plasmid pBR322 (SUTCLIFFE 1978), to the left terminal 2320 nucleotides of Ad12 DNA (SUGISAKI et al. 1980; BAS et al. 1981), and to the left and right terminal sequences of Ad2 DNA (GINGERAS et al. 1982) as indicated. The designations used are the following: 8:2 two common octanucleotides, 10:1 one common decanucleotide, etc. Compare legend to Fig. 17 for relative lengths of cellular sequences

6 Computer Analyses of Cellular Sequences at Junction Sites

The cellular nucleotide sequences linked to the left or right terminus of adenovirus DNA were computer analyzed for the presence of common features that might explain the mechanism of insertion. The following parameters were included in the analyses.

6.1 Sequence Homologies Including Patch Homologies to Ad2 and/or Ad12 DNA

The data in Fig. 16 list such homologies found between cellular junction sequences as listed (v-c, viral-cellular sequence or c-v, cellular-viral sequence in direction of reading), and the left terminal *Hind*IIIG fragment of Ad12 DNA (2320 bp) (SUGISAKI et al. 1980; BOS et al. 1981), of the left terminal 38% of Ad2 DNA, or the sequence comprised by the right terminal EcoRI fragments F, D, E, and C of Ad2 DNA (GINGERAS et al. 1982). As expected from previous analyses on the occurrence of patch homologies (Gahlmann et al. 1982), such homologies are very abundant. They comprise sequence identities of up to 14 nucleotides. Homologies shorter than eight nucleotides have not been tabulated, as they occur very frequently. The longer the viral and cellular sequences included in the analysis, the more abundant is the number of sequence homologies detected. It is obvious that such sequence homologies could theoretically be used in directing recombination events. In some instances, they may actually have been used (Fig. 9b). In other cases, no homologies have been found. Thus, patch homologies represent a very sizeable repertoire that can potentially become important in directing recombination events. A purely statistical treatment of sequence homologies is obviously meaningless. As revealed by the abundance of sequence homologies, their occurrence is statistically predicted, and they could be useful as signals in recombination events.

6.2 Sequence Homologies to the Nucleotide Sequence of the Cloning Vector pBR322

The entire nucleotide sequence of pBR322 was included in this comparison, mainly to rule out cloning artifacts in the sense that longer nucleotide sequences from the plasmid vector could have recombined with the cellular junction sequence. The data presented in Fig. 16 provide no evidence whatsoever for this possibility. Patchy homologies occur at rates comparable to those found between cellular junction sequences and adenoviral DNA. It is concluded that cloning artifacts have not occurred with the junction sequences analyzed. This possibility had already been ruled out by recloning and sequencing of the original unoccupied cellular sequence that had remained unaltered in cell lines HE5 and CLAC1 (Gahlmann et al. 1982; Stabel and Doerfler 1982; Gahlmann and Doerfler 1983). It is also apparent from the data presented that patch homologies between cellular junction sequences are about equally frequent to the sequence of the prokaryotic vector as to the sequences of the human adenoviruses. Thus, patch homologies appear to be a common. These homologies can, however, be utilized to stabilize recombination complexes (cf. Fig. 11, 12).

6.3 Structural Peculiarities

The computer programs used in the analyses were not the most sophisticated ones. Complex structures with multiple mismatches permitted may not have been detected. Only in occasional cases (CLAC3, HE5 internal link, CBA-12-1-T) were stemmed loops or palindromic sequence arrangements discovered. Their functional significance is unknown. A stemmed loop has also been described at a viral-cellular junction in a mutant adenovirus ge-

	HE5 vc	HE5 cv	CLAC3 vc	CLAC3 cv	CLAC1 vc	CLAC1 cv	SYREC vc	SYREC cv	T1111(2) vc	T1111(2) cv
HE5 vc		8:5	-	8:1	8:2	-	8:4	9:1 8:1	-	8:1
HE5 cv			8:1	-	-	8:2	9:1 8:1	8:4	8:1	-
CLAC3 vc				-	-	-	-	-	-	-
CLAC3 cv					-	-	-	-	-	-
CLAC1 vc						-	10:1 9:1 8:10	8:1	-	8:1
CLAC1 cv							8:1	10:1 9:1 8:10	8:1	-
SYREC vc								9:2 8:5	-	8:3
SYREC cv									8:3	-
T1111(2)vc										8:1
T1111(2)cv										

Fig. 17. Computer-aided comparison of all cellular junction sequences with each other. The lengths in nucleotide pairs of the cellular DNA sequences are the following: HE5 401, CLAC3 82, CLAC1 529, SYREC 304, T1111 (2) 170. Thus, strictly speaking, comparisons of frequencies of patch homologies have to take into account the relative nucleotide numbers of each cellular sequence

nome that carries an insertion of cellular DNA (Shimizu et al. 1983). There is at present no proven way to determine the stability and/or functional significance of such structures. One must therefore refrain from overinterpreting these findings.

6.4 Some of the Viral-Cellular Junction Sequences Have Been Analyzed for the Existence of Open Reading Frames

Several open reading frames comprising shorter or longer sequences have been found at some of the junction sites. We hesitate to ascribe biological significance to these results, as we have not yet completely analyzed the RNA transcribed from junction sequences in the cell lines investigated. These analyses have just been initiated (Gahlmann and Doerfler 1983).

6.5 Comparison of All Cellular Junction Sequences

Lastly, all cellular junction sequences were compared with each other (Fig. 17). Again, the matrix reveals only short sequence homologies at frequencies that may not exceed values statistically expected. Moreover, there are no striking similarities or identities of nucleotide sequences when these patch homologies are compared with each other. Hence, a cellular nucleotide sequence common to all junction sites involving adenoviral DNA integration does not exist. Allowing for maximal flexibility of the recombination mechanism in eukaryotic cells, patch homologies as abundantly found may be utilized in stabilizing recombination complexes, but it cannot be claimed that such homologies are absolutely essential, since they are not always found.

7 Main Conclusions

Obviously, we are far from understanding the mechanism of recombination in molecular terms that explains the insertion of foreign (viral) DNA into the genome of mammalian cells. For a number of reasons, it will be very important to elucidate this mechanism. In fact, one may have to accept the possibility that more than one functional pathway for insertion exists.

From the data collected in this review, we shall try to deduce a few general features which may help to formulate new experimental approaches. In evaluating these general features, we are aware of the fact that we have analyzed highly selected, clonal lines of adenovirus-transformed cells or adenovirus-induced tumors. In these cells the site of insertion may reflect in a complex way the mode of selecting for transformed cells with certain growth properties. Proximity of viral promoters and/or enhancers to certain cellular sequences and vice versa may bestow upon the carrier cell of such inserted genomes growth advantages that have led to the selection of these cells for further analyses. Thus, the cells selected for analyses may not necessarily reflect the entire gamut of recombinatorial possibilities between viral and host DNA.

What are the most important general findings gleaned from extensive analyses of host-viral DNA junction sites?

1. The insertion sites have been studied in a series of transformed or tumor cell lines which were usually initiated by adenovirions, not by transfection with viral DNA. Exceptions have been noted (Deuring et al. 1981; Westin et al. 1982; Shimizu et al. 1983). The mode of introducing foreign DNA into the host cell may be of importance in guiding or selecting the mechanism of recombination. Usually adenovirus DNA has been linked to cellular DNA via the terminal viral sequences. This finding suggested a mechanism that somehow gave special significance to the termini of viral DNA. A circular intermediate, in particular one that was formed by the interaction of the terminal protein molecules, appeared conceivable. On the other hand, in some instances the recombination event had also involved internal viral DNA sequences (Sambrook et al. 1974; Vardimon and Doerfler 1981). Thus, one type of mechanism may not exclusively apply; there may be alternative ways of inserting foreign DNA.

2. It has been previously pointed out (Doerfler 1982) that the colinear insertion of the nearly intact genome of Ad12 and the insertion of partly deleted Ad2 genomes may be the consequence of the degree of permissivity

of the interaction between virus and host cells. Ad12 infects hamster and mouse cells abortively; Ad2 replicates in hamster cells. In some of the Ad12 tumors analyzed, fragments of viral DNA were integrated. Moreover, cell lines carrying multiple copies of nearly intact Ad12 genomes contained, in addition, nonstoichiometric sets of fragments of parts of the viral genome.

3. In considering a model that emphasizes the functional role of the terminal viral protein in the recombination event, this function could be exerted in a circular, as well as a linear, intermediate. In fact, a circular structure may never be formed, but the two viral DNA termini may contact the same cellular site in immediate succession, the second terminus perhaps only after endonucleolytic cleavage of cellular DNA has been effected.

4. In cell line HE5, it was striking that insertion of Ad2 DNA did not lead to the deletion of a single cellular nucleotide at the site of linkage. In cell line CLAC1, the original cellular sequence was unaltered up to the site of junction. Thus, deletion of cellular DNA cannot be the rule in inserting foreign DNA.

5. Moreover, repetitions of cellular nucleotides at the sites of linkage between adenoviral and cellular DNA were never observed. This result ruled out the possibility of a transposonlike mechanism, as has been reported for the integration of the retroviral proviruses (Majors and Varmus 1981).

6. Deletions of viral nucleotides of varying lengths (2, 45, 174, 8/10, 64, 9) were seen at all junctions analyzed. An invasive model of recombination (cf. Fig. 11) would explain these microdeletions, but other mechanisms are also possible. In cell line HE5, hamster cell DNA was linked to nucleotide 11 on the left terminus and to nucleotide 9 on the right terminus of the integrated Ad2 genome (Gahlmann and Doerfler 1983). In the mouse tumor CBA12-1-T, cellular DNA was connected to nucleotide 10 on the left terminus of Ad12 DNA (Schulz and Doerfler, manuscript in preparation). In two variants of adenovirus type 16 (Ad16), the viral DNA carried a reduplication of left terminal DNA sequences on the right terminus of the Ad16 genome. In the DNA of both variants, the 8 right terminal nucleotides were deleted (Hammarskjöld and Winberg 1983). It was therefore tempting to speculate that a hot spot of recombination might exist around nucleotides 8 to 10 at the adenoviral DNA termini. These nucleotides are located just in front of the essential consensus sequence at which viral DNA replication is initiated (Tolun et al. 1979).

7. We have discussed at length the hypothetical role that patch homologies or combinations of patches could play in the recombination event. A stabilizing function for the recombination complex has been envisaged. We should like to emphasize, however, that such patch homologies have not always been found and they may not be essential in some integration events. Perhaps they occur on one terminus of viral DNA only and suffice there to stabilize the complex. Furthermore, one has to consider patch homologies of up to 14 nucleotides in length and certainly combinations of nucleotide patches involving even longer stretches (cf. Fig. 9b) as naturally occurring which could then be exploited by the mechanisms of recombination. This concept would explain the large number of different recombina-

tion sites actually found. In the sequences analyzed, we have noted two types of patch homologies; one type existing between neighboring cellular and preserved viral DNA sequence, another one between the deleted viral nucleotides and the cellular sequence replacing them.

8. Another interesting phenomenon is the insertion of nucleotides of unknown origin at the internal junction of Ad2 DNA in cell line HE5 (cf. Fig. 6). Is this insertion haphazard or does it point to a complex recombination event involving other (cellular?) sequences?

9. Adenoviral DNA integration has been observed in unique or repetitive cellular sequences. It is always possible that the repetitive cellular sequences were generated as a consequence of the insertion event or were due to another viral function. This reservation obviously does not hold for the data obtained with cell line CLAC1, since in that case integration had occurred into repetitive cellular DNA (STABEL and DOERFLER 1982).

10. We have just begun to study the expression as cytoplasmic RNA of cellular junction sequences in normal and adenovirus-transformed cells (GAHLMANN and DOERFLER 1983). These investigations will have to be carried out in more detail in the future.

11. The mechanism of adenoviral DNA insertion is unknown. In any event, this mechanism is not like the integration of bacteriophage λ DNA into the bacterial chromosome at one highly specific site, since all cellular junction sequences are different (cf. Fig. 17). Moreover, the mechanism of adenoviral DNA insertion does not resemble that of retroviral proviruses or of transposons. There must be yet another mechanism or other mechanisms of recombination in mammalian cells.

Comparisons with the integration of bacteriophage λ DNA are interesting also, in that λ DNA can integrate at low frequency at nonspecific sites. Several of these sites have been sequenced (ROSS and LANDY 1983). The highly significant finding has been reported that at presumably nonspecific sites the internal hexanucleotide TTTATA of the highly specific "attachment site" GCTTTTTTATACTAA sequence (LANDY and ROSS 1977) has been preserved. Thus, even part of this highly specific sequence can be recognized. A hexanucleotide is consequently sufficient to direct recombination mechanisms. This hexanucleotide is certainly a less conspicuous signal than some of the patch combinations we have observed at sites of adenoviral DNA integration. The data on sequence peculiarities at nonspecific sites of λ DNA insertion tend to caution one toward a purely statitistical interpretation of patchy homologies and their possible function. Highly specific proteins that have the capacity to recognize short sequence identities and utilize them are probably involved in recombination.

12. In searching for signals to direct the recombination event, certain DNA structures may also play a role (cf. Figs. 3b, 6, 15). Future research may have to be directed, in particular, toward specific chromatin structures in the cell that may predispose for foreign DNA insertion. Such investigations are at present possible only to a limited extent.

13. The sequence data on the SYREC2 DNA molecule (cf. Fig. 13) at the site of junction between Ad12 DNA and human cell DNA (DEURING and DOERFLER 1983) and on the Ad7 – human cell DNA recombinant

(Shimizu et al. 1983) provide unequivocal proof for the occurrence of recombinants between adenoviral and cellular DNA in productive infection (Burger and Doerfler 1974).

14. A comparison of many of the features of adenoviral DNA insertion with those derived from investigations on the SV40 (Stringer 1982) or polyoma virus system (Hayday et al. 1982; Mendelsohn et al. 1982) reveal striking similarities. The recombination mechanisms of a mammalian cell – there may in fact be more than one exclusive way of inserting foreign DNA – may not differentiate among different foreign DNA molecules. This assumption would rather tend to belittle the role that the terminal adenoviral protein might have in the insertion process. Since viral-cellular recombinations with simian virus 40, polyoma virus, and adenovirus DNAs have many features in common, it is conceivable that the terminal adenoviral protein has a guiding function in recombination and that the actual recombination is catalyzed by cellular functions. It is, however, remarkable that adenovirus DNA frequently recombines via its termini, whereas SV40 DNA or polyoma virus DNA recombine at random sites of the viral sequence.

Acknowledgments. We thank Petra Böhm for much help with the list of references and for typing this manuscript. Research in the authors' laboratory was supported through SFB74-C1 of the *Deutsche Forschungsgemeinschaft*.

References

Benton WD, Davis RW (1977) Screening λgt recombinant clones by hybridization to single plaques in situ. Science 196:180–182

Bos JL, Polder JL, Bernards R, Schrier PI, Van den Elsen PJ, Van der Eb AJ, Van Ormondt H (1981) The 2.2 kb E1b mRNA of human Ad12 and Ad5 codes for two tumor antigens starting at different AUG triplets. Cell 27:121–131

Botchan M, Topp W, Sambrook J (1976) The arrangement of simian virus 40 sequences in the DNA of transformed cells. Cell 9:269–287

Burger H, Doerfler W (1974) Intracellular forms of adenovirus DNA. III. Integration of the DNA of adenovirus type 2 into host DNA in productively infected cells. J Virol 13:975–992

Cook JL, Lewis AM Jr (1979) Host response to adenovirus 2-transformed hamster embryo cells. Cancer Res 39:1455–1461

Deuring R, Doerfler W (1983) Proof or recombination between viral and cellular genomes in human KB cells productively infected by adenovirus type 12: structure of the junction site in a symmetric recombinant (SYREC). Gene 24

Deuring R, Klotz G, Doerfler W (1981a) An unusual symmetric recombinant between adenovirus type 12 DNA and human cell DNA. Proc Natl Acad Sci USA 78:3142–3146

Deuring R, Winterhoff U, Tamanoi F, Stabel S, Doerfler W (1981b) Site of linkage between adenovirus type 12 and cell DNAs in hamster tumor line CLAC3. Nature 293:81–84

Doerfler W (1968) The fate of the DNA of adenovirus type 12 in baby hamster kidney cells. Proc Natl Acad Sci USA 60:636–643

Doerfler W (1969) Nonproductive infection of baby hamster kidney cells (BHK21) with adenovirus type 12. Virology 38:587–606

Doerfler W (1970) Integration of the deoxyribonucleic acid of adenovirus type 12 into deoxyribonucleic acid of baby hamster kidney cells. J Virol 6:652–666

Doerfler W (1982) Uptake, fixation and expression of foreign DNA in mammalian cells: The organization of integrated adenovirus sequences. Curr Top Microbiol Immunol 101:127–194

Doerfler W, Lundholm U, Hirsch-Kauffmann M (1972) Intracellular forms of adenovirus deoxyribonucleic acid. I. Evidence for a deoxyribonucleic acid-protein complex in baby hamster kidney cells infected with adenovirus type 12. J Virol 9:297–308
Doerfler W, Lundholm U, Rensing U, Philipson L (1973) Intracellular forms of adenovirus DNA. II. Isolation in dye buoyant density gradients of a DNA-RNA complex from KB cells infected with adenovirus type 2. J Virol 12:793–807
Eick D, Doerfler W (1982) Integrated adenovirus type 12 DNA in the transformed hamster cell line T637: Sequence arrangements at the termini of viral DNA and mode of amplification. J Virol 42:317–321
Eick D, Stabel S, Doerfler W (1980) Revertants of adenovirus type 12-transformed hamster cell line T637 as tools in the analysis of integration patterns. J Virol 36:41–49
Eick D, Kemper B, Doerfler W (1983) Excision of amplified viral DNA at palindromic sequences from the adenovirus type 12-transformed hamster cell line T637. The EMBO J 2:1981–1986
Esche H, Siegmann B (1982) Expression of early viral gene products in adenovirus type 12-infected and -transformed cells. J Gen Virol 60:99–113
Esche H, Schilling R, Doerfler W (1979) In vitro translation of adenovirus type 12-specific mRNA isolated from infected and transformed cells. J Virol 30:21–31
Fanning E, Doerfler W (1976) Intracellular forms of adenovirus DNA. V. Viral DNA sequences in hamster cells abortively infected and transformed with human adenovirus type 12. J Virol 20:373–383
Gahlmann R, Doerfler W (1983) Integration of viral DNA into the genome of the adenovirus type 2-transformed hamster cell line HE5 without loss or alteration of cellular nucleotides. Nucleic Acids Res 11:7347–7361
Gahlmann R, Leisten R, Vardimon L, Doerfler W (1982) Patch homologies and the integration of adenovirus DNA in mammalian cells. EMBO J 1:1101–1104
Gallimore PH (1974) Interactions of adenovirus type 2 with rat embryo cells: Permissiveness, transformation and in vitro characteristics of adenovirus-transformed rat embryo cells. J Gen Virol 25:263–273
Gingeras TR, Sciaky D, Gelinas RE, Bing-Dong J, Yen CE, Kelly MM, Bullock PA, Parsons BL, O'Neill KE, Roberts RJ (1982) Nucleotide sequences from the adenovirus 2 genome. J Biol Chem 257:13475–13491
Groneberg J, Doerfler W (1979) Revertants of adenovirus type 12-transformed hamster cells have lost part of the viral genomes. Int J Cancer 24:67–74
Groneberg J, Chardonnet Y, Doerfler W (1977) Integrated viral sequences in adenovirus type 12-transformed hamster cells. Cell 10:101–111
Groneberg J, Sutter D, Soboll H, Doerfler W (1978) Morphological revertants of adenovirus type 12-transformed hamster cells. J Gen Virol 40:635–645
Gutai MW, Nathans D (1978) Evolutionary variants of simian virus 40: cellular DNA sequences and sequences at recombinant joints of substituted variants. J Mol Biol 126:275–288
Hammarskjöld ML, Winberg G (1980) Encapsidation of adenovirus 16 DNA is directed by a small DNA sequence at the left end of the genome. Cell 20:787–795
Hammarskjöld ML, Winberg G (1983) Further characterization of the packaging sequence of Ad16 (Abstr). Gulbenkian Workshop on the Molecular Biology of Adenoviruses, Sintra, Portugal
Hayday A, Ruley HE, Fried M (1982) Structural and biological analysis of integrated polyoma virus DNA and its adjacent host sequences cloned from transformed rat cells. J Virol 44:67–77
Hohn B (1979) In vitro packaging of λ and cosmid DNA. Methods Enzymol 68:299–309
Hohn B, Murray K (1977) Packaging recombinant DNA molecules into bacteriophage λ particles in vitro. Proc Natl Acad Sci USA 74:3259–3263
Holliday R (1964) A mechanism for gene conversion in fungi. Genet Res 5:282–304
Ibelgaufts H, Doerfler W, Scheidtmann KH, Wechsler W (1980) Adenovirus type 12-induced rat tumor cells of neuroepithelial origin: Persistence and expression of the viral genome. J Virol 33:423–437

Johansson K, Persson H, Lewis AM, Pettersson U, Tibbetts C, Philipson L (1978) Viral DNA sequences and gene products in hamster cells transformed by adenovirus type 2. J Virol 27:628–639

Kemper B, Garabett M (1981) Studies on T4 head maturation. 1. Purification and characterization of gene-49-controlled endonuclease. Eur J Biochem 115:123–131

Ketner G, Kelly TJ Jr (1976) Integrated simian virus 40 sequences in transformed cell DNA: analysis using restriction endonucleases. Proc Natl Acad Sci USA 73:1102–1106

Kimura T, Sawada Y, Shinawawa M, Shimizu Y, Shiroki K, Shimojo H, Sugisaki H, Takanami M, Uemizu Y, Fujinaga K (1981) Nucleotide sequence of the transforming early region E1b of adenovirus type 12 DNA: structure and gene organization, and comparison with those of adenovirus type 5 DNA. Nucleic Acids Res 9:6571–6589

Kuhlmann I, Doerfler W (1982) Shifts in the extent and patterns of DNA methylation upon explantation and subcultivation of adenovirus type 12-induced hamster tumor cells. Virology 118:169–180

Kuhlmann I, Achten S, Rudolph R, Doerfler W (1982) Tumor induction by human adenovirus type 12 in hamsters: loss of the viral genome from adenovirus type 12-induced tumor cells is compatible with tumor formation. EMBO J 1:79–86

Kruczek I, Doerfler W (1982) The unmethylated state of the promoter/leader and 5′ regions of integrated adenovirus genes correlates with gene expression. EMBO J 1:409–414

Landy A, Ross W (1977) Viral integration and excision: structure of the lambda att sites. Science 197:1147–1160

Majors JE, Varmus HE (1981) Nucleotide sequence at host-proviral junctions for mouse mammary tumor virus. Nature 289:253–258

Maxam AM, Gilbert W (1977) A new method for sequencing DNA. Proc Natl Acad Sci USA 74:560–564

Maxam AM, Gilbert W (1980) Sequencing end-labeled DNA with base-specific chemical cleavages. Methods Enzymol 65:499–560

Mendelsohn E, Baran N, Neer A, Manor H (1982) Integration site of polyoma virus DNA in the inducible LPT line of polyoma-transformed rat cells. J Virol 41:192–209

Mizuuchi K, Kemper B, Hays J, Weisberg RA (1982) T4 endonuclease VII cleaves Holliday structures. Cell 29:357–365

Ortin J, Scheidtmann KH, Greenberg R, Westphal M, Doerfler W (1976) Transcription of the genome of adenovirus type 12. III. Maps of stable RNA from productively infected human cells and abortively infected and transformed hamster cells. J Virol 20:355–372

Roberts RJ, Sciaky D, Gelinas RE, Jiang B-D, Yen CE, Kelly MM, Bullock PA, Parsons KE, O'Neill KE, Gingeras TR (1982) Information content of adenovirus-2 genome. Cold Spring Harbor Symp Quant Biol 47:1025–1037

Robinson AJ, Younghusband HB, Bellet AJD (1973) A circular DNA-protein-complex from adenoviruses. Virology 56:54–69

Ross W, Landy A (1983) Patterns of λ int recognition in the regions of strand exchange. Cell 33:261–272

Ruben M, Bacchetti S, Graham F (1983) Covalently closed circles of adenovirus 5 DNA. Nature 301:172–174

Sambrook J, Botchan M, Gallimore P, Ozanne B, Pettersson U, Williams J, Sharp PA (1974) Viral DNA sequences in cells transformed by simian virus 40, adenovirus type 2 and adenovirus type 5. Cold Spring Harbor Symp Quant Biol 39:615–632

Sambrook J, Greene R, Stringer J, Mitchison T, Hu S-L, Botchan M (1979) Analysis of the sites of integration of viral DNA sequences in rat cells transformed by adenovirus 2 or SV40. Cold Spring Harbor Symp Quant Biol 44:569–584

Sawada Y, Fujinaga K (1980) Mapping of adenovirus 12 mRNAs transcribed from the transforming region. J Virol 36:639–651

Schick J, Baczko K, Fanning E, Groneberg J, Burger H, Doerfler W (1976) Intracellular forms of adenovirus DNA: Integrated form of adenovirus DNA appears early in productive infection. Proc Natl Acad Sci USA 73:1043–1047

Schirm S, Doerfler W (1981) Expression of viral DNA in adenovirus type 12-transformed cells, in tumor cells, and in revertants. J Virol 39:694–702

Sharp PA, Gallimore PH, Flint SJ (1974) Mapping of adenovirus 2 RNA sequences in lytically infected cells and transformed cell lines. Cold Spring Harbor Symp Quant Biol 39:457–474

Shimizu Y, Yoshida K, Ren C-S, Fujinaga K, Rajagopalan S, Chinnadurai G (1983) Hinf family: a novel repeated DNA family of the human genome. Nature 302:587–590

Southern EM (1975) Detection of specific sequences among DNA fragments separated by gel electrophoresis. J Mol Biol 98:503–517

Stabel S, Doerfler W (1982) Nucleotide sequence at the site of junction between adenovirus type 12 DNA and repetitive hamster cell DNA in transformed cell line CLAC1. Nucleic Acids Res 10:8007–8023

Stabel S, Doerfler W, Friis RR (1980) Integration sites of adenovirus type 12 DNA in transformed hamster cells and hamster tumor cells. J Virol 36:22–40

Stringer JR (1981) Integrated simian virus 40 DNA: Nucleotide sequences at cell-virus recombinant junctions. J Virol 38:671–679

Stringer JR (1982) DNA sequence homology and chromosomal deletion at a site of SV40 DNA integration. Nature 296:363–366

Starzinski-Powitz A, Schulz M, Esche H, Mukai N, Doerfler W (1982) The adenovirus type 12 mouse-cell system: permissivity and analysis of integration patterns of viral DNA in tumor cells. EMBO J 1:493–497

Sugisaki H, Sugimoto K, Takanami M, Shiroki K, Saito I, Shimojo H, Sawada Y, Uemizu Y, Uesugi S, Fujinaga K (1980) Structure and gene organization in the transforming *Hind*III-G fragment of Ad12. Cell 20:777–786

Sutcliffe JG (1978) Complete nucleotide sequence of the *Escherichia coli* plasmid pBR322. Cold Spring Harbor Symp Quant Biol 43:77–90

Sutter D, Doerfler W (1979) Methylation of integrated viral DNA sequences in hamster cells transformed by adenovirus 12. Cold Spring Harbor Symp Quant Biol 44:565–568

Sutter D, Doerfler W (1980) Methylation of integrated adenovirus type 12 DNA sequences in transformed cells is inversely correlated with viral gene expression. Proc Natl Acad Sci USA 77:253–256

Sutter D, Westphal M, Doerfler W (1978) Patterns of integration of viral DNA sequences in the genomes of adenovirus type 12-transformed hamster cells. Cell 14:569–585

Tikchonenko TI (to be published) Molecular biology of S16 (SA7) and some other simian adenoviruses. In: Doerfler W (ed) The molecular biology of adenoviruses 2. Curr Top Microbiol Immunol 110

Tiemeier D, Enquist L, Leder P (1976) Improved derivative of a phage λ EK2 vector for cloning recombinant DNA. Nature 263:526–527

Tolun A, Aleström P, Pettersson U (1979) Sequenced of inverted terminal repetitions from different adenoviruses: demonstration of ϕ conserved sequences and homology between SA 7 termini and SV40 DNA. Cell 17:705–713

Tyndall C, Younghusband HB, Bellet AJD (1978) Some adenovirus DNA is associated with the DNA of permissive cells during productive or restricted growth. J Virol 25:1–10

Vardimon L, Doerfler W (1981) Patterns of integration of viral DNA in adenovirus type 2-transformed hamster cells. J Mol Biol 147:227–246

Visser L, Van Maarschalkerweerd MW, Rozijn TH, Wassenaar ADC, Reemst AMCB, Sussenbach JS (1979) Viral DNA sequences in adenovirus-transformed cells. Cold Spring Harbor Symp Quant Biol 44:541–550

Visser L, Reemst ACMB, Van Mansfeld ADM, Rozijn TH (1982) Nucleotide sequence analysis of the linked left and right hand terminal regions of adenovirus type 5 DNA present in the transformed rat cell line 5RK20. Nucleic Acids Res 10:2189–2198

Westin G, Visser L, Zabielski J, Van Mansfeld ADM, Pettersson U, Rozijn TH (1982) Sequence organization of a viral DNA insertion present in the adenovirus type-5-transformed hamster line BHK268-C31. Gene 17:263–270

Wilson JH, Berget PB, Pipas JM (1982) Somatic cells efficiently join unrelated DNA segments end-to-end. Mol Cell Biol 2:1258–1269

Yasue H, Ishibashi M (1982) The oncogenicity of avian adenoviruses. III. In situ hybridization of tumor line cells localized a large number of a virocellular sequence in few chromosomes. Virology 116:99–115

Subject Index

The Molecular Biology of Adenoviruses 3

Edited by Walter Doerfler

1984. (Current Topics in Microbiology and Immunology, Volume 111)
ISBN 3-540-13138-8
In preparation

Contents:

Springer-Verlag
Berlin
Heidelberg
New York
Tokyo